FUNDAMENTALS OF SOIL MECHANICS FOR SEDIMENTARY AND RESIDUAL SOILS

FUNDAMENTALS OF SOIL MECHANICS FOR SEDIMENTARY AND RESIDUAL SOILS

Laurence D. Wesley

WILEY

JOHN WILEY & SONS, INC.

For general information about our other products and services, please contact our Customer Care Department within the United States at (800) 762-2974, outside the United States at (317) 572-3993 or fax (317) 572-4002.

Wiley also publishes its books in a variety of electronic formats. Some content that appears in print may not be available in electronic books. For more information about Wiley products, visit our web site at www.wiley.com.

Library of Congress Cataloging-in-Publication Data:

Wesley, Laurence D.
   Fundamentals of soil mechanics for sedimentary and residual soils / Laurence D. Wesley.
        p. cm.
   Includes bibliographical references and index.
   ISBN 978-0-470-37626-3 (cloth)
   1. Residual materials (Geology) 2. Soil mechanics. 3. Sediments (Geology)
I. Title.
   TA709.5.W47 2010
   624.1'5136—dc22

                                                                        2009009721

ISBN 978-0-470-37626-3

Printed in the United States of America

10 9 8 7 6 5 4 3 2 1

CONTENTS

9   SHEAR STRENGTH OF SOILS   185

PREFACE

This book was originally planned with the title *Soil Mechanics for Geotechnical Engineers*. I mention this because its target readership is indeed geotechnical engineers, including those who teach and train geotechnical engineers, and students aiming to become geotechnical engineers. Its name was changed to the present title following discussions with my publisher; together we agreed that the current title is preferable as it conveys more clearly the distinctive content of the book.

My first aim in writing this book has been to give equal coverage to residual soils and sedimentary soils. I have believed for a long time that there is a need for such a book, because many graduates are leaving universities throughout the world without even hearing of residual soils, let alone having any understanding of their properties. This is despite the fact that in not a few cases, the universities from which they graduate are surrounded by residual soils on every side, as far as the eye can see.

All graduates should have a basic knowledge of residual soils, first, because they are likely to encounter residual soils from time to time during their working life and, second, because there are important characteristics of residual soils that do not fit into the conventional concepts or the "theoretical framework" of classical soil mechanics. The application of these concepts to residual soils can produce quite misleading conclusions. A prime example is the use of stress history, and the *e*-log *p* graph associated with it, as an explanation of soil behavior. Stress history and the concepts of normal consolidation and overconsolidation have little or no relevance to residual soils.

Having said the above, I do not think that the differences between residual soils and sedimentary soils are such that residual soils should be covered as an alternative to or an extension of conventional soil mechanics. The most basic fundamentals of soil mechanics, such as the principle of effective stress and the Mohr-Coulomb failure criterion, are equally applicable to both groups. The important characteristics of residual soil behavior can easily be incorporated into conventional soil mechanics teaching. I have tried to do that in this book. There are no chapters dedicated specifically to residual soils. Material has simply been included on residual soils throughout the book, wherever their properties deviate significantly from those of sedimentary soils. I guess I am hoping that this book will give a push (or at least a gentle nudge) for the inclusion in "mainstream" soil mechanics of those aspects of residual soil behavior that ought to be there.

I would add also that the supposed differences between residual and sedimentary soils are perhaps not as wide as is often imagined. The e-log p graph for soil compressibility is a case in point. Because of my experience with residual soils, I have been pushing for many years for the use of a linear pressure scale (rather than a log one) for interpreting the one-dimensional compression behavior of soils only to discover in recent years that Professor Janbu of the Norwegian University of Science and Technology has been urging this over a much longer period of time on the basis of his experience with sedimentary soils. The reasons for using a linear pressure scale are almost as compelling for sedimentary soils as for residual soils.

My second aim with this book is to emphasize concepts and principles rather than methods. My experience in mentoring and training graduate engineers has been that they have a strong command of methods but a rather weak grasp of concepts and principles. This is not surprising; engineers want to get on with designing and building things and have a "mental predisposition" toward methods rather than concepts. This can easily lead to a "handbook" or "recipe" approach to design. This might be acceptable in some branches of civil engineering but is decidedly unsatisfactory in geotechnical engineering. Nature rarely produces the tidy situations that are amenable to such an approach. Unfortunately, the advent of the computer has added to this emphasis on methods rather than concepts and principles.

In keeping with the above aims I have concentrated on the properties of undisturbed soils.

As far as possible, I have avoided presenting conceptualized or idealized versions of soil behavior, especially those derived from the study of remolded soils, and presented only the results of actual tests on undisturbed soils. Idealizations certainly have their place and are inevitable in design situations. However, idealizations are only appropriate when the limitations or approximations associated with them are clearly understood and taken account of.

If I was to dedicate this book to anyone, I think it would have to be Professor Nilmar Janbu, mentioned above. The following quotations (Janbu, 1998) highlight what he has been saying for many years:

It is very surprising, to say the least, to observe all the efforts still made internationally in studying remoulded clays. If the aim of such research is practical application, it is obviously a total waste of money.

...it remains a mystery why the international profession still uses the awkward e-log p plots, and the incomplete and useless coefficient Cc which is not even determined from the measured data, but from a constructed line outside the measurements...

Both statements are highly relevant to sedimentary soils. They are even more relevant to residual soils, and I hope in writing this book I have been adequately mindful of them.

REFERENCE

Janbu, N. *Sediment deformation*. Bulleting 35, Norwegian University of Science and Technology, Trondheim, Norway, 1998.

ACKNOWLEDGMENTS

I am indebted to a large number of people who have been my teachers, mentors, and valued colleagues since I first encountered soil mechanics some 50 years ago. I cannot acknowledge them all by name, but I especially wish to mention the following.

Professor Peter Taylor, of Auckland University, who introduced me to soil mechanics in my undergraduate course and supervised my Master of Engineering thesis. Professor Taylor is a very gifted teacher and researcher, and I owe him a great deal. He also reviewed an early draft of this book and provided me with some very helpful comments.

Ir Zacharias, Ir Soedarmanto, and Ir Soelastri and all the engineers and technicians with whom I worked during my time at the Institute for Soil and Highway Investigations in Bandung, Indonesia. The eight years I spent at the Institute were probably the most formative influence in my understanding of the behavior of natural soils, especially residual and volcanic soils. Muljono Purbo Hadiwidjoyo, an engineering geologist at the Indonesian Geological Survey in Bandung, and Yunus Dai, a soil scientist at the Indonesian Soil Research Centre in Bogor. They have been outstanding figures in their respective fields and the knowledge they shared with me was invaluable in my early years in Indonesia as I sought to come to grips with local geology and soil conditions.

Russell Bullen and the staff of the New Zealand Ministry of Works Central Laboratories in Wellington, with whom I worked for five years between two spells in Indonesia.

Professors Alan Bishop and Peter Vaughan and the staff of Imperial College, London. I learned a great deal from discussions and interaction with these two professors as well as with two PhD colleagues, Richard Pugh and Mamdouh Hamza.

Dr. David Hight, formerly of Imperial College and currently of Geotechnical Consulting Group, London, with whom I have enjoyed many fruitful discussions.

Alan Pickens, Terry Kayes, and Tim Sinclair and all the staff of the Auckland consulting firm Tonkin and Taylor with whom I worked for nearly 11 years. I learned a lot during this time from a wide range of geotechnical projects in both New Zealand and Southeast Asia but also from the experience and wisdom of those around me.

Michael Dobie, Regional Manager for Asia Pacific, Tensar International, who shared with me his expertise in geogrid reinforced earth and who reviewed various sections of this book.

Professor Michael Pender, Dr. Tam Larkin, and Dr. John St George, colleagues in the geotechnical group at the University of Auckland, where I taught for the last 18 years of my career. I am especially grateful to Michael Pender, who encouraged me in my move from the consulting world to lecturing. My thanks also go to Michael Pender and John St. George for reviewing and providing helpful comments on parts of this book.

Professors Ramon Verdugo, Claudia Foncea, Ricardo Moffat, and Leonart Gonzalez, of the Geotechnical Team in the Civil Engineering Department of the University of Chile in Santiago, where I have been a visiting lecturer over the past five years. I appreciate very much the warm welcome they have given me into their "circle" and I have enjoyed the many stimulating discussions, both technical and philosophical, we have engaged in during lunch and coffee breaks.

Finally, thanks to my wife, Barbara, and my children, especially Kay, for their tolerance and support during the rather large portion of my "retirement" that has been devoted to writing this book.

CHAPTER 1

SOIL FORMATION, COMPOSITION, AND BASIC CONCEPTS

1.1 WEATHERING PROCESSES, SEDIMENTARY AND RESIDUAL SOILS

The word *soil* is used in soil mechanics to mean any naturally formed mineral material that is not rock. It thus covers all loose material ranging in particle size from clay through silt and sand to gravel and boulders. The main focus of soil mechanics is the material at the fine end of the range, particularly clay and silt and to a lesser extent sand.

Soils are formed by the physical and chemical weathering of rock. **Physical weathering** may be one of two types. First, there is disintegration—caused primarily by wetting and drying or by freezing and thawing in cracks in the rock. Second, there is erosion—caused by the action of glaciers, water, or even wind. These processes produce a range of particles of varying sizes which are still composed of the same material as the parent rock. Sand and silt particles produced by physical weathering generally consist of single rock minerals, rather than combinations of these, as is the case in their parent rock or in gravel-sized material. It is important to recognize that no matter how fine the particle size of the material produced by physical weathering may be, it can never have the properties of clay because the chemical conversion needed to form true clay particles is not present.

Chemical weathering processes are much more complex and involve chemical changes to the mineral content of the parent rock caused by the action of percolating water, oxygen, and carbon dioxide. The minerals of which rock is composed are converted into a very different group of materials known as clay minerals. Well-known members of this group are kaolinite, illite, and montmorillonite, but less well known clay minerals of considerable importance in volcanic areas are halloysite and allophane. Clay mineral particles are generally crystalline in form and are of colloidal size, that is, they are less than 0.002 mm. These minerals give soil the properties of cohesion and plasticity, which are the distinctive characteristics of clay.

The nature of the clay mineral produced in any given situation is dependent on both the parent rock and the weathering environment, in particular the local climate, whether the site is well drained, and whether the percolating water is acidic or alkaline. For example, kaolinite is formed from feldspar by the action of water and carbon dioxide. Quartz is one of the minerals most resistant to weathering, so that soils weathered from granite tend to have a substantial proportion of coarse quartz particles within a matrix of finer material. Weathering is most intense in warm, wet climates and least intense in cold, dry climates. In the wet tropics, weathering can extend to many tens of meters below the ground surface. The chemical processes involved in weathering are complex and not of direct interest or concern to geotechnical engineers; it is the properties of the end product that are of paramount interest.

Apart from the direct physical and chemical processes that convert rock to soil, there are other processes that transport soil particles and redeposit them in lakes and the ocean. This process is illustrated in Figure 1.1. The soil formed directly from the chemical weathering process is called a **residual soil**. It remains in place directly above and in contact with its parent rock. Rainfall erodes some of this residual soil and transports it via streams and

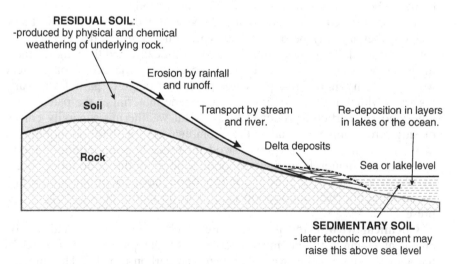

Figure 1.1 Soil formation processes.

rivers to eventually end up in lakes or the sea where it is redeposited as sediment at the bottom of the lake or sea. This process may continue for many thousands or millions of years, and the soils undergo a great deal of compression, or "consolidation," as additional layers are deposited above them. In this way the soil can build up to a great thickness. Soils formed in this way are termed **sedimentary soils** or **transported soils**.

Once formed, sedimentary soils undergo further changes due to the weight of overburden material above them as well as natural hardening, or "aging," processes. Seeping water influences these processes, possibly by providing chemical cementing agents that tend to bond the soil particles together or by dissolving some materials or chemicals present in the soil, a process known as leaching. In many situations, compression from the weight of overlying material combined with chemical cementing processes converts the soil into a sedimentary rock. Sandstones and clay stones (or mudstones) are formed in this way.

These sedimentary soils may be uplifted later by tectonic movement, so that in many parts of the world today they exist far from the sea or lake where they were formed and well above sea level. Once they are uplifted in this way, the erosion cycle from rain and streams begins all over again, and the thickness of the soil decreases.

1.2   CLAY MINERALS

Clay minerals are a very distinctive type of particle that give particular characteristics to the soils in which they occur. The most well known clay minerals are kaolinite, illite, and montmorillonite. These have a crystalline structure, the basic units of which are termed a silica tetrahedron and an alumina octahedron. These units combine to form sheet structures, which are usually represented graphically in the form shown in Figure 1.2. The actual clay mineral particles are formed by combinations of these basic sheets, which form multilayered "stacks," as indicated in Figure 1.2. The nature of the bonds between the sheets has a very important influence on the properties of the whole particle.

Kaolinite particles have a basic structure consisting of a single sheet of silica tetrahedrons and a single sheet of alumina octahedrons. These combined sheets are then held in a stack fairly tightly by hydrogen bonding. Illite particles have a basic structure made up of a central alumina sheet combined with silica sheets above and below. The combined sheets are in turn linked together by potassium ions sandwiched between them. This is a fairly weak form of bond. Montmorillonite is made up of the same basic unit as illite, but the form of bond between these basic units is different. Water and exchangeable cations provide this bond, which is a much weaker bond than that in illite particles.

This special structure means that these clay particles are not inert, as are rock particles. The term "active" is used to describe clay minerals, meaning

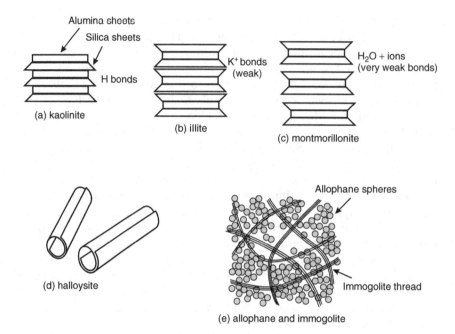

Figure 1.2 Schematic of clay minerals.

they are capable of swelling and shrinking by taking in water or losing it, depending on the environment surrounding them. Kaolinite is of relatively low activity, illite of medium activity, and montmorillonite of high activity. In general, the higher the activity of the clay, the less favorable the engineering properties of the soil. Montmorillonite clays are of relatively low strength as well as being highly compressible and often cause problems with foundations because of shrinkage or swelling. On the other hand, kaolinite clays, because of their low activity, have relatively good engineering properties. We can note that there are some engineering situations where high activity is desirable, as, for example, in some water-retaining structures where a low-permeability, highly plastic material is required as a barrier to seepage.

In addition to these common clay minerals, there are two rather unusual minerals often found in clays derived from the weathering of volcanic material. These are halloysite and allophane/immogolite. Although distinct, allophone and immgolite normally occur together. These minerals are formed from the weathering of volcanic "ash," a loose silt- or sand-sized material produced by volcanic eruptions, especially andesitic eruptions. Unlike rocks and other volcanic material such as coarse pyroclastic deposits (material produced by explosive events) and lava flows, volcanic ash particles do not have a crystalline structure and for this reason undergo a different and unique weathering process that leads to the formation of halloysite and allophane. The form of these materials is illustrated diagrammatically in Figure 1.2. Both are of much smaller particle size than the three well-known clay minerals already described and have a less well developed crystalline structure. The allophone and immogolite diagram is based on an electron micrograph of Wada (1989). A detailed account of the geotechnical properties of allophone clays is given by Wesley (2002).

Halloysite consists of cylindrical "rolls" some of which are properly formed and others are mere fragments. Allophane and immogolite normally occur together; the allophane particles are essentially spherical in shape, while immogolite has the form of long threads that weave between the allophane spheres. While the well-known clay minerals kaolinite, illite, and montmorillonite consist of particles somewhat less than 0.002 mm (2 μm) in size, the spherical allophane particles are approximately one thousand times smaller. Neither halloysite nor allophane/immogolite is of high activity. Allophane/immogolite soils are very unusual, being characterized by extremely high water contents and Atterberg limits (described in chapter 3). Despite this they have remarkably good engineering characteristics.

Current understanding of the weathering of volcanic ash is that it involves the following sequence, at least in wet tropical areas where the weathering process is intense:

Ash $\longrightarrow$ allophane/immogolite $\longrightarrow$ halloysite $\longrightarrow$ kaolinite $\longrightarrow$ sesqui-

oxides $\longrightarrow$ laterite

The progress of weathering from kaolinite to sesqui-oxides involves the leaching out of the silica-based minerals and the concentration of aluminum and iron compounds known as sesqui-oxides. With time, these compounds act as cementing agents, forming "concretions," and the material becomes a nonplastic, sandy gravel. This material is known as laterite, and the weathering process that produces it is termed laterization. This laterization process can occur in both volcanic and nonvolcanic soils, although it appears to be more common in the former. In cooler climates, the weathering may not progress as far as the formation of laterite.

1.3 INFLUENCE OF TOPOGRAPHY ON WEATHERING PROCESSES

Topography has a strong influence on the weathering process and thus on the type of soil formed (see Figure 1.3). This is especially true in the wet tropics. In hilly and mountainous areas, the soil is well drained with seepage tending to occur vertically downward. This leads to the formation of low-activity minerals, especially kaolinite. In volcanic areas, the minerals allophane and halloysite are likely to be formed initially, leading with time to the formation of kaolinite, as noted above. Soils containing these minerals generally have good engineering properties.

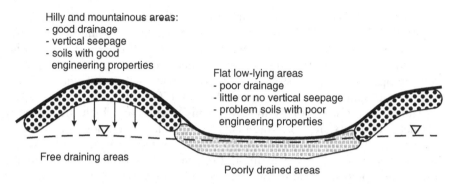

Figure 1.3 Influence of topography on clay mineral formation.

In wide, flat, areas, drainage of any sort is much more limited, and the weathering process is quite different. It tends to produce montmorillinite and associated highly active minerals. This is particularly the case in wet tropical areas that have distinct wet and dry seasons. Clays of this sort are called vertisols by soil scientists because the cyclic wetting and drying process and associated surface cracking tend to cause movement of water (and soil) in both the upward and downward direction close to the surface. These soils are often termed black clays by geotechnical engineers and generally have poor or undesirable engineering properties.

1.4 FACTORS GOVERNING THE PROPERTIES OF SEDIMENTARY AND RESIDUAL SOILS

The title of this book possibly suggests that sedimentary and residual soils are quite distinct materials with different mechanical or physical properties. There is important, but limited, truth to this suggestion. Many of the fundamental principles that govern soil behavior, in particular the principle of effective stress (Chapter 4), the laws governing seepage (Chapter 7), and the Mohr–Coulomb shear strength failure criterion (Chapter 9), are equally applicable to both groups. The stability concepts governing foundation design, earth pressure, and slope stability (Chapters 11–14) are also of universal applicability. However, there are some aspects of residual and sedimentary soil behavior that are different and can only be appreciated if we have a sound understanding of the processes by which these two soil groups are formed. At the risk of being repetitive, we will therefore give further consideration to the formation process of the two groups and the influence this has on their properties.

Residual soils are the direct product of the weathering of their parent rock and are generally more closely related to characteristics of their parent rock than is the case with sedimentary soils. They often exhibit a property known as "structure"; that is, the particles are packed together or even bonded together in a way that forms a soil "skeleton" having characteristics quite different from those of a simple collection of individual particles.

Sedimentary soils undergo various additional processes beyond the initial physical and chemical weathering of the parent rock and subsequent transport and redeposition. In particular, they undergo compression from the weight of the layers deposited above them, making them stronger or harder. In some situations the load on the soil may later be reduced as a result of subsequent geological uplift and erosion processes. Soils that have not been subjected to stresses greater than those currently acting on them are termed **normally consolidated** soils, while those that have had higher loads on them sometime in the past are called **overconsolidated** soils. The sequence of stresses to which the soil has been subjected since its formation is termed its **stress history**. Figure 1.4 is an attempt to illustrate the processes involved in the formation of the two groups.

While stress history has been an important concept (for sedimentary soils) in soil mechanics since its inception, it has been increasingly recognized that other factors, especially the cementation and hardening effects that occur with time (aging), are equally important. This means that structure may be just as important in sedimentary soils as in residual soils and is becoming an increasingly important focus of research.

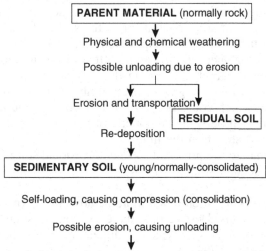

Figure 1.4 Formation factors influencing properties of sedimentary and residual soils.

The formation of both soils is complex, but two important factors lead to a degree of uniformity and predictability with sedimentary soils that is absent from residual soils:

(a) The sorting processes that take place during erosion, transportation, and deposition of sedimentary soils tend to produce homogeneous deposits. Coarse particles get deposited in one place and finer particles in a different place.

(b) Stress history is a prominent factor in determining the behavioral characteristics of sedimentary soils and leads to a convenient division of these soils into normally consolidated and overconsolidated materials.

The absence of these two factors with residual soils means that they are generally more complex and less capable of being divided into tidy categories or groups.

It is worth noting at this point that with time the processes forming these two soil groups tend to have an opposite effect on their properties, as illustrated in Figure 1.5. The weathering of rock tends to make the rock less dense and steadily reduces its strength. Solid rock contains essentially no void space whereas soils often contain a similar or greater volume of voids than solid particles. With some soils, solid material makes up as little as 20 percent of the volume of the soil. The term "void ratio" is used to define the volume of void space in a soil and is the ratio of the volume of voids to the volume of solid material (i.e., the soil particles).

With sedimentary soils, the compression of the soil from the weight of material above it together with the aging effect makes it denser and harder. Figure 1.5b shows a graph illustrating the way the void space in the soil steadily decreases as the weight of material above it increases. If the load on it is reduced, which could occur as a result of tectonic uplift followed by erosion, there will be some "rebound" of the soil with a small increase in void content, as shown in the figure.

We can summarize the principal aspects of residual soils that distinguish them from sedimentary soils as follows:

1. Residual soils are often much more heterogeneous (nonuniform) than sedimentary soils. Despite this, there are some residual soils that are just as homogeneous as typical sedimentary soils. Tropical red clays are often in this category.

2. Some residual soils, especially those of volcanic origin, may contain very distinctive clay minerals not found in sedimentary soils and which strongly influence their behavior.

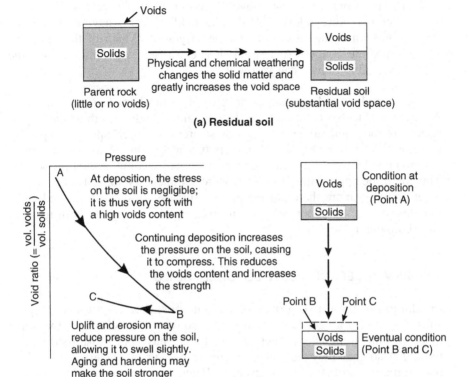

Figure 1.5 Formation processes and density of residual and sedimentary soils.

3. Some residual soils are not strictly particulate, that is, they do not consist of discrete particles. They may appear to consist of individual particles, but when disturbed or subjected to shear stress, these particles disintegrate and form an array of much smaller particles.

4. Stress history is not a significant formative influence on the properties of residual soils.

5. Some behavioral frameworks based on the study of sedimentary soils, especially the logarithmic plot used to express compressibility (Chapter 8), may not be helpful to an understanding of the behavior of residual soils.

6. Empirical correlations between soil properties developed from the study of sedimentary soils may not be valid when applied to residual soils.

7. The pore pressure state above the water table (Chapter 4) is of considerably more relevance to understanding the behavior of residual soils than is the case with sedimentary soils. Much of the action of immediate concern to geotechnical engineers takes place above the water table rather than below it.

The discipline soil mechanics developed in northern Europe and North America, and its basic concepts evolved from the study of sedimentary soils, which are the dominant soil type in these areas. It is perhaps not surprising therefore that few textbooks or university courses on soil mechanics even mention residual soils, let alone give an adequate account of their properties. This might be appropriate if all soils were sedimentary soils. This is not the case; probably more than half the earth's surface consists of residual soils. For this reason, this book attempts to give equal coverage to both residual and sedimentary soils.

1.5 REMOLDED, OR DESTRUCTURED, SOILS

In addition to these two groups of natural soils, there is a third group of soils that are no longer natural and are therefore of much less importance to geotechnical engineers. These are soils that have been disturbed and/or remolded so that they no longer retain important characteristics of their undisturbed in situ state. This group includes soils prepared by sedimentation from an artificial slurry, which are often used to investigate sedimentary soil behavior, and also compacted soils.

The term **destructured** is frequently used these days to designate these soils and has a slightly different meaning from the term **remolded**. The term "destructured" means that the soil has been manipulated in such a way that bonds between particles or any other structural effects are destroyed, but the particles themselves are not altered. "Remolding" is a somewhat vague term but is generally taken to mean that the soil has been thoroughly manipulated, and any special characteristics associated with its undisturbed state are no longer present. With residual soils, thorough remolding may well completely destroy some particles as well as destroy the structure of the material.

The properties of remolded soils are thus not governed by any form of structure, as is normally the case with most undisturbed soils, regardless of whether they are residual or sedimentary. Compacted clays may be an exception to this statement to a small extent, as it is possible that the compaction process does create some form of structure. For example, compacted clays tend to have a higher permeability in the horizontal direction than in the vertical direction, due to the horizontal layering effect produced by the compaction method. They may also have a greater stiffness in the vertical direction than the horizontal direction.

REFERENCES

Wada, K. 1989. Allophane and imogolite. In J. B. Dixon and S. B. Weed (eds.), *Minerals in Soil Environments* (2nd ed.). SSSA Book Series No 1., pp. 1051–1087. Madison, WI: Soil Science Society of America.

Wesley, L. D. 2003. Geotechnical characterization and behaviour of allophane clays. In *Proc. International Workshop on Characterisation and Engineering Properties of Natural Soils*, Vol. 2, Singapore, December 2002. Leiden: A. A. Balkema, pp. 1379–1399.

CHAPTER 2

BASIC DEFINITIONS AND PHASE RELATIONSHIPS

2.1 COMPONENTS OF SOIL

Unlike most other engineering materials such as steel or concrete, soil is not a single uniform material. It normally consists of two materials but not infrequently it may consist of three materials. These materials are called phases, and soil is referred to as a two- or three-phase material. The phases are:

1. Soil particles (solids)
2. Water
3. Air

The soil particles tend to be interlocked to some extent and form what is often referred to as the **soil skeleton**, as shown in Figure 2.1.

Understanding and formulating soil behavior involve primarily an understanding of the roles played by these phases and the interaction between them, especially in relation to the stresses acting on the soil. Most soils encountered by geotechnical engineers contain water only in the void space between the particles. Such soils are called **fully saturated**. Soils in which air is also present are termed **partially saturated** or **unsaturated**. This book deals primarily with fully saturated soils, especially clays, and unless stated otherwise it can be assumed that the soils described are fully saturated. The only exception is compacted soil, which normally contains a small volume of air, in the vicinity of 5–10 percent of the total volume.

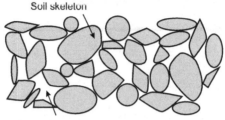

Soil skeleton

Voids containing water and/or air

Figure 2.1 The soil skeleton.

2.2 PHASE RELATIONSHIPS

A number of definitions and terms are used to describe the properties and relative proportions of the three phases that make up the soil. Figure 2.2 shows the three phases "lumped" together for definition purposes and for ease of working out relationships between them.

Table 2.1 lists the terms used for defining the mass, weight, and volume of the phases and the relationships between them. The units commonly used for each term (or property) are also listed. It should be noted that there are fixed limits to the values of some of the properties, depending on how they have been defined. For example, the porosity must range between 0 and 1, as it is related to the total volume. Void ratio, on the other hand, does not have specific limits, as it is related to the volume of solids rather than the total volume. The degree of saturation can only range between 0 and 100 percent.

Mention should be made of an additional parameter that we will use from time to time in this book. This is the submerged unit weight of the soil, defined as $\gamma' = \gamma - \gamma_w$, where γ_w is the unit weight of water. The submerged unit weight is thus the effective weight of the soil when it is beneath the water level in the ground. It takes account of the buoyant influence of the water.

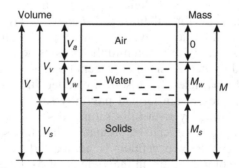

Figure 2.2 Diagrammatic representation of the three phases of soil.

Table 2.1 Phase Definitions and Relationships between Them

Term	Usual Symbol	Definition	Unit	Normal Range
Density (bulk density)	ρ	Total mass/total volume (M/V)	kg/m^3 gm/cm^3 $tonne/m^3$	1500–2100 1.5–2.1 1.5–2.1
Unit weight	γ	Total weight/total volume $(Mg/V = W/V)$	kN/m^3	15–21
Water content	w	Mass water/mass soil (M_w/M_s)	%	15–80 (can be > 100)
Dry density	ρ_d	Mass solids/total volume (M_s/V)	kg/m^3 gm/cm^3 $tonne/m^3$	1200–800 1.2–1.8 1.2–1.8
Dry unit weight	γ_d	Weight solids/total volume $(M_s g/V = W_s/V)$	kN/m^3	12–18
Particle density	ρ_s	Mass of solids/volume of solids (M_s/V_s)	kg/m^3 gm/cm^3	2600–2750 2.6–2.75 2.6–2.75
Specific gravity	G_s	Particle density/water density (ρ_s/ρ_w)	Dimensionless	
Void ratio	e	Void volume/solid volume (V_v/V_s)	Dimensionless	0.3–2.0 (can be >8)
Porosity	n	Void volume/total volume (V_v/V)	Dimensionless	0.2–0.6
Degree of saturation	S_r	Water volume/void volume (V_w/V_v)	%	0–100
Air voids	a_v	Volume of air/total volume (V_a/V)	%	0–20

Three of these parameters can be measured directly, namely the unit weight, water content, and specific gravity. The remaining parameters, such as dry unit weight, void ratio, degree of saturation, or air voids, can be calculated from these.

The property with the smallest range is the specific gravity, which is normally between 2.6 and 2.75. Occasionally, soils rich in iron are found, and the specific gravity may then exceed 2.8.

Some elementary relationships between the above parameters can easily be identified. For example the relationship between the mass of solids and the total mass, or between dry density and bulk density, can be established as follows:

$$M = M_s + M_w = M_s + wM_s \quad (\text{since} \quad w = M_w/M_s)$$
$$= M_s(1 + w)$$

Hence $M_s = M/1 + w$, and $W_s = W/1 + w$.

If we divide both sides by the volume of the soil V, then we obtain

$$\rho_d = \rho/1 + w \text{ and } \gamma_d = \gamma/1 + w$$

Other relationships are most easily worked out by drawing the phase diagram in a slightly different form in which the volume of solids is taken as unity, as shown in Figure 2.3. With this starting point the void volume is e, the void ratio, and the remaining quantities can be determined using the definitions given in Table 2.1.

We can then write down a number of other relationships from the definitions of the various parameters. We can obtain the relationship between water content, specific gravity, void ratio, and degree of saturation as

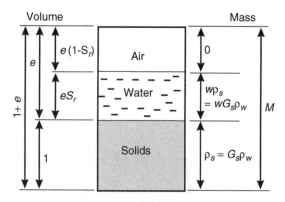

Figure 2.3   The three phases, taking the volume of the solids as unity.

follows:

$$\text{Mass of water} = wG_s\rho_w = (\text{volume of water})\rho_w = (eS_r)\rho_w$$

so that $wG_s = eS_r$. The bulk density can be expressed as

$$\rho = \frac{G_s(1+w)}{1+e}\rho_w$$

and the unit weight as

$$\gamma = \frac{G_s(1+w)}{1+e}\gamma_w$$

The air voids are related to void ratio, water content, and specific gravity as follows:

$$a_v = \frac{e - eS_r}{1+e} = \frac{e - wG_s}{1+e}$$

2.3 EXAMPLES IN USE OF PHASE RELATIONSHIPS

Example 1

Measurements made on an undisturbed sample of fully saturated clay give the following results:

Height $= 80\,\text{mm} = 8.0\,\text{cm}$
Diameter $= 63\,\text{mm} = 6.3\,\text{cm}$
Weight $= 425.0\,\text{g}$
Weight after oven drying $= 275.2\,\text{g}$

Determine the unit weight, water content, void ratio, and specific gravity.

We can note before continuing that the units used in geotechnical engineering vary from country to country and are seldom entirely consistent. Soil samples are almost invariably "weighed" on scales that record in grams or kilograms, so strictly speaking what is being determined is mass rather than weight. Many countries, especially those that have historically used the metric system, continue to use grams and kilograms in preference to newtons or kilonewtons, without worrying whether they are dealing in weight or mass. Countries that have adopted the International System (SI) version of the metric system avoid the use of centimeters because of a system preference for meters or millimeters. This is rather unfortunate because the ideal units for laboratory work are grams and centimeters, since $1\,\text{cm}^3$ of water weighs $1\,\text{g}$, which makes weight-to-volume conversions very simple and avoids the use of powers of 10 that arise when millimeter or meter is used. In general, the SI version of the metric system will be used in this book. The only exception will be the occasional use of centimeter for laboratory calculations.

We will continue with our example:

The volume of the sample is given as

$$V = \frac{\pi (63)^2}{4} 80 = 249{,}379 \, \text{mm}^3 = 249.4 \, \text{cm}^3$$

The bulk density is therefore equal to

$425/249{,}379 = 1.704 \times 10^{-3} \text{g/mm}^3 = 1.704 \, \text{g/cm}^3 = 1704 \, \text{kg/m}^3$

and the unit weight is

$1704 \times 9.81 \, \text{N/m}^3 = 1704 \times 9.81/1000 \, \text{kN/m}^3 = 16.72 \, \text{kN/m}^3$

The weight of water is $425 - 275.2 = 149.8 \, \text{g}$

so that the water content is $149.8/275.2 = 0.544 = 54.4\%$.

The volume of water is

$149.8 \times 1 \, \text{cm}^3 = 149.8 \, \text{cm}^3 =$ volume of voids (since soil is fully saturated)

and the volume of solids is $249.4 - 149.8 = 99.6 \, \text{cm}^3$.

The void ratio is given as $149.8/99.6 = 1.504$.

The particle density is $275.2/99.6 = 2.76 \, \text{g/cm}^3$.

so that the specific gravity is $2.76 \, \text{g/cm}^3 / 1.00 \, \text{g/cm}^3 = 2.76$.

It is often helpful in carrying out calculations of this sort to use a table, as shown below, and to start by filling in the known values and working from these to fill in the unknowns. The known values are shown in bold. From these figures we can easily fill in the remaining unknowns as calculated above. Once the mass and volume of each component is established, it is easy to calculate any other parameters we may wish to know.

	Mass (g)	Volume (cm^3)
Air	**0**	**0**
Water	149.8	149.8
Solids	**275.2**	99.6
Total	**425.0**	**249.4**

Example 2

To monitor the properties of a compacted clay fill, a small excavation is made in the surface and the volume of the hole is measured. It is found to be $0.30\,\text{m}^3$. The excavated soil is retained and weighed immediately to give a weight of 506.3 kg. It is then put in an oven and dried to give a dry weight of 386.2 kg. The specific gravity is measured and found to be 2.69. Determine the following:

- Unit weight
- Dry unit weight
- Water content
- Void ratio
- Degree of saturation
- Air voids

We will use a table and fill in the known quantities (shown in bold), and proceed from there to determine the unknowns.

	Mass (kg)	Volume (m^3)
Air	**0**	0.0363
Water	120.1	0.1201
Solids	**386.2**	0.1436
Total	**506.3**	**0.30**

The mass of water is determined as $506.3 - 386.2 = 120.1$ kg. The volume of water is $120.1/1000 = 0.1201\,\text{m}^3$ (since $1\,\text{m}^3$ of water "weighs" 1000 kg). The particle density is $2.69 \times 1000 = 2690\,\text{kg/m}^3$. Thus the volume of the solids is $386.2/2690 = 0.1436\,\text{m}^3$. The volume of air is therefore $0.30 - (0.1201 + 0.1436) = 0.0363\,\text{m}^3$. We can note also that the volume of the voids equals $0.0363 + 0.1201 = 0.1564\,\text{m}^3$.

We can now calculate the parameters listed above:

Bulk density $= 506.3/0.30\,\text{kg/m}^3 = 1687.7\,\text{kg/m}^3$
Unit weight $= 1687.7 \times 9.81/1000\,\text{kN/m}^3 = 16.56\,\text{kN/m}^3$
Water content $= 120.1/386.2 = 0.3110 = 31.1\%$
Void ratio $= (0.0363 + 0.1201)/0.1436 = 1.089$
Degree of saturation $= 0.1201/0.1564 = 0.768 = 76.8\%$
Air voids $= 0.0363/0.30 = 0.121 = 12.1\%$

Example 3

An embankment is to be constructed using clay excavated from a nearby "borrow" source. In its natural state in this borrow source, the clay is fully saturated with a unit weight of $17.35\,\text{kN/m}^3$ and a water content of 41.5 percent. After excavation it is transported and water is added to it before it is compacted to form the embankment. The embankment is to have a volume of 75,000 m³, with specified properties of dry unit weight equal to 10.40 kN/m³ and water content of 48.5 percent.

Determine the following:

(a) Specific gravity of the soil
(b) Volume of water (in liters) to be added to each cubic meter of soil excavated from borrow source in order to achieve required water content for compaction
(c) Air voids in compacted soil
(d) Volume of soil to be excavated from borrow source

There are several ways of carrying out the calculations required to answer the above questions. The method we will use here is to determine all of the properties of the soil in both its initial state and final states, setting these out in tables, as we did in the previous examples.

Properties in the borrow area (water content 41.5%):

	Weight (kN)	**Volume (m³)**
Air	**0**	**0**
Water	5.09	0.519
Solids	12.26	0.481
Total	**17.35**	**1.00**

Weight of solids $W_s = W/1 + w = 17.35/1.415 = 12.26$ kN
Weight of water $= 17.35 - 12.26 = 5.09$ kN
Volume of water (volume of voids) $= 5.09/9.81 = 0.519$ m³
Volume of solids $= 1.00 - 0.519 = 0.481$ m³

The properties in the compacted fill in the embankment (water content 48.5%) are as follows:

	Weight (kN)	**Volume (m³)**
Air	**0**	0.078
Water	5,044	0.514
Solids	**10,400**	0.408
Total	15.444	**1.00**

Weight of water $= 10.40 \times 0.485 = 5.044 \, \text{kN}$
Total weight $= 5.044 + 10.40 = 15.444 \, \text{kN}$
Volume of water $= 5.044/9.81 = 0.514 \, \text{m}^3$
Volume of solids $= 10.4/25.49 = 0.408 \, \text{m}^3$
Volume of air $= 1.00 - (0.514 + 0.408) = 0.078 \, \text{m}^3$

We can now answer the questions above:

(a) Specific gravity of the soil: The particle "density" (particle unit weight) equals $12.26/0.481 = 25.49 \, \text{kN/m}^3$. Therefore the specific gravity is $25.49/9.81 = 2.60$.

(b) Volume of water from each cubic meter excavated from the borrow source to raise the water content from 41.5 percent to 48.5 percent: Each cubic meter from the borrow contains 12.26 kN of solids and 5.09 kN of water. The required weight of water per cubic meter is $12.26 \times 0.485 = 5.95 \, \text{kN}$.

Therefore the water to be added to each cubic meter equlas $5.95 - 5.09 = 0.86 \, \text{kN.} = 0.86 \times 1000 \, \text{N} = 860 \, \text{N} = 860/9.81 \, \text{kg} = 87.7 \, \text{kg}$.

The volume of water per cubic meter is 87.7 liters (since 1 liter of water "weighs" 1 kg).

(c) Air voids in the compacted soil are $0.078 \, \text{m}^3 = 7.8\%$.

(d) The volume of soil to be excavated from the borrow source.

This is perhaps the calculation that creates the most difficulty or confusion, although it should not do so. The key point to understand is that during earthworks (that is, during excavation, transport, and recompaction), no solid material is lost or gained. Some water may be lost from evaporation or may be added as necessary, and the compacted soil will normally contain some air voids, even though in its original state it is fully saturated. Thus the water and air in the soil may change but the solid material remains constant. We therefore need to focus on the solid material to determine the change in volume that occurs.

In the borrow area, the solids are "packed" in at $12.26 \, \text{kN/m}^3$, while in the compacted fill they are packed in at $10.40 \, \text{kN/m}^3$. The soil particles are therefore packed in more densely in the original ground than in the fill and will occupy less volume. Thus $1 \, \text{m}^3$ of the compacted fill would occupy $10.40/12.26 = 0.848 \, \text{m}^3$ in its original state in the borrow area. The volume to be excavated will therefore be $75,000 \times 0.848 = 63,600 \, \text{m}^3$.

2.4   MEASUREMENT OF BASIC PROPERTIES

2.4.1   Bulk Density

The bulk density γ can be measured either in the field or in the laboratory and consists simply of determining the weight of a known volume of soil.

Laboratory Methods   To determine the unit weight (or density) in the laboratory, an undisturbed sample of the material is needed. An undisturbed sample means one taken from the ground in such a way that it retains its in situ (i.e., in-place) characteristics. Undisturbed samples are possible with clay but not with sand or gravel. The procedure is then very straight-forward:

(a) The sample is trimmed to have known dimensions so that its volume can be calculated.
(b) The sample is weighed.

The ratio of weight to volume gives the unit weight.

Field Methods   The simplest method in the field is to use a sampling device or a "core cutter" to obtain directly a sample of known dimensions. This is then weighed to obtain the unit weight. Other, less direct methods can be used when sampling is not possible or convenient. Neat holes with a regular shape can be excavated, their volume measured, and the soil excavated retained and weighed. The ratio of weight to volume gives the unit weight.

2.4.2   Water Content

This is normally determined by weighing a soil sample before and after it has been dried in an oven at a temperature between 105 and 110 °C. Normal practice is to leave the soil sample in the oven overnight

Weight of wet soil and container $= W_1$
Weight of dry soil and container $= W_2$
Weight of water $= W_1 - W_2$
Weight of container $= W_3$
Weight of dry soil (solids) $= W_2 - W_3$
Water content $w = (W_1 - W_2)/(W_2 - W_3)$

2.4.3   Solid Density and Specific Gravity

A special container known as a pycnometer (or specific gravity bottle) is used. Its special feature is that its volume can be determined to a high

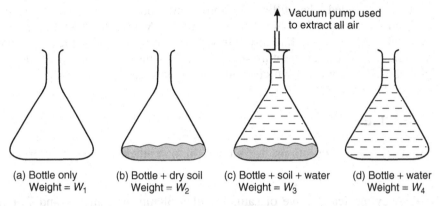

Figure 2.4 Procedure for determining the specific gravity of soil particles.

degree of accuracy. The test, illustrated in Figure 2.4, involves the following steps:

1. The pycnometer is weighed empty (W_1).
2. A sample of the soil being investigated is placed in the bottle and the bottle is weighed (W_2).
3. The bottle (with the soil still in it) is filled with water and weighed (W_3).
4. The pycnometer is filled with water and weighed (W_4).

Mass of soil $= W_2 - W_1$

Mass of water when bottle is full of water $= W_4 - W_1$

Mass of water when bottle is filled with water plus soil $= W_3 - W_2$

   Therefore

Mass of water displaced by the soil $= (W_4 - W_1) - (W_3 - W_2)$

   and

Volume occupied by soil $= \dfrac{(W_4 - W_1) - (W_3 - W_2)}{\rho_w}$

Density of particles $= \dfrac{\text{Mass of soil}}{\text{Volume of soil}} = \dfrac{(W_2 - W_1)\,\rho_w}{(W_4 - W_1) - (W_3 - W_2)}$

The specific gravity is defined by

$$G_s = \frac{\rho_s}{\rho_w} = \frac{W_2 - W_1}{(W_4 - W_1) - (W_3 - W_2)}$$

Care is needed when filling the bottle with both soil and water to ensure that no air is trapped within the soil particles. A vacuum extraction procedure is used to ensure that any trapped air is removed. The measurements are influenced by temperature and corrections must be made if the test is not carried out at a known controlled temperature.

EXERCISES

Answers to the questions are shown in parentheses in bold after each question.

1. A cylindrical sample of saturated clay 38 mm in diameter and 76 mm long has a mass of 154.0 g. After oven drying, the mass is 107 g. For this soil determine the following:
   (a) Water content (**43.7%**)
   (b) Bulk density (**1.79 g/cm³**)
   (c) Dry density (**1.24 g/cm³**)
   (d) Porosity (**0.55**)
   (e) Void ratio (**1.2**)
   (f) Specific gravity (**2.73**)
2. An oven-dried sand sample contained in a water-tight cylinder has a bulk density of 1700 kg/m³ and a specific gravity of 2.8.
   Determine the void ratio. (**0.647**)
   Sufficient water is added to the sample to give a degree of saturation of 60 percent.
   Determine the water content. (**13.9%**)
   Further water enters the sample until it becomes fully saturated.
   Determine the bulk density and the water content. (**2093 kg/m³, 23.1%**)
3. Determine the unit weight and the dry unit weight of a soil having a void ratio of 1.15 and a specific gravity of 2.67 assuming:
   (a) Soil is fully saturated (**17.43 kN/m³, 12.18 kN/m³,**)
   (b) Degree of saturation is 68 percent (**15.75 kN/m³, 12.18 kN/m³**)
4. During an earthworks operation a cylindrical sample of the compacted soil is obtained using a core cutter (a cylindrical sampler). Measurements on the sample give the following:
   Length: 12 cm (120 mm)
   Diameter: 10 cm (100 mm)
   Mass when sampled: 1688.0 g
   Mass after oven drying: 1285.6 g
   Specific gravity: 2.71

Determine the following:

Unit weight of soil $(17.6\,\text{kN/m}^3, = 1.79\,\text{g/cm}^3)$

Dry unit weight $(13.4\,\text{kN/m}^3, = 1.32\,\text{g/cm}^3)$

Water content (31.3%)

Void ratio (0.99)

Degree of saturation (86%)

Air voids (7%)

5. A motorway project includes construction of a compacted earth fill embankment of $50,000\,\text{m}^3$ final volume. Material for the embankment is to be taken from a borrow pit, where the soil is found to have the following average properties:

   Bulk density $1940.3\,\text{kg/m}^3$

   Water content 29.5%

   Specific gravity 2.69

   The soil is mechanically excavated and transported in trucks which, fully loaded, carry 6 tonnes $(6000\,\text{kg})$ of soil each. The soil is dumped on the embankment, spread, and broken up, after which a sprinkler adds water until the water content is 38.0 percent. The material is mixed by rotary disc and compacted by sheepsfoot rollers until a dry density of $1291\,\text{kg/m}^3$ is achieved. Determine the following:

   (a) Dry density, void ratio, and degree of saturation of undisturbed material in borrow area $(\gamma_d = 1498\,\text{kg/m}^3, e = 0.795, S_r = 100\%)$

   (b) Number of truck loads required to construct embankment $(13,933)$

   (c) Volume of excavation in borrow pit $(43,082\,\text{m}^3)$

   (d) Volume of water (liters) to be added per truck load assuming losses from evaporation are negligible $(349\,\text{liters})$

6. The properties of a compacted soil are investigated by excavating a hole in it and measuring the volume of the hole and the weight and properties of the excavated soil. The following data were obtained:

   Volume of hole: $0.027\,\text{m}^3$

   Weight of soil when excavated: $46.41\,\text{kg}$

   Weight of soil after oven drying: $34.01\,\text{kg}$

   Specific gravity of soil: 2.70

   For this soil determine the following:

   —unit weight, dry unit weight, water content, void ratio, degree of saturation, percent air voids $(\gamma = 16.9\,\text{kN/m}^3, \gamma_d = 12.4\,\text{kN/m}^3, w = 36.5\%, e = 1.000, S_r = 98\%, a_v = 7.4\%)$

7. Soil is excavated from a borrow area to form an embankment with a volume of $60,000\,\text{m}^3$. In its natural state in the borrow area the soil is fully saturated with a unit weight of $16.2\,\text{kN/m}^3$ and a water content

of 62.1 percent. After drying and compaction in the embankment the soil is to have a dry unit weight of 13.3 kN/m³ and a water content of 33.0 percent. Determine the following:

(a) Specific gravity of soil **(2.77)**

(b) Volume of excavation in borrow area **(79,849.3 m³)**

(c) Weight (in kilograms) of water to be removed by drying (evaporation) per tonne (1000 kg) of soil excavated from borrow area **(145 kg)**

8. Soil with a water content of 56.5 percent and a specific gravity of 2.72 is found after compaction to have an air void of 9 percent. Determine the dry unit weight and the unit weight of the soil. $(\gamma_d = 9.5\,\text{kN/m}^3, \gamma = 14.9\,\text{kN/m}^3)$

CHAPTER 3

BASIC INDEX TESTS, SOIL CLASSIFICATION AND DESCRIPTION

3.1 GENERAL

The properties of soils measured in soil mechanics can be divided into two broad groups. First, there are properties that give a general picture of the soil and its expected characteristics but which are not used directly in analytical design procedures. Most of the properties described in Chapter 2 and listed in Table 2.1 belong in this group. They are valuable in providing an indication of likely engineering properties of the soil. Second, there are properties that are used directly for design purposes. These are primarily parameters governing the strength, compressibility, and permeability of the soil. In this chapter, further properties belonging to the first group are described. As we shall see later, some of these properties can be related to design parameters by means of empirical correlations.

For engineering purposes soils are divided into two main categories with two subgroups in each category, as shown in Figure 3.1. Coarse-grained soils consist of gravel and/or sand and are commonly referred to also as granular materials or noncohesive soils. Fine-grained soils consist of silt and/or clay and are often referred to also as cohesive soils.

3.1.1 Gravel and Sand

These consist of rock fragments of various sizes and shapes. Gravel particles usually consist of rock fragments but may occasionally consist of single minerals. Sand particles normally consist of single minerals, frequently quartz. In some cases there may be only one size of particle present,

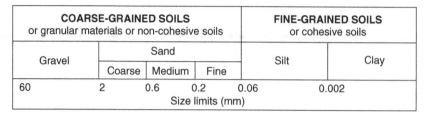

Figure 3.1 Principal soil groups and their particle size limits.

in which case the material is described as "uniform." In other cases a whole range of particle sizes from boulders down to fine sand may be present and the material is described as well graded.

3.1.2 Clay

Clay consists of very small particles and possesses the properties of cohesion and plasticity, which are not found in sands or gravels. **Cohesion** refers simply to the fact that the material sticks together, while **plasticity** is the property that allows the material to be deformed without volume change or rebound and without cracking or crumbling.

3.1.3 Silt

This is an intermediate material lying between clay and fine sands. Silts are less plastic than clays (strictly speaking, true silts hardly possess the property of plasticity at all) and more permeable and display the distinctive properties of "quick" behavior and dilatancy, which are not found in clays. **Quick behavior** refers to the tendency of silt to liquify when shaken or vibrated, and **dilatancy** refers to its tendency to undergo volume increase when deformed.

The meanings given to the terms gravel, sand, silt, and clay in soil mechanics are essentially the same as the meanings given to them in normal everyday usage.

3.2 PARTICLE SIZE AND ITS ROLE IN INFLUENCING PROPERTIES

Soil behavior always depends to some extent on particle size and it is logical to use this as a starting point for grouping or classifying soils. The particle size limits commonly adopted for each group are shown in Figure 3.1. With **coarse-grained materials,** it is found that engineering properties are closely related to particle size and hence particle size is the dominant criterion used in evaluating and classifying these soils.

With **fine-grained soils**, however, it is found that there is no longer a direct relationship between properties and particle size. This is because their properties are influenced by both the **size** and **composition** of their particles. For this reason, other methods have been devised for evaluating and classifying them, in particular tests known as Atterberg limits. These measure the property called plasticity, which we will describe shortly.

It is important to understand the difference between the terms "clay" and "clay fraction" or "silt" and "silt fraction." **Clay** is a descriptive term applied to fine-grained soils which behave as clays, that is, they have cohesion and plasticity, are not dilatant, and do not contain a noticeable amount of coarse material. **Clay fraction** is the proportion by weight of the particles in the soil finer than 0.002 mm. Similarly, **silt** is a descriptive term and **silt fraction** is the proportion of the material between 0.002 and 0.06 mm. It is possible for a soil to consist entirely of particles finer than 0.002 mm but still not be a clay. Rocks can be ground, either artificially or naturally, by glacial action, to produce "rock flour" that consists entirely of clay-sized particles but which does not have the properties of plasticity and cohesion needed to make it qualify as a clay. It would behave as a silt. Many clays found in nature have clay fractions in the range of only 15–50 percent.

Soils normally consist of a range of particle sizes and do not necessarily fall into one of the size categories listed above. To portray the particle size composition of a soil, a graph termed a particle size distribution curve (or grading curve) is used, as illustrated in Figure 3.2. The graphs show the proportion of the soil finer than any particular particle size.

Some soils contain particles ranging from coarse gravel through sand and silt to clay. These materials are termed **well graded**. Other soils are made up of particles that are nearly all the same size. These are termed **uniformly graded** or **poorly graded**. A third category, not commonly found, may be made up of a range of particle sizes from which a particular size range is missing. These are termed **gap graded**, though they may also be called **poorly graded**. Examples of the three types are given in Figure 3.2. Note that the gap-graded material contains no particle sizes over the horizontal section of the graph, that is, between about 0.03 and 0.3 mm.

3.2.1   Measurement of Particle Size

The particle sizes of coarse-grained soils are easily measured using sieves. The finest sieve that can conveniently be manufactured and used in practice happens to correspond to the particle size boundary between fine sand and silt, which as we have seen above is 0.06 mm. To measure the particle size of silts and clays, it is normal practice to use a sedimentation process. The process is not difficult, though a little tedious, and involves a number of practical calibrations and corrections. For this reason, only an outline will be given here; the method involves the following principles and steps:

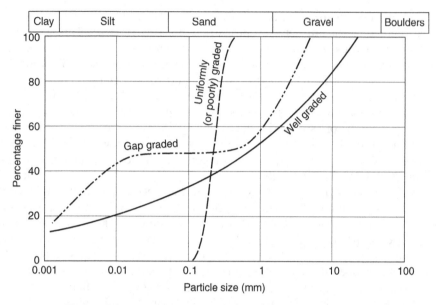

Figure 3.2   Particle size distribution (or grading) curves.

1. The method makes use of Stokes's law, which relates the velocity at which particles settle through a liquid under gravity to their diameter and the viscosity of the liquid.
2. A known weight of soil is thoroughly mixed with water to form a uniform suspension and then is allowed to stand so that particles settle out. The largest particles settle out first and the smallest settle out last.
3. The weight of sediment in the suspension at any time is directly related to the density of the suspension, so that the concentration of sediment present at a particular time can be determined simply by measuring the density (or the specific gravity) of the suspension.
4. The specific gravity of the suspension can be determined in two ways—either by using a hydrometer or by using a pipette and extracting a small sample of the suspension.
5. From Stokes's law and the specific gravity of the suspension, the proportion of particles smaller than a specific size can be determined each time a measurement is made. This enables a particle size curve of the type shown in Figure 3.2 to be drawn for fine-grained soils.

*There are a number of approximations involved in this procedure. In particular, clay-sized particles tend to be plate shaped rather than spherical, so their rates of settlement will not be strictly in accordance with the assumptions made. However, it is usually found that for mixed-grained soils containing both coarse and fine particles the particle size curve for the coarse-grained fraction obtained by sieving connects fairly well with that for the fine-grained fraction obtained by sedimentation.

3.3 PLASTICITY AND ATTERBERG LIMITS

The distinctive property of clay is plasticity, and hence some form of test that measures or evaluates this property is a useful indicator of the nature of the clay. The tests devised for this purpose in the formative years of soil mechanics are the **Atterberg limits**, and they have become firmly established as very valuable indicators of the properties of a clay or silt. They are a measure of the water content range over which the soil is plastic. The way clay passes through stages of plasticity is illustrated in Figure 3.3.

If the clay is initially fairly dry, it is likely to be hard and have the properties of a solid. If its water content is increased, it will start to soften and enter a stage termed semisolid. With continuing increase in water content it will further soften and eventually become truly plastic. In its plastic state clay can be deformed or molded without cracking or breaking and without undergoing volume change. With further increases in water content the soil will eventually become soft enough to be more liquid than solid. The water contents that mark the boundaries between these stages are termed the **shrinkage limit**, the **plastic limit**, and the **liquid limit**, as indicated in Figure 3.3. By far the most important of these are the plastic limit and the liquid limit, which are known as the Atterberg limits after the man who devised them.

3.3.1 Determination of Atterberg Limits

The liquid limit (LL) is determined using a rather unusual device, illustrated in Figure 3.4. It consists of a shallow cup that can be raised and dropped by turning a handle.

The test procedure is as follows:

1. The soil is thoroughly mixed with water to form a paste with a consistency judged to be a little dryer than the liquid limit.

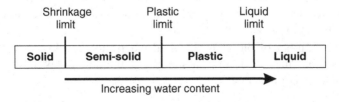

Figure 3.3 Solid, plastic, and liquid stages of a soil and the boundaries between them.

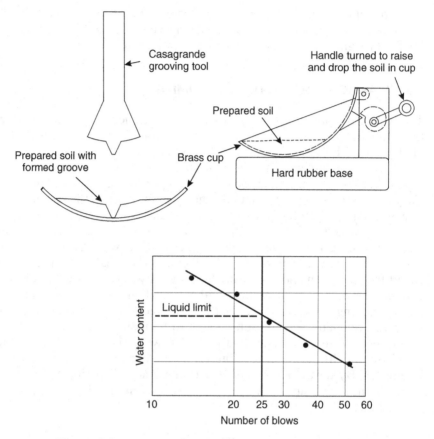

Figure 3.4   Liquid limit apparatus and plot of test results.

2. A portion of this soil is then placed in the cup of the apparatus and leveled with a spatula.

3. A groove is formed in the soil using a tool of special shape and dimensions as shown. This divides the soil into two halves separated by a gap of 2 mm.

4. The handle of the device is then turned; each rotation raises the cup by 1 cm and drops it. The effect of repeated drops (or blows) is to make the two halves of the soil flow toward each other and gradually close the gap separating them.

5. The number of blows is recorded when the gap closes over the specified distance of 13 mm, a sample of the soil is taken, and its water content is measured.

The LL is defined as the water content at which the gap closes over a distance of 13 mm with 25 blows. It is not practical to adjust the water content to give this exact result, so a series of tests is carried out over a range of water contents on each side of the LL. Most standards require that there be at least two points both above and below the LL. A graph of number of blows versus water content is then plotted and the water content at 25 blows read from it, as indicated in Figure 3.4. The LL is this water content in percent rounded to the nearest whole number.

The plastic limit (PL) is defined as the water content at which a thread of the soil can be rolled to a diameter of 3 mm but no smaller. Rolling to a smaller diameter causes the thread to crumble because of lack or plasticity. At higher water contents the thread can be rolled to a smaller diameter, and at lower water contents it will crumble before reaching 3 mm.

The procedure for the test is as follows:

1. The soil is prepared with a water content judged to be slightly wetter than the plastic limit.
2. A small portion of this soil is taken and rolled on a glass plate using hand or finger pressure to form a thread with a diameter as small as possible.
3. If the thread becomes smaller than 3 mm in diameter, the soil is wetter than the PL.
4. The thread is gathered up and reformed into a ball and rolled again into a thread. Each time it is rolled it will lose moisture, so that eventually it will crumble at 3 mm. When this happens, the thread is placed in a moisture content tin and covered to prevent evaporation.
5. This procedure is continued until a reasonable weight of threads in the tin is reached.
6. The moisture content of the threads is determined. When rounded to the nearest whole number, this is the PL.

The following points should be noted with respect to the Atterberg limits

(a) Standards for Atterberg limit tests normally require that the test be carried out only on material finer than 0.425 mm (425 μm). This does not affect most clays, as particles of this size are not usually present. It is usually possible by visual inspection to determine whether significant material of this size is present; if only a very small percentage is present, it may be acceptable to ignore it, since the work involved in sieving the soil to remove the coarse material is considerable.

(b) In general it is desirable to carry out the Atterberg limit tests without any predrying of the soil. This is because predrying can alter the properties of some soils. In particular soils containing the clay minerals halloysite and/or allophane are likely to undergo irreversible changes when air dried or oven dried. Clays containing a high proportion of allophane can change from moderately plastic clays to nonplastic silty sands when oven dried.

A third parameter is defined from the LL and PL, namely the plasticity index (PI). This is the water content range over which the soil is plastic, that is, PI $= $ LL $-$ PL. Based on the Atterberg limits, Casagrande (1948) developed a chart known as the plasticity chart, which is shown in Figure 3.5. The plasticity chart is intended as a means of dividing or classifying fine-grained soils into groups expected to have similar engineering properties. The PI is plotted against the LL, as seen in Figure 3.5, so a soil can be represented by a single point on this chart. There is a dividing line separating soils with clay characteristics from those with silt characteristics; this line is known as the A-line. Casagrande established the A-line by conducting Atterberg limit tests on a wide range of soils at the same time examining their engineering characteristics. This enabled him to establish the line separating the two fine-grained soils—silt and clay. On the basis of this chart Casagrande developed a classification system for fine-grained soils, which we will describe in detail in Section 3.5. In addition to a division based on the A-line, Casagrande also used a vertical division at LL = 50 to divide clays and silts into two further groups—those of high LL and those of low LL, as indicated in Figure 3.5.

Regardless of whether or not it is used for systematic classification of soils, the plasticity chart is a very useful indicator of the likely properties of fine-grained soils. The Atterberg limit values by themselves are not particularly reliable as indicators of soil behavior but become very useful when plotted on the plasticity chart. Soils that plot well above the A-line generally have poor engineering properties. They are likely to be of high compressibility and low shear strength and display shrink/swell behavior.

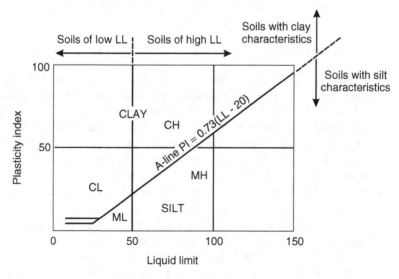

Figure 3.5 Plasticity chart.

On the other hand, soils that plot below the A-line tend to have the opposite characteristics and be good engineering materials.

Along with this general trend will be another trend dependent on the LL. For a given position in relation to the A-line, the higher the LL the less desirable will be the engineering characteristics. These are general trends followed by most soils, though there are exceptions. The position the soil occupies on the plasticity chart can be regarded as a good indicator of the intrinsic properties of the soil.

3.4   LIQUIDITY INDEX OF CLAY AND RELATIVE DENSITY OF SAND

In addition to indicating the intrinsic nature of the soil, the Atterberg limits, when compared with the natural water content of the soil, give a valuable indication of the natural state of the soil in the ground. The parameter used for this purpose is the **liquidity index** (LI), which expresses the water content of the soil in relation to the PL and LL. It is defined by the following relationship and illustrated in Figure 3.6:

$$LI = \frac{w - PL}{LL - PL} = \frac{w - PL}{PI}$$

where w = natural water content of the soil.

A similar index is used with sand, known as the relative density (D_r) or density index (I_D). This relates its in situ density to two reference density states and is defined as follows:

$$D_r (= I_D) = \frac{e_{max} - e}{e_{max} - e_{min}}$$

where e_{max} and e_{min} are respectively the void ratios of the sand in its loosest and densest state and e is the natural void ratio of the sand.

This index is also illustrated in Figure 3.6. Relative density values of zero or unity (100 percent) therefore indicate very loose and very dense states, respectively. Standard tests have been devised to measure these density states. The densest state is determined by a vibration process, while the loosest state is determined by simply pouring the sand into a container using a technique that will produce its loosest state.

While the liquidity index (LI) and the relative density (D_r) index are serving similar purposes, they have been defined in different ways. The LI of clay is measured from the compact (solid) state toward the noncompact (liquid) state, but the D_r of sand is measured from the noncompact (loose) state toward the compact (dense) state.

The term liquidity index is somewhat misleading and needs further explanation. As an index of liquidity it is only valid for soils that are fully remolded. At their PL (LI = 0) remolded soils are "solid" while at their LL (LI = 1) they are "liquid." With undistorted soils this is not the case.

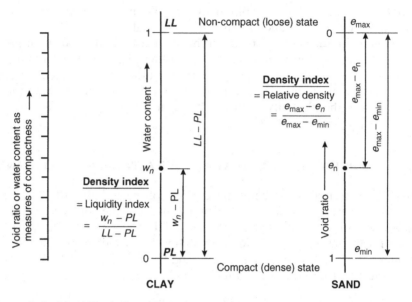

Figure 3.6 Liquidity index of clay and relative density of sand (compactness indexes).

It is not unusual to encounter natural (undisturbed) clays having values of LI greater than unity but which are still firm or hard materials. This is particularly likely with residual soils, where the weathering process often results in a relatively strong, porous structure containing a large proportion of entrapped water. When remolded, the structure collapses, the water is released, and the material becomes very soft or essentially liquid.

For undisturbed soils it is therefore more appropriate to think of the LI as a measure of the compactness or "denseness" of the soil rather than its liquidity. At their PL, soils are in a relatively compact state, whereas at their LL, they are in an "open," noncompact state. Despite this ambiguity in significance, the LI it is still a very valuable parameter for undisturbed soils. A firm to stiff clay with a LI equal to or greater than 1 is highly likely to suffer severe loss of strength when disturbed or remolded in any way. Excavation and compaction of such soils will destroy their structure and severely weaken them. This loss of strength on remolding is termed sensitivity and is described in the next section.

3.5 SENSITIVITY, THIXOTROPY, AND ACTIVITY OF CLAYS

The **sensitivity** (S_t) of a clay is defined as follows:

$$S_t = \frac{\text{Undisturbed (undrained) shear strength}}{\text{Remolded (undrained) shear strength, at the same water content}}$$

We have not yet described the shear strength of soils; at this stage it is sufficient to accept that the undrained shear strength is a simple measure of the shear strength of the soil.

Most clays have a sensitivity between 1 and about 4, but values much higher than this are not uncommon. A soil may have an undrained shear strength of 100 kPa, which would be a reasonable value for many soils but may also have a LI of 1. This means that when remolded its undrained shear strength is likely to be only about 2 kPa. It would thus have a sensitivity of 50. Some clays, especially those found in parts of Norway, have sensitivities greater than 100 and are called **quick clays**.

Thixotropy is the term used to describe a tendency in some clays to regain strength after they have lost strength due to remolding. It is rare for a soil to regain all of its original strength and with many clays very little regain in strength occurs. Thixotropy is not an important characteristic of clay because it is of little practical significance.

Another property of clay that is of some significance is termed **activity**, and is defined as follows.

$$\text{Activity} = \frac{\text{Plasticity index}}{\text{Clay fraction}}$$

This is a measure of the plasticity of the clay-sized particles. Activity values less than 0.75 are considered low, values between 0.75 and 1.25 are normal, and values greater than 1.25 indicate clays of high activity. There is a fairly close correlation between clay mineral type and activity. Kaolinite, halloysite, and allophane are of low activity; illite is of medium, or normal, activity; and montmorillonite (or smectite) is of high activity.

3.6 SYSTEMATIC CLASSIFICATION SYSTEMS

Geotechnical engineers, in order to undertake design work, gather together as much information as possible on the nature and properties of the soils they are dealing with. This information comes from a variety of sources. Some comes from site investigation work, especially borehole drilling and the log (record) made of the materials encountered during drilling, some from descriptions made by laboratory staff when opening and examining samples in the laboratory, and some from the results of laboratory tests. It is highly desirable that everyone involved in this process uses consistent terminology in describing soil characteristics or properties. Only in this way can information be reliably conveyed from one person to another. For this purpose, systematic classification and description systems have been devised.

The terms classification and description are used somewhat loosely in soil mechanics. For convenience and consistency they are best given the following meanings:

Classification refers to the identification of the soil itself, that is, what its composition and intrinsic properties are.

Description refers to the state in which the soil exists in the ground, that is, whether it is hard or soft, loose or dense, and any other properties reflecting its undisturbed state in the ground.

Classification systems used in soil mechanics, such as the Unified Soil Classification System, refer primarily to the material itself; they make only passing reference to describing the state in which the material exists in the ground.

Descriptive systems, which are used in logging borehole cores, investigation pits, or exposures in cuttings, are systems that enable accurate accounts to be given of the state of the material in situ.

It seems logical to keep these two aspects distinct when undertaking soil classification and/or description. In some situations, for example, when borrow sources are being investigated or when only disturbed samples are available, classification of the material itself is possible (and relevant), but description of its undisturbed state is not possible. We will consider soil classification first.

3.6.1 Unified Soil Classification System

The classification system in most common use around the world today is the **Unified Soil Classification System (USCS)**. This was developed by Casagrande during World War II, primarily for use in the evaluation of soils for airfield construction (Casagrande, 1948). The almost universal adoption of the USCS, with or without minor modifications to suit local conditions, is not surprising. It is based on sound principles and is very much superior to any other systems in use. The system provides for the use of group symbols (CH, GW, etc.) to classify or name soils but can be used equally well without the group symbols. The use of a rigid system of symbols tends to create rather narrow, artificial limits to the classification process. The process and basis of classification according to the USCS is illustrated in Figure 3.7. Classification can be done either visually or on the basis of laboratory tests.

The first step is to place soil into either the coarse- or the fine-grained category. Soils with more than 50 percent smaller than 0.06 mm belong in the **fine-grained group**, while those with less than 50 percent smaller than 0.06 mm belong in the **coarse-grained group**. As well as being the smallest size that can be measured using a sieve, 0.06 mm is also the smallest size that can be seen with the naked eye. Further subdivisions are then made within each group, but the basis of these subdivisions is different, depending on whether the soil belongs in the coarse- or fine-grained group. Since the properties of coarse-grained soils are governed primarily by their particle size, this is retained as the sole basis for further subdivision into sand

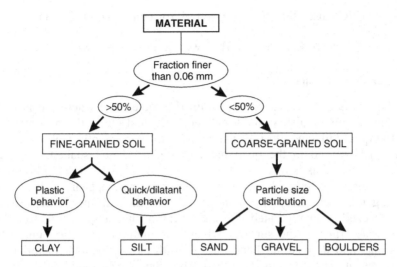

Figure 3.7   Classification of soils according to the USCS.

or gravel. With fine-grained soils, the division into silt or clay is made on the basis of either behavioral characteristics or Atterberg limits. These reflect both particle composition and particle size. Details of the further subdivisions are as follows.

Coarse-Grained Soils

Coarse-grained soils are subdivided into sand, gravel, or boulders according to their particle size distribution and the boundaries given in Figure 3.1. An estimate is made of the relative proportions of its principal constituents and an appropriate name given to it; for example:

SANDY GRAVEL—a material consisting mainly of gravel but containing a significant amount of sand.

In addition to the name given to the soil, further descriptive information should be given. This should generally cover the following:

(a) Maximum particle size
(b) Grading, that is, whether material is well graded, poorly graded, or uniform
(c) Shape of grains—angular, rounded, elongated, etc.
(d) Hardness of grains
(e) Fines content and whether the fines are silty or clayey
(f) Color
(g) Geological origin and dominant mineral or rock type (if relevant or useful)

The USCS uses the following letters for coarse-grained soils:

S—sand, **G**—gravel, **W**—well graded, **P**—poorly graded

Fine-Grained Soils

If the soil is fine grained, it is examined to determine whether it is a silt or a clay. To distinguish visually between silts and clays, the best test to use is the quick/dilatancy test. A pat of soft soil (sufficiently wet to be almost sticky) is placed in the open palm of the hand and shaken, or vibrated horizontally. This is most effectively done by tapping the hand holding the soil against the other hand. With silt, water will appear on the surface, giving it a wet, shiny appearance, and will then disappear if the sample is squeezed or manipulated. During vibration the sample tends to collapse and water runs to the surface; when it is manipulated, the sample tends to dilate and draw water back into it. With clays, these characteristics are not present. The division into silt or clay can also be made on the basis of Atterberg limits, as illustrated in Figure 3.5.

An appropriate name is then given to the soil, for example, as follows:

Sandy clay—a soil which is predominantly clay but which contains significant sand

Silty clay—a soil which behaves primarily as clay but also displays some tendency toward silt behavior (It is debatable whether the use of this terminology is strictly in accordance with the USCS, but it is much more sensible than trying to force an intermediate material into categories to which it does not belong.)

In addition to the name of the soil, other descriptive information should be given, generally covering the following:

(a) Plasticity—low, medium, or high (see note below).
(b) Presence of coarse material, that is, sand or gravel content
(c) Color
(d) Geological origin (if useful or relevant)

The USCS uses the following letters for fine-grained soils:

C—clay, **M**—silt, **H**—high LL, **L**—low LL

3.6.2 Additional Notes Regarding Classification

1. Field and Laboratory Classification: Classification can be based on either visual examination or laboratory tests. In practice most classification is done visually in the field, but there may be occasions when reference to laboratory tests is possible and beneficial. With coarse-grained soils, the grading curve is used, and with fine-grained soils the plasticity chart is used.

2. Plasticity: As implied above, the most important property of a clay or silt is its plasticity, so that in describing or classifying a fine-grained soil some indication of its plasticity is desirable. There is some confusion in soil mechanics as to what is meant by a highly plastic soil. The following is a reasonable statement on plasticity as used in soil mechanics:

> A *highly plastic soil is one that can be molded or deformed over a wide moisture content range without cracking or showing any tendency to volume change. It also shows no trace of quick or dilatant behavior.*

Hence, to evaluate plasticity in the field, it is necessary to remold the soil over a range of moisture contents and manipulate it to check whether it behaves in a plastic manner. The dry strength of the material is also a good guide to plasticity. Highly plastic clays will become rock hard when dry, while those of low plasticity can be crumbled in the fingers.

With respect to the use of Atterberg limits for indicating plasticity, it should be recognized that neither the LL nor the PL is very useful on its own. In general, the PI is a better indication of plasticity than the LL, but ideally the PI and LL need to be examined together by plotting them on the plasticity chart. A soil with a high LL (above 50) which plots above the A-line is a highly plastic material, but a soil with the same LL which plots well below the A-line is unlikely to be a highly plastic material.

There is a tendency to think the H and L letters used by the USCS mean high plasticity and low plasticity. The British version of the USCS assumes this. This usage is not strictly correct, unless the soil has already been identified as a clay. In the case of silts the L and the H should simply be taken to refer to low LL and high LL, respectively, since silts by definition can never be of "high plasticity."

3. Grading: Gravels and sands should be described as well graded (i.e., a good representation of all particle sizes from largest to smallest) or poorly graded (the opposite) according to their particle size distribution curves. Poorly graded materials may be further divided into uniformly graded (i.e., most particles about the same size) and gap graded (i.e., absence of one or more intermediate sizes).

4. Peat and Organic Soils: The term peat is used to describe soils that consist of nearly 100 percent organic matter. Such a material can generally be identified by its appearance; it will be seen to consist of fibrous matter, that is, decomposed leaves and branches. Soils which do not consist of nearly 100 percent organic matter but which contain a high organic content should be described as organic clays (or highly organic clays) rather than peat. When the organic material is in a fibrous state, its presence is easily detected, but when it is amorphous, its presence may be more difficult to establish. Dark color is usually an indication of the presence of organic matter, but it is not a totally sure guide. The organic content should be described using terms such as *slightly*, *moderately*, and *highly*, with the terms *fibrous* and *amorphous* used to indicate its nature.

5. Color: Color is not normally an important characteristic, since there is no systematic relationship between color and engineering properties. However, color may be of some significance, especially for indicating changes of material, although this may not necessarily be the case.

6. Inconsistencies: Some soils do not fit easily into classification systems and should not be forced to. Soils derived from volcanic ash, especially those containing allophane, are examples. These soils normally plot well below the A-line on the plasticity chart, but their behavior is not really that of silt. They do not normally display the quick or dilatant behavior associated with silt, although neither do they appear to be have the plasticity of normal clays. They are probably best called clayey silt (or a silty clay).

3.6.3 Description of In situ (Undisturbed) Characteristics of Soil

The procedures outlined above refer only to the material itself; they do not include information on the state in which it exists in the ground. This section outlines the information and procedures for covering this aspect.

Coarse-Grained Soils

The most important property describing the in situ state of coarse-grained soils is their relative density, so this should be the main focus of interest in describing the state of the material in the ground. The **relative density** refers to the "denseness," or degree of compactness, of the material in the ground, as indicated earlier in Figure 3.6. The terms **loose, medium dense, dense, and so on**, are used to describe this property. Table 3.1 is a guide relating descriptive terms to SPT or CPT values. (The meaning of these terms is described later in chapter 10 of this book).

Table 3.1 Density Index (of Relative Density) Correlations

Descriptive Term	Density Index (D_r)	SPT "N" Value (blows/300 mm)	CPT Value (MPa)
Very dense	>85	50	>20
Dense	65–85	30–50	12–20
Medium dense	35–65	10–30	4–12
Loose	15–35	4–10	1.6–4
Very loose	<15	<4	–1.6

For visual classification, a simple field assessment can be made using the terms loosely packed and tightly packed, based on the following guide:

Loosely packed—can be removed from exposures by hand or removed easily by shovel.

Tightly packed—requires pick for removal, either as lumps or as disaggregated material.

Fine-Grained Soils

The most important property describing the in situ state of a fine-grained soil is its **strength**, or "consistency." Table 3.2 is a guide to the terms used to designate this together with indicative properties.

It is very useful also to give an indication of the sensitivity of fine-grained soils, that is, the loss of strength which occurs when the soil is disturbed or remolded. For all soil types it is useful also to record any significant structural features such as the presence or absence of bedding planes, faults, fissures, or discontinuities.

The complete description of an undisturbed soil is probably best given in the following order, although alternative orders are acceptable:

1. The **soil name**, made up of the **primary constituent** with the **secondary constituent** as a qualifying adjective.
2. Any other information on **composition**, including:
   - Presence of a third constituent or extraneous matter
   - For fine-grained soils, an indication of plasticity, or dilatancy
   - For coarse-grained soils, an indication of grading, that is, well graded or poorly graded, maximum particle size, angularity or particles, and so on

Table 3.2 Guidelines for Undrained Shear Strength of Cohesive Soils

Undrained Shear Strength (kPa)	Descriptive Term	Diagnostic Features
<12	Very soft	Easily exudes between fingers when squeezed
12−25	Soft	Exudes with difficulty between fingers when squeezed
25−50	Firm	Penetrated by thumb with moderate effort
50−100	Stiff	Can be indented by thumb pressure
100−200	Very stiff	Can be indented by thumb nail
200−500	Hard	Difficult to indent by thumb nail

3. Information on **in situ state**, covering:
   For coarse-grained soils: **relative density (denseness)**
   For fine-grained soils; **stiffness (consistency) and sensitivity**
4. Any **in situ structural features**, that is, bedding, fissures, discontinuities, and so on
5. Color
6. Any other information such as geological origin

Examples

Sand: fine to medium sand only, fairly uniform but with occasional shells, medium dense, dark grey

Clay: high plasticity, homogeneous, firm to stiff, dark grey

Silt, clayey: containing traces of peat, low plasticity, moderately sensitive, firm, light grey

Silty sand: gravelly, sand sizes from coarse to fine, about 20 percent hard angular gravel particles 15 mm maximum size, fines are low plasticity, dense, grey (alluvial)

As indicated earlier, classification should be thought of as two distinct steps. The first is to identify the nature or composition of the material itself, and the second is to describe the state in which it exists in the ground. The descriptive sequence suggested above reflects this process and endeavors to place the most important items first and the least important at the end. However, the sequence is not of paramount importance and other sequences are acceptable. In some cases, for example, when bulk samples are taken from borrow pits, only the first step, that is, classification, may be involved.

3.7 CLASSIFICATION OF RESIDUAL SOILS

Various attempts have been made over the years to develop classification systems specific to residual soils. This has been done because of the particular properties of residual soils and the supposed inadequacy of existing systems to take account of these properties. None of these systems has been very successful, and none has found wide acceptance. This is not surprising, as the nature and properties of residual soils vary so widely that it is unrealistic to think that a single system could adequately cover all types. Rather than focusing on or trying to implement these systems, it is more useful to have a general understanding of residual soil properties and the factors to take account of in their evaluation or "classification." The following brief comments may be helpful in this respect.

3.7.1 Parent Rock

It is always useful to know the parent rock of any residual soil, although there is not necessarily a consistent link between parent rock and the properties of soil derived from it. The environment in which the weathering takes place can significantly influence the resulting soil.

3.7.2 Usefulness of Existing Systems

The central feature of the USCS for fine-grained soils, namely the plasticity chart, is still a very good indicator of the intrinsic properties of fine-grained residual soils. Soils that plot well below the A-line generally have good engineering properties and those that plot well above it have poor engineering properties. Figure 3.8 shows the position occupied on the plasticity chart by three well-known residual soil groups. The first of these is a group of soils originally identified in soil mechanics literature as "black cotton" soils (found in parts of India) but also called simply black clays, or "vertisols," a soil science term. These soils are rich in the clay mineral montmorillonite and are the source of many swelling and shrinkage problems affecting building foundations.

The second group is comprised of tropical red clays, normally made up of a large proportion of the clay mineral halloysite. These are often ideal clays for construction purposes, being of high strength, low compressibility, and low shrink/swell potential. They are also usually very suitable for earth works as their natural water content tends to be close to their PL (although this cannot be deduced from their position on the plasticity chart).

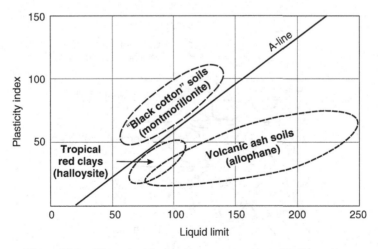

Figure 3.8 Three residual soil groups on the plasticity chart.

The third group is made up of volcanic ash soils, which normally contain a large proportion of the clay mineral allophane (in conjunction with immogolite). These are very unusual soils, as they generally have good engineering properties despite extremely high natural water contents and Atterberg limits. As mentioned earlier, despite plotting well below the A-line, they do not clearly exhibit the properties of silt.

Thus, plotting the Atterberg limits of residual soils on the plasticity chart is just as useful as it is with sedimentary soils but should be used primarily as a guide to likely engineering properties rather than as a means of rigid classification of the soil.

3.7.3   Classification of Weathering Profile

Little (1969) proposed a system for classifying the degree of weathering of a soil/rock profile which has rightly found wide use in practice. It is illustrated in Figure 3.9. The profile is made up of six categories, ranging from fresh rock to a true soil. Little stated that the classification system he proposed was intended to cover the residue resulting from the weathering of igneous rocks in the humid tropics. It was in no way intended to be a classification system for residual soils other than for this purpose. Its limitation to igneous rocks in the humid tropics is important and should be recognized. The weathering of basic rocks such as basalt or andesite is

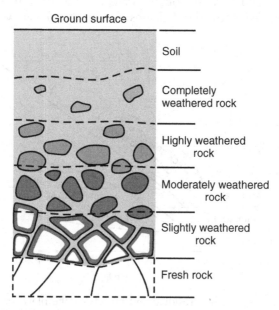

Figure 3.9   Classification of the weathering profile (adapted from Little, 1969).

significantly different in that the transition from rock to soil occurs in a very narrow zone which cannot be divided up into six categories, as is the case with weathered granites or other acidic rocks.

The weathering of sedimentary rocks, such as sandstones or mudstones, is also quite different to that of igneous rocks. The sequence of layers produced by the weathering of these materials is more likely to reflect the properties of each individual layer of the parent material than the degree of weathering. Little's system is therefore very useful for the purpose for which it was intended, but it does not provide any comparative information on the nature of the top horizon, that is the true soil layer, which is often of greatest interest to the engineer. The top horizon will be very different depending on the parent rock and the weathering environment.

3.7.4   Importance of Mineralogy and Structure

Two properties that tend to give residual soils their distinctive characteristics are their mineralogical content and their structure. Despite what has been said in the introduction to this section about the impossibility of devising an adequate classification system for residual soils, the author made an attempt at a "classification" system based on these two properties (Wesley, 1988; Wesley and Irfan, 1997). However, this system is not intended to be a rigid classification system; rather it is a means of grouping together residual soil types that can be expected to have similar engineering properties. It is in no way intended to displace the USCS and should be used in parallel with that system. The following is a brief summary of its main features.

(a) Three main groups are created on the basis of mineralogical content as follows:

Group A: Soils without a strong mineralogical influence.

Group B: Soils strongly influenced by common clay minerals, the principal group being those containing montmorillonite.

Group C: Soils strongly influenced by clay minerals essentially found only in residual soils. Subgroups within this group are (1) allophane, (2) halloysite, and (3) sesquioxides (laterite)

(b) Further subdivisions of these groups are made on the basis of structure, which is considered in two categories as follows:

Macrostructure or discernible structure: This includes all features discernible to the naked eye, such as layering, planes of weakness, fissures, pores, presence of unweathered or partially weathered rock, and relic structures.

Microstructure or nondiscernible structure: This includes interparticle bonding or cementation, clustering of particles, and so on.

REFERENCES

Casagrande, A. 1948. Classification and identification of soils. *Trans. ASCE,* Vol. 113, pp. 901–930.

Little, A.L. 1969. The engineering classification of residual tropical soils. *Proceedings, Specialty Session on Engineering Properties of Lateritic Soils.* Seventh International Conference on Soil Mechanics and Foundation Engineering, Mexico City, pp. 1–10.

Wesley, L. D. 1988. Engineering classification of residual soils. In *Proceedings 2nd International Conference Geomechanics in Tropical Soils*, Singapore, pp. 73–84.

Wesley, L. D. and T. Y. Irfan. 1997. The classification of residual soils. In G. E. Blight (ed.), *Mechanics of Residual Soils*. Rotterdam: A. A. Balkema.

CHAPTER 4

STRESS AND PORE PRESSURE STATE IN THE GROUND

4.1 VERTICAL STRESS IN THE GROUND

Consider a site where the ground is level. If we drill a hole in the ground, we will normally find that at some point the soil will become wetter and water will start to flow into the hole. If we leave the hole for some time, the water level will rise and then become stable. The depth at which this occurs is known as the groundwater level or, more commonly, the **water table**. This is illustrated in Figure 4.1.

We can determine the stress state on the soil element shown at depth D. We will assume that the unit weight of the soil is γ and is the same above and below the water table. This will normally be the case if the soil is clay in a temperate climate but may not be the case in a dry climate or if the soil is a coarse silt or sand.

The vertical stress on the element is given by $\sigma_v = \gamma D$, where σ_v is the total vertical stress on the soil element. It is called the **total stress** because it is the stress resulting from the total weight of material above it.

We can also determine a second stress acting at depth D. This is the pressure in the water contained in the void space (or pore space) of the soil. This pressure is called the pore water pressure or just the pore pressure. This void space is totally interconnected so that in this situation, with a level water table, the pore pressure is hydrostatic and is given by:

$$u = \gamma_w (D - H_w) \text{ where } u \text{ is the pore pressure and}$$
$$\gamma_w \text{ is the unit weight of water.}$$

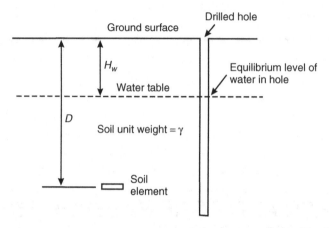

Figure 4.1   Water table and stress state in the ground.

The difference between these two stresses is called the **effective stress**, that is,

$$\sigma' = \sigma_v - u = \gamma D - \gamma_w(D - H_w)$$

where σ' is the effective vertical stress.

This relationship between total stress, pore pressure, and effective stress applies generally within the soil mass, not just in the vertical direction, and is normally written simply as

$$\sigma' = \sigma - u \tag{4.1}$$

Equation 4.1 is undoubtedly the most important equation in soil mechanics, as it expresses a concept known as the **principle of effective stress**, which will be covered more fully in Chapter 6. This principle states that soil behavior is governed by effective stress and not by total stress. Deformation, compression, or changes in strength of the soil only occur as a result of changes in effective stress, not changes in total stress, although changes in the latter may well cause changes in the former. To predict how soil will behave in any particular situation, it is essential to consider how the effective stresses are changing. This will become clear in later chapters.

4.2   PORE PRESSURES ABOVE WATER TABLE AND SEASONAL VARIATIONS

Although it might be thought that the pore pressure above the water table will be zero, this is not normally the case. The pore spaces between the soil particles act as fine tubes and water is drawn into this space, or retained in it, by capillarity or surface tension forces. In a fine-grained soil, made up entirely of clay-sized particles (smaller than 0.002 mm), the effective pore size will be about 20 percent of this, which is 0.0004 mm. The theoretical capillary rise in such a material would be about 75 m. With most fine-grained soils water will thus not drain out of their void space under gravity forces. It is only when the material is of fine sand size that water will begin to drain out of the pore space from gravity forces alone. Even in fine sand, drainage will be limited, and considerable water will still remain in the void space. Only in course sands and gravels will water drain almost completely from the pore space.

Measurements in the field have shown that many clays remain fully saturated for many meters or tens of meters above the water table, and only in the top 1 or 2 m is the soil less than fully saturated. This partial saturation occurs not because of gravity drainage downward, but because of water loss by evaporation at the surface. The pore pressure and saturation state in relation to the water table are illustrated in Figure 4.2. The pore pressure shown is the hydrostatic or equilibrium state, the pore pressure being negative above the water table and positive below it. No seepage flow occurs in this state.

The depth of the partially saturated zone is governed by the particle size of the soil and the climatic conditions. In areas with temperate or wet climates, the depth of the partially saturated zone in clay is unlikely to be more than 1 or 2 m below the surface. However, for dry climates this depth could be meters or tens of meters.

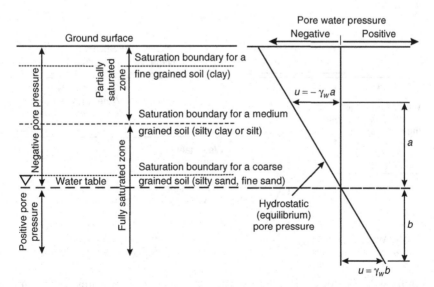

Figure 4.2 Pore pressure and saturation states in relation to the water table.

The pore pressure state above the water table tends not to receive much attention in soil mechanics, possibly because it is not of great importance with sedimentary soils. However, with residual soils, the pore pressure state above the water table is of considerable importance for two reasons. First, the water table is often deep and the zone of prime interest to geotechnical engineers is above the water table. Second, the high permeability of many residual soils means that seasonal changes in pore pressure can be very significant and govern the behavior of the soil. For these reasons, the pore pressure state above the water table, and the way it is influenced by climate seasons, is described in some detail in the following sections.

4.2.1   Case A: Coarse-Grained Soils

Case A is the simplest case and is illustrated in Figure 4.3. Sand (or gravel) acts like a "reservoir" into which water flows under gravity during wet weather, causing the water level to rise and from which water is lost by evaporation during dry weather with an accompanying drop in the water table. Below the water table the soil is fully saturated and the pore pressure state is hydrostatic. Above the water table the soil has a very low (almost zero) degree of saturation, and the pore pressure is essentially zero. Seasonal changes extend at least to the bottom of the sand layer. In many situations, the water table in the sand may be governed by external controls, such as the proximity of a river, rather than by the influence of weather at the surface. We should note in passing that with coarse-grained soils the soil properties governing the way the water table changes with seasons are the permeability and the porosity (storage capacity) of the soil. These are described in Chapter 7.

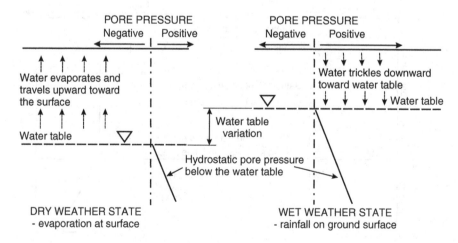

Figure 4.3   Seasonal variation in water table and pore pressures: coarse-grained soil.

4.2.2 Case B: Low-Permeability Clays

Low-permeability clays are at the opposite end of the spectrum to coarse-grained soils. Because of their low permeability the seasonal influence is not deep and may not reach the water table, as illustrated in Figure 4.4.

During dry weather, water is lost by evaporation at the surface, causing the pore pressure to become more negative and water to flow toward the surface. The ground shrinks and a zone close to the surface may become partially saturated. During wet weather, the pore pressure is zero (or very slightly positive) at the surface, and water will enter the soil and seep downward. The soil will gradually absorb water and swelling will occur. The equilibrium state shown in Figure 4.4 is a transient state that can be expected to occur briefly during the year when the seasons are changing. Most heavily overconsolidated sedimentary clays, such as London clay, belong in case B.

4.2.3 Case C: Medium- to High-Permeability Clays

Case C is intermediate between the above two and is illustrated in Figure 4.5. The seasonal influence now extends beyond the water table, which rises in the summer and falls in the winter. The pore pressure is

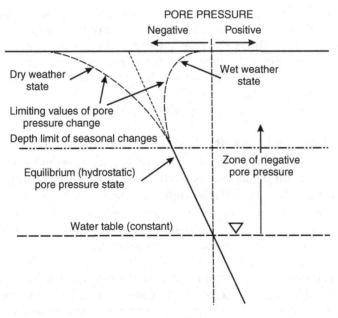

Figure 4.4 Seasonal variation in water table and pore pressures: low-permeability clay.

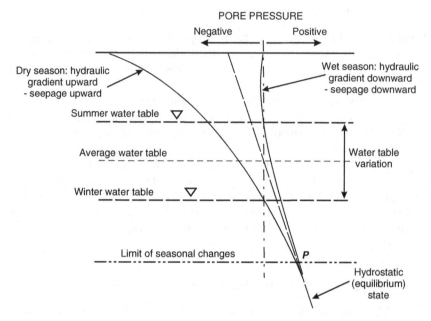

Figure 4.5 Seasonal variation in water table and pore pressure: moderate- to high-permeability clay.

hydrostatic with respect to the average position of the water table, which will occur from time to time, especially midway between the wet and dry seasons. At other times the pore pressure will not be hydrostatic, either below or above the water table. The term "hydraulic gradient," used in the figure, can be ignored at this stage. It governs the direction and rate of seepage flow and is explained fully in Chapter 7.

We should note the following with respect to cases B and C:

(a) There is a limit to the depth influenced by seasonal effects.
(b) The governing parameters are the permeability and compressibility of the soil or, in their combined form, the coefficient of consolidation. These are described fully in Chapter 8.
(c) The pore pressure is not necessarily hydrostatic below (or above) the water table, except transiently in the "average" situation between the extremes caused by the seasonal influence.

The pattern of behavior illustrated in Figures 4.4 and 4.5 is essentially theoretical, based on the seepage and consolidation principles presented in Chapters 7 and 8. Field evidence confirming this pattern has been obtained from measurements on a number of sites, such as, for example, Kenny and Lau (1984), Urciuoli (1998), and Pun and Urciuoli (2008). Pun and

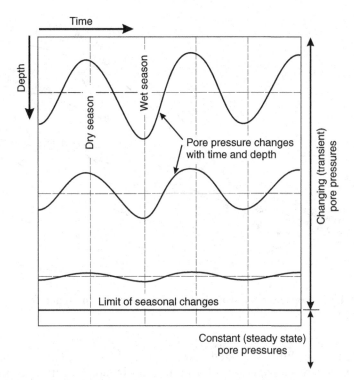

Figure 4.6   Decline of seasonal pore pressure variation with depth. (Adapted from Pun and Urciuoli, 2008)

Urciuoli (2008) installed piezometers at three different depths at a site in Italy and recorded pore pressures over a 10-year period. These show regular seasonal changes which progressively decrease with depth; they are shown conceptually in Figure 4.6.

This is the same behavior as portrayed in Figure 4.5 presented in a different form. We can note that the behavior in Figure 4.6 is not related to any particular water table position. The water table could be below the limit of seasonal changes or could be fluctuating within the zone of seasonal changes.

4.3   HILL SLOPES, SEEPAGE, AND PORE PRESSURES

In the previous section we considered only level sites and assumed no horizontal movement of water. In most natural situations this will not be the case. The ground will be sloping and the water table will also be sloping. This means water will be continually seeping through the ground in the "downhill" direction. Figure 4.7 illustrates the situation for a clay slope

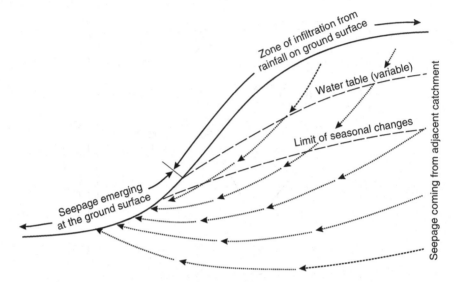

Figure 4.7 Seepage in a hillside coming from direct rainfall and from an adjacent catchment.

assumed to be fully saturated, except possibly for a shallow zone at the surface. Where seepage is occurring in a horizontal direction, especially in man-made structures such as earth dams, the term phreatic surface is often used in place of water table.

Water enters the upper part of the slope from two possible sources. One is surface rainfall, and the other is seepage from an adjacent rainfall catchment. Water emerges again at the lower part of the slope and the valley floor and contributes to the flow in the streams and rivers normally found in valleys. The dotted arrows indicate the likely seepage pattern. Although this is an accurate portrayal of the general shape of the seepage pattern in a hill slope, it is not strictly correct near the ground surface. The slope will be subject to the same seasonal influences described in the previous section for level sites, which will result in continuous changes to the seepage pattern, at least close to the surface. Near the ground surface, the flow will fluctuate between upward and downward, depending on the weather (or the season), in the same way as for level sites. As for level sites, there will be a lower limit to the zone of seasonal changes, as indicated in the figure.

4.4 SIGNIFICANCE OF THE WATER TABLE (OR PHREATIC SURFACE)

The significance of the water table (or phreatic surface) in fine-grained soils should be clearly understood. In particular, the following points should be noted:

(a) The water table is not a boundary line below which the soil is fully saturated and above which it is unsaturated or partially saturated.

(b) It is not a boundary below which seepage is occurring and above which there is no seepage.

(c) It is simply a line of zero (atmospheric) pore pressure, below which pore pressures are positive and above which they are negative.

(d) The water table therefore does not constitute a discontinuity in the seepage pattern. Seepage occurs in a continuous fashion above and below the water table according to the same physical laws (provided the soil remains fully saturated). These laws are described later in Chapter 7.

(e) If the soil consists of coarse material, such as sand or gravel, then the situation is different, as water will drain out of the sand above the water table. In this special case the water table will be the upper limit of the seepage zone.

(f) To understand the seepage situation in any slope, it is not sufficient to measure only the depth of the water table. Measurements of pore pressure need to be made at a number of locations over a period of time to get an accurate picture.

4.5 HORIZONTAL STRESS IN GROUND

There are situations in geotechnical engineering when it is necessary, or at least useful, to know the magnitude of the horizontal stress in the ground. Deep excavations and tunneling are examples of such situations. A number of factors affect the value of the horizontal stress, and there are no simple ways of knowing what the value may be in any particular situation. It is affected by the type of soil, the way the soil has been formed, and any further stresses the soil has been subjected to after formation.

The relationship between the horizontal stress and the vertical stress in a natural soil deposit is usually expressed as a value of K_o, called the **coefficient of earth pressure at rest**:

$$K_o = \sigma_h'/\sigma_v' \tag{4.2}$$

The terms "at rest" or the "K_o state" are used to describe a situation where the soil is restrained in the horizontal direction so that no lateral deformation can occur. This is the case with a sedimentary soil formed by deposition in a marine or lake environment. Because sedimentation occurs over a wide area, no significant horizontal deformation can occur. Vertical compression occurs because of the stress from the increasing thickness of the deposit.

However, the term K_o can also be used somewhat loosely to indicate the horizontal stress in the ground even if this stress is caused by factors such

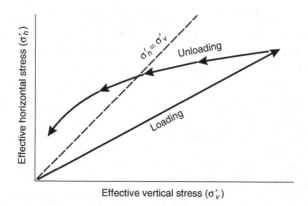

Figure 4.8   Relationship between horizontal and vertical effective stresses during loading and unloading of a soil in a K_o condition (no horizontal deformation).

as tectonic movement, in which case the condition of zero lateral movement no longer applies

The way in which the horizontal stress relates to the vertical stress in this situation is illustrated in Figure 4.8. As the vertical load increases, the horizontal stress also increases, but at a much lower rate. The relationship is close to linear and the value of K_o in this situation is usually between about 0.3 and 0.5. This is in reasonable accordance with expectations from elastic theory, which predicts the following relationship between K_o and Poisson's ratio:

$$K_o = \frac{\upsilon'}{1 - \upsilon'} \text{ where } \upsilon' \text{ is Poisson's ratio} \tag{4.3}$$

Poisson's ratio for soils is normally between about 0.2 and 0.4, giving K_o values from 0.25 to 0.67. Experimental measurements on sand and normally consolidated clay suggest the following relationship between K_o and the angle of shearing resistance (friction angle) ϕ' of the material. An explanation of the angle of shearing resistance is given in Chapter 9:

$$K_o = 1 - \sin \phi', \text{ where } \phi' \text{ is the friction angle} \tag{4.4}$$

In some situations the soil layers may undergo up-lift as a result of tectonic forces followed by erosion. The vertical stress on the soil is reduced and the soil becomes overconsolidated, as described in Chapter 1. During reduction in vertical stress the soil does not behave elastically and the horizontal stress does not decrease in proportion to the decrease in the vertical stress. Initially the decrease in horizontal stress is much less than the decrease in vertical stress, as indicated by the unloading line in Figure 4.8. As unloading continues, the horizontal stress may become equal to the vertical stress and then exceed the vertical stress. Measurements in heavily overconsolidated clays have indicated K_o values as high as 3.

In residual soils the situation is quite different and less predictable, and there is very little information on K_0 values. It is possible to make some comments based on intuition and an understanding of the weathering process. Generally, the weathering process that forms residual soils is accompanied by some loss of material and a reduction in the stiffness and strength of the material. Removal of material tends to relieve any "locked-in" stresses in the soil or its parent material. There may well be horizontal stresses in the parent material caused by formation processes.

Vaughan and Kwan (1984) suggest a theoretical method of analysis to investigate the possible influence of weathering on the horizontal stress. They postulate a decrease in stiffness and strength of the material as weathering progresses and analyze the influence of this on the stress state, making use mainly of elastic theory concepts. In summary, their analysis suggests that the influence of the initial horizontal stress in the parent rock disappears quite early in the weathering process, and the stress state tends towards the "at-rest" K_o value given by simple elastic theory, expressed in Equation 4.3 Because Poisson's ratio is likely to be quite low, especially in weakly cemented material, the value of the horizontal stress is likely to be low.

In contrast to the above, if the principal effect of the weathering process is the removal of cementing material and the release of active clay minerals, the opposite trend may occur. The active clay minerals may take in water and swell, causing an increase in horizontal stress. This may well be the case with some mudstones and shales. Figure 4.9 illustrates the way the parameter K_o can vary with time in sedimentary and residual soils.

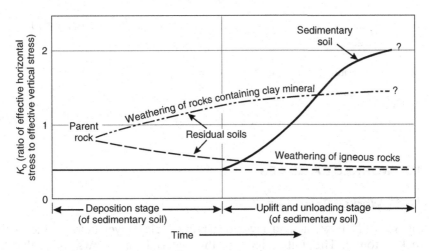

Figure 4.9 Possible changes in K_0 with time in sedimentary and residual soils.

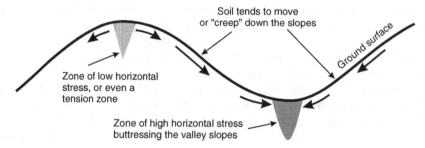

Figure 4.10   Influence of topography on the horizontal stress state in the ground.

There are other factors that influence the magnitude of the horizontal stress in both sedimentary and residual soils and possibly lead to situations where the horizontal stress may exceed the vertical stress. For example, tectonic movement may be forcing one land block against another, inducing very large horizontal stresses. Also, at the base of valleys the natural tendency for soil on the slope to slide toward the bottom means that horizontal stresses are likely to be higher at the base of valleys where the soil is buttressing the soil on the slopes. This is illustrated in Figure 4.10.

Similarly, at the crest of slopes the horizontal stress is likely to be very low, and in some cases a zone of tension may develop where K_o would be zero or even tending to be negative.

Returning to Figure 4.1 and the stress state on the soil element at depth D, we can calculate the horizontal stresses as follows:

$$\text{Total vertical stress } \sigma_v = \gamma D$$

$$u = \gamma_w(D - H_w)$$

$$\sigma'_v = \sigma_v - u = \gamma D - \gamma_w(d - H_w)$$

$$\text{Horizontal effective stress } \sigma'_h = K_o \sigma'_v$$

$$\text{Total horizontal stress } \sigma_h = \sigma'_h + u$$

Thus

$$\sigma_h = K_o \sigma_v' + u \tag{4.5}$$

4.6   WORKED EXAMPLES

4.6.1   Worked Example 1

Figure 4.11 shows a sand layer overlying a clay layer below which solid rock is found. The unit weights of the materials are shown. Note that the unit weight of the sand is less above the water table than below it. This is

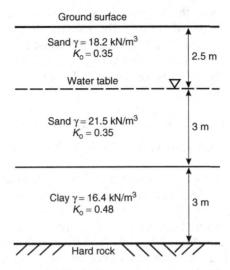

Figure 4.11 Soil conditions for worked example 1.

because in a coarse material such as sand some water drains from the void space above the water table. The sand above the water table is thus partially saturated. Determine the total stress, pore pressure, and effective stress in both the vertical and horizontal direction throughout the soil profile.

The stress changes within the soil are linear within each unit, so we only need to determine the values at each change depth, which in this case are 2.5, 5.5, and 8.5 m.

At 2.5 m:

Vertical total stress $\sigma = 2.5 \times 18.2 = 45.5\,\text{kPa}$

Pore pressure $u = 0$

Vertical effective stress $= 45.5\,\text{kPa}$

Horizontal effective stress $= 0.35 \times 45.4 = 15.9\,\text{kPa}$

Horizontal total stress $= 15.9 + 0 = 15.9\,\text{kPa}$

At 5.5 m:

Vertical total stress $\sigma = 2.5 \times 18.2 + 3 \times 21.5 = 110.0\,\text{kPa}$

Pore pressure $u = 3 \times 9.8 = 29.4\,\text{kPa}$

Vertical effective stress $= 110.0 - 29.4 = 80.6\,\text{kPa}$

Horizontal effective stress within sand $= 0.35 \times 80.6 = 28.2\,\text{kPa}$

Horizontal total stress within sand $= 28.2 + 29.4 = 57.6\,\text{kPa}$

We must now note that the value of K_o is different in the clay from that in the sand, so that there is a discontinuity in the horizontal stress at the boundary between the sand and clay. We must therefore also determine the horizontal stress in the clay as well as in the sand. We thus have two values for horizontal stresses, one marginally above the boundary and one marginally below it:

Horizontal effective stress within clay $= 0.48 \times 80.6 = 38.7\,\text{kPa}$

Horizontal total stress within clay $= 38.7 + 29.4 = 68.1\,\text{kPa}$

We can continue in this manner at 8.5 m and complete the calculation. It is normally convenient to carry out these calculations using a table such as Table 4.1.

The complete picture is presented in Figure 4.12. It should be noted that there cannot be any discontinuities in the vertical stress, whether this is expressed as the total stress or the effective stress, because there cannot be static equilibrium within the soil mass if there are any discontinuities in vertical stress state. The situation with regard to the horizontal stresses is different; there can be sudden changes at soil boundaries (as at 5.5 m in the example) because the condition of horizontal equilibrium is still satisfied.

4.6.2 Worked Example 2

Figure 4.13 shows a submerged surface in a lake or river. The initial depth of water is only 1 m but during a flood rises by a maximum of 3 m. Determine the vertical total stress and vertical effective stress at point P initially and during the flood peak.

Initially:

Total (vertical) stress $= 1 \times 9.8 + 3 \times 20.5 = 71.3\,\text{kPa}$
Pore pressure $= 4 \times 9.8 = 39.2$
Effective (vertical) stress $= 32.1\,\text{kPa}$

Table 4.1 Stresses in Worked Example 1

Depth (m)	Vertical Total Stress (kPa)	Pore Pressure (kPa)	Vertical Effective Stress (kPa)	Horizontal Effective Stress (kPa)	Horizontal Total Stress (kPa)
Surface	0	0	0	0	0
2.5	45.5	0	45.5	15.9	15.9
5.5 (within the sand)	110.0	29.4	80.6	28.2	57.6
5.5 (within the clay)	110.0	29.4	80.6	38.7	68.1
8.5	159.2	58.8	100.4	48.2	107.0

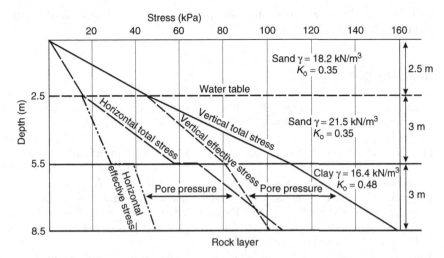

Figure 4.12 Stresses throughout the soil profile analyzed in worked example 1.

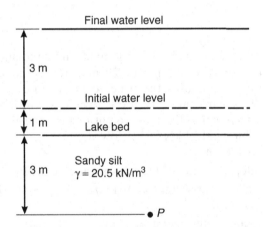

Figure 4.13 Estimation of stress state beneath a submerged surface in a river or lake

During flood:

Total stress $= 4 \times 9.8 + 3 \times 20.5 = 100.7$ kPa
Pore pressure $= 7 \times 9.8 = 68.6$
Effective stress $= 32.1$ kPa

We see from these calculations that while there is a large increase in the total stress at point P there is no change in effective stress. This is to be expected in this situation as the change in water level produces identical increases in the total stress and the pore pressure so the effective stress remains constant.

REFERENCES

Kenny, T. C. and K. C. Lau. 1984. Temporal changes of groundwater pressure in a natural slope of non fissured clay. *Canadian Geotechnical Journal*, Vol. 21, No. 1, pp. 138–146.

Pun, W. K. and G. Urciuoli. 2008. Soil nailing and subsurface drainage for slope stabilisation. *Proceedings Tenth International Symposium on Landslides and Engineered Slopes*. Vol.1, pp. 85–125. London: Taylor & Francis Group.

Urciuoli, G. 1998. Pore pressures in unstable slopes constituted by fissured clay shales. In A. Evangelista and L. Picarelli (eds.), *Proceedings Second International Symposium on the Geotechnics of Hard Soils—Soft Rocks*, Vol. 2, pp. 1177–1185. Rotterdam: A. A. Balkema.

Vaughan, P. R. and C. W. Kwan. 1984. Weathering, structure and in situ stress in residual soils. *Geotechnique*, Vol. 34, No. 1, pp. 43–59.

EXERCISES

1. A layer of sand extends from the ground surface to an indefinite depth. The water table is at a depth of 2.5 m. Groundwater extraction from a neighboring site causes the water table to drop to a depth of 5 m. Calculate the change in vertical effective stress in the sand layer at a depth of 8 m. Assume the unit weight of the sand is $18 \, \text{kN/m}^3$ above the water table, and $21 \, \text{kN/m}^3$ below it.
   ($\Delta \sigma' = 17.0 \, \text{kN/m}^3$)

2. A soil profile, from the surface downwards, consists of the following:
   (a) 4 m of clay, unit weight $= 16 \, \text{kN/m}^3$ (above and below the water table), $K_o = 0.5$
   (b) 8 m of sand, unit weight $= 21 \, \text{kN/m}^3$, $K_o = 0.3$
   (c) hard rock from 8 m
   The water table is at a depth of 2 m.
   Estimate the following:
   (i) The total vertical stress, the pore pressure, and the effective vertical stress throughout the layers.
   (ii) The effective horizontal stress and the total horizontal stress throughout the layers.

Draw graphs illustrating your results [one graph for (i) and one graph for (ii)].

Depth (m)	σ_v	u	σ'_v	σ'_h	σ_h
0	0	−19.6	19.6	9.8	−9.8
2	32	0	32	16	16
4 (clay)	64	19.6	44.4	22.2	41.8
4 (sand)	64	19.6	44.4	13.3	32.9
12	232	98.1	133.9	40.2	138.3

(Answer: stresses in kPa)

CHAPTER 5

STRESSES IN THE GROUND
FROM APPLIED LOADS

5.1 GENERAL

We now move on to look at stresses in the ground arising from applied loads, in particular loads from buildings or other structures at the ground surface. Such structures are often supported on surface foundations, commonly known as spread footings or "shallow foundations." Our interest in knowing the stress in the soil caused by such loads is primarily so that we can estimate the likely settlement of the foundation. Any increase in vertical stress in the ground will result in compression of the soil layer in accordance with the laws that govern the compressibility of the soil, covered in later chapters.

Figure 5.1 shows a surface foundation. The load on the foundation applies a stress to the soil at the surface. This stress "spreads out" with depth, or distance from the foundation, so that as the depth increases, the load is resisted by a wider area of soil, and the resulting stress decreases with depth. In order to estimate the settlement at the surface, we need to calculate the stress change throughout the layer and apply Equation 5.1 to a series of sublayers to give us the compression of each layer. The procedure is covered in detail in Chapter 8. In this chapter we are only concerned with the stress changes caused by various shapes of surface load.

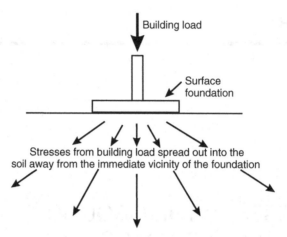

Figure 5.1   Stress distribution in the ground beneath a surface foundation.

5.2   ELASTIC THEORY SOLUTIONS FOR STRESSES BENEATH LOADED AREAS

Solutions for the stress increases have been obtained using elastic theory for a range of footings of varying shapes. The simplest solution is that for a point load, which was derived by Boussinesq in 1885. The expression for the vertical stress σ_z is given by

$$\sigma_z = \frac{Q}{z^2} \left\{ \frac{3}{2\pi} \left(\frac{1}{1 + (r/z)^2} \right)^{5/2} \right\} \tag{5.1}$$

where

z is the depth below the surface, r is the horizontal (radial) distance from the line of action of the force, and
Q is the magnitude of the point load.

This can be written as $\sigma_z = I_\sigma(Q/z^2)$, where I_σ is the influence factor and is a function of r/z.

The value of I_σ is shown graphically in Figure 5.2 followed by graphs for other foundation shapes. The graph is self-explanatory and provides a simple method for determining the stress increase at any point in the soil resulting from the application of a point load at the surface. Although in practice loads will not be point loads, it is informative to examine the stress pattern in the ground resulting from a point load, as similar patterns result from loads acting on finite areas. We will determine the increase in the vertical stress resulting from a point load at the surface at depths $z = 1, 2, 3$ m and radius $r = 0, 1, 2$ m. The value of the point load will be taken as 100 kN.

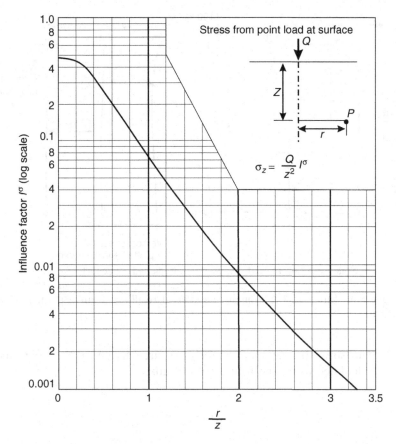

Figure 5.2   Stresses in the ground from a point load at the surface.

Table 5.1   Stresses in ground from a point load at the surface

	$r = 0$		$r = 1\,M$		$r = 2\,m$	
z	r/z	σ_z	r/z	σ_z	r/z	σ_z
$1\,m$	0	**47.7**	1.0	**8.44**	2.0	**0.85**
$2\,m$	0	**11.9**	0.5	**6.83**	1.0	**2.11**
$3\,m$	0	**5.31**	0.333	**4.10**	0.667	**2.10**

The units are not important, as the same pattern is obtained irrespective of the units. The analysis is given in Table 5.1 and is illustrated in Figure 5.3.

Elastic theory solutions for line loads, circular loaded areas, and square loaded areas are given in Figures 5.4, 5.5 and 5.6, respectively. These graphs are also self-explanatory, although the following points should be noted.

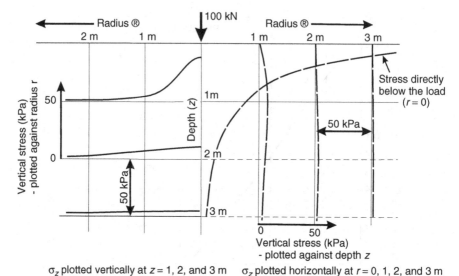

Figure 5.3 Stress distribution in the ground resulting from a point load at the surface.

1. The stress beneath the center of a circular loaded area can also be calculated using the following formula:

$$\sigma_z = q \left[1 - \left\{ \frac{1}{1 + (R/z)^2} \right\}^{3/2} \right] \qquad (5.2)$$

   where R is the radius of the loaded area and z is the depth.

2. The graph for a line load is appropriate for many building perimeter foundations, which tend to be very long and narrow.

3. The solution in Figure 5.6 is for a point below one corner of the loaded area and is given in terms of m and n, where these are defined as $m = B/Z$ and $n = L/Z$, where B and L (which are interchangeable) are the length and width of the foundation. To find the stress at other points within or external to the loaded area requires a composite analysis, as illustrated in the following example.

Figure 5.7 shows two rectangular loaded areas when the point of interest lies within and outside the limits of the loaded area.

Consider first a point within the loaded area as shown in Figure 5.7a. To determine the stress at point P, we divide the total area into segments having corners above the point in question, as shown and labeled 1–4. We

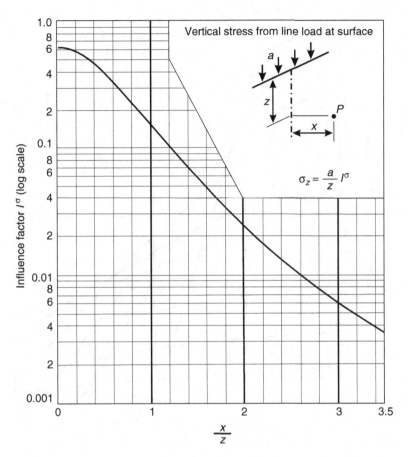

Figure 5.4   Stresses in the ground from a line load at the surface.

then apply the solution to each segment in turn. The values of L and B vary for each segment. This gives us the stress coming from each of the four areas, and by summing these we obtain the total stress increment coming from the complete area, that is, $\sigma_p = \sigma_1 + \sigma_2 + \sigma_3 + \sigma_4$, where σ_p is the stress coming from the total area and $\sigma_1, \ldots, \sigma_4$ are the stresses coming from the four sub-areas 1–4.

For points outside the loaded area, such as point P in Figure 5.7b, a slightly more subtle procedure is needed involving adding and subtracting segments. The loaded area in this case is $ABCD$, and the point at the surface directly above P is the point O. To handle this situation, a new area is created, including the original area and extending to the point O as shown. This new area is subdivided into the three smaller areas shown. We can now determine the stress increase at point P coming from each area that has a corner above point P. To determine the stress due to the actual loaded

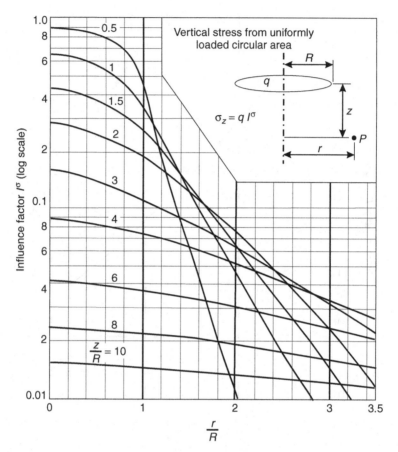

Figure 5.5 Stresses in the ground from a circular loaded area at the surface.

area, we need to subtract or add the appropriate areas so that only the influence of the original area remains. In this case the necessary subtractions and additions are as follows:

$$\sigma_{ABCD} = \sigma_{OHCE} - \sigma_{OHBF} - \sigma_{OGDE} + \sigma_{OGAF}$$

where σ_{ABCD} to σ_{OGAF} are the stresses coming from the areas $ABCD$ to $OGAF$, respectively.

Subtracting areas $OHBF$ and $OGDE$ from the total area means that the area $OGAF$ has been subtracted twice; hence area OGAF needs to be added again.

The solutions presented in Figures 5.2 and 5.4–5.6 are based on elastic theory and are for uniform pressure distributions on the loaded area and a deep homogeneous layer. This implies that the foundations are infinitely flexible. In practice this will not normally be the case. Most building

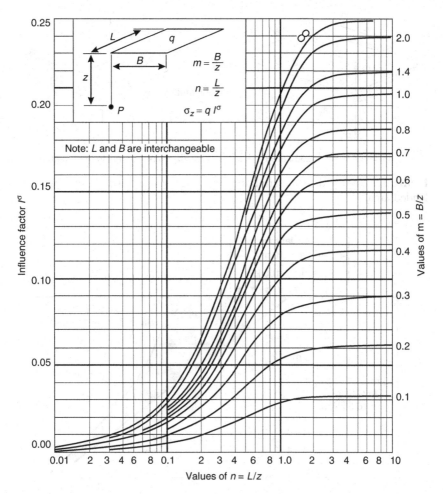

Figure 5.6 Stresses in the ground from a rectangular loaded area at the surface.

foundations are made of reinforced concrete and will be quite rigid. The foundations most likely to qualify as flexible are those of large steel storage tanks. The base of such tanks is normally made of welded steel plates, which can be expected to deform readily. Despite the fact that many foundations are rigid, the use of the charts still provides estimates of sufficient accuracy for practical purposes. Foundations with complex shapes can be handled using Figure 5.6 by subdividing the shape into a number of rectangles.

Use of the above charts gives us the stress arising from the applied load at the surface. The total stress in the soil will be the sum of the initial stress from the weight of the soil itself plus the additional stress coming from the foundation load. This is illustrated in Figure 5.8, which shows a circular tank constructed on top of a 15-m-thick clay layer overlying hard

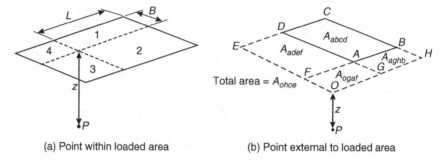

(a) Point within loaded area   (b) Point external to loaded area

Figure 5.7   Method for determining stress below points within and external to a loaded area.

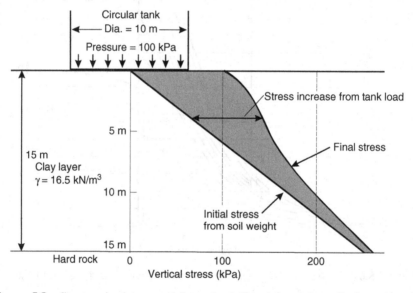

Figure 5.8   Stresses in the ground from the self-weight of the soil plus a foundation load.

rock. The initial total stress increases linearly with depth. The additional stress, determined using the graphs in Figure 5.5, is the same as the applied pressure (100 kPa) at the surface and steadily decreases with depth, though not in a linear relationship.

REFERENCES

Boussinesq, J. 1885. *Application des Potentiels a L'Etude de L'Equilibre et du Mouvement des Solides Elastiques*. Paris: Gauthier-Villars.

EXERCISES

1. Figure 5.9 shows a rectangular area 40 m by 60 m applying a uniform pressure of 50 kPa on the ground surface. The site consists of stiff clay down to a depth of 20 m below which hard rock is found. Using elastic theory, find the increase in vertical stress at a depth of 15 m below points A, B, and C shown in Figure 5.9.
   (**A: 43.0 kPa, B: 12.25 kPa, C: 0.4 kPa**)

2. Four masses, each of 50 tonnes, apply vertical forces, which may be considered as point loads, at the corners of a square 10 m by 10 m on the ground surface. Find the increase in vertical stress at depths of 7 and 14 m:
   (a) Beneath one of the loads
   (b) Beneath the center of the square.

Depth	7 m	14 m
Beneath any load	5.42 kPa	2.27 kPa
Beneath center of square	3.30 kPa	2.70 kPa

3. Compare the vertical stress increase σ_z in the ground below the center of a uniformly loaded circular area (intensity q, radius a) with that beneath a point load $Q = \pi a^2 q$. Do this by plotting graphs of σ_z/q against z/a. (In other words, investigate the difference in the stresses beneath a surface load when the load is either a point load or a uniform stress on a circular area.)
   Determine the value of z/a at which the two stresses differ by 5 percent. For greater values of z/a the load on a circular area may be considered to act at a point, with little loss of accuracy in settlement analysis.
   (**$z/a = 5$ approximately**)

4. (a) At a level site where uniform clay ($\gamma = 18\,\text{kN/m}^3$) is found to a depth of 40 m, an excavation 30 m² and 1 m deep is made with

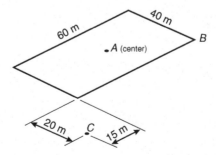

Figure 5.9   Stresses from a loaded area (exercise 1).

vertical sides. Calculate the change in vertical stress 10 m below the original ground level at points (i) beneath the corner and (ii) beneath the center.

(Corner: 4.5 kPa, Center: 15.1 kPa)

(b) A circular oil tank 28 m in diameter with a total weight of 180 MN is to be erected centrally in the excavated area. Assuming the contact stress to be uniform, find the increase in vertical stress at the same points and hence determine *the final vertical stresses*.

(Final σ_v: Corner 197.4 kPa, Center 410 kPa)

CHAPTER 6

PRINCIPLE OF EFFECTIVE STRESS

6.1 THE BASIC PRINCIPLE

In Chapter 4 we introduced the basic concept of effective stress and its relationship to total stress and pore pressure, as given by Equation 4.1 for a fully saturated soil, namely, $\sigma' = \sigma - u$, where σ' is the effective stress, σ is the total vertical stress acting on the plane, and u is the pore pressure.

To understand the theoretical validity of this expression, consider a "plane" $Y-Y$ through a fully saturated soil, as shown in Figure 6.1. The plane is selected to pass only through the void space and the points of contact between the particles and not through the particles themselves. Although $Y-Y$ is shown somewhat curved in Figure 6.1, in practice this will not be the case. In fine-grained soils, especially clays, the particle size is so small that deviations from a true plane are negligible.

The total vertical stress acting on the plane $= \sigma$

The pore pressure in the soil $= u$

If the average vertical component of the inter-particle contact force is P and the average contact area between the particles (projected onto the plane $Y-Y$) is A, then the stress per unit area acting on the soil skeleton is given by NP, where N is the number of interparticle contacts per unit area. The total stress per unit area is given by $\sigma = NP + u(1 - NA)$, since the area on which the pore pressure acts equals $1 - NA$.

Now NP is the stress per unit area acting on the soil skeleton given as σ' and NA is the total interparticle contact area per unit area given as a. Hence we can write

$$\sigma = \sigma' + u(1 - a)$$

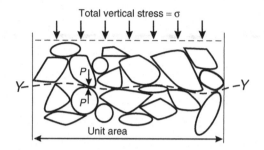

Figure 6.1 Total and effective stress in a soil.

so that

$$\sigma' = (\sigma - u) + ua \qquad (6.1)$$

The effective stress is thus not exactly equal to $\sigma - u$ but depends on the value of a, the contact area. In general, $a < 0.5\%$, so that the equation $\sigma' = \sigma - u$ is a very good approximation.

It can be shown, however, that the deformation of the soil skeleton is independent of the area a. Consider again the stress state between two particles, as shown in Figure 6.2.

The actual stress state is shown in Figure 6.2a; this can be considered to be made up of the two components shown in Figure 6.2b. The pressure u which acts over the whole particle will not cause any distortion of the particle and thus no distortion of the soil skeleton, although there may be a very small volume change. The stress causing distortion or compression of the soil skeleton is thus the component $P - uA$. If we define σ'_c as the stress causing deformation of the soil skeleton, then σ'_c is obtained by summation of the terms $P - uA$. Therefore

$$\sigma'_c = (P - uA)N$$
$$= NP - uNA$$
$$= \sigma' - au$$

where N is the number of contacts per unit area.

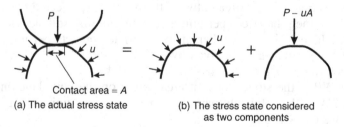

(a) The actual stress state

(b) The stress state considered as two components

Figure 6.2 Stress acting on the soil skeleton (after Bishop, 1964).

Inserting the value of σ' from Equation 6.1 above we obtain

$$\sigma'_c = (\sigma - u) + ua - au$$

Thus

$$\sigma'_c = \sigma - u \qquad (6.2)$$

This analysis shows that, although the interparticle stress depends on the value of the contact area a, the volume change of the soil due to deformation of the soil skeleton depends only on the difference $\sigma - u$. If compression of the soil particles themselves is taken into account, then there will be some additional, very small volume change. Analysis of the way the effective stress relates to the shear strength of the soil also shows that the use of Equation 4.1 is an approximation that is dependent on the ratio of the interparticle contact area to the total cross-sectional area but that for normal engineering situations the shear strength is controlled only by the difference $\sigma - u$, that is, by Equation 4.1

The important assumptions or limitations in the effective stress equation $\sigma' = \sigma - u$ are therefore as follows:

(a) For estimating volume change, it is accurate provided the compressibility of the soil particles themselves is negligible compared to the compressibility of the soil skeleton.

(b) For estimating changes in shear strength it is correct provided the interparticle contact area is negligible compared to the total cross-sectional area.

These limitations are strictly of theoretical interest only and for practical purposes in geotechnical engineering can be ignored. The concept of effective stress, and its expression in the equation $\sigma' = \sigma - u$, is the most fundamental principle on which soil mechanics rests and has stood the test of time since its formulation by Karl Terzaghi in the 1920s. The principle of effective stress (for fully saturated soils) can be summarized as follows:

- In general, the stress acting on a soil consists of two components—the total stress and the pore water pressure. The difference between these is called the effective stress, and this is the stress which acts on the soil "skeleton," that is, the array of solid particles of which the soil is composed.

- Soil only experiences and responds to effective stresses, not to total stresses. Soil will only undergo volume change, or a change in strength, if the effective stress acting on it changes in some way. Changes in total stress on their own will not affect the soil.

6.2 APPLIED STRESSES, DRAINED AND UNDRAINED BEHAVIOR

The most common situation in geotechnical engineering involving the application of stress to soil is the construction of a foundation or an embankment, as illustrated in Figure 6.3. In this case the applied stress is a vertical normal stress, which will induce both normal and shear stresses within the soil. Beneath the center of a wide load, the induced stresses will be predominantly vertical and horizontal normal stresses, while at the edges of the loaded area shear stresses will tend to be predominant. Normal stresses can be resisted by either the soil skeleton or the pore water in the soil or distributed between the two. Shear stresses can only be resisted by the soil skeleton itself.

The way in which the soil responds to and resists the applied loads depends on the type of soil and the rate at which the load is applied. If the soil is of low permeability and the load is applied quickly, the applied stress is likely to be resisted by both the soil skeleton and the pore water. This means the applied stress will induce a change in the pore pressure in the soil. This situation is referred to as **undrained behavior**, implying that during the load application no movement of water occurs in the soil and the water content remains constant. Pore pressures in the soil and the position of the water table change during undrained loading. With time, the pore pressures induced by the load application will drain away and the pore pressure state and water table will revert to the equilibrium position that existed prior to the load application. The soil will undergo a volume decrease. This process is termed consolidation and is fully described in Chapter 8.

On the other hand, if the soil is of high permeability, as would be the case with sand, then any induced pore pressures will drain away as rapidly as the load is applied. This is referred to as **drained behavior**. In this situation there is no change in the water table or in pore pressures in the ground as the load is applied.

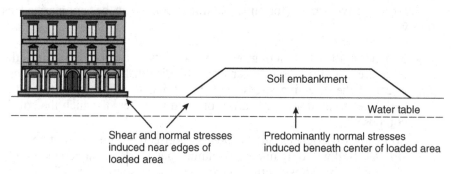

Figure 6.3 Stresses in ground from applied surface loads.

The terms **drained** and **undrained** are widely used in soil mechanics and should be clearly understood. The terms do not refer to the presence or absence of drains, although drains can influence whether the soil behaves in a drained or undrained manner. They refer only to whether or not any movement of water occurs in the soil. Whether the soil behaves in a drained or undrained manner depends primarily on two factors:

(a) The rate at which the load is applied relative to the permeability and compressibility of the soil. The higher the rate of load application in relation to the permeability of the soil, the more likely it is that behavior will be undrained and vice versa.

(b) The opportunities that exist for water to drain from the soil as the load is applied, in other words, the presence or not of escape routes within the soil and the proximity of drainage surfaces or open boundaries to which the water can escape.

6.3   PORE PRESSURE CHANGES UNDER UNDRAINED CONDITIONS

When an all-round pressure is applied to a saturated soil, from which no drainage is permitted, the proportion of stress carried by the pore water and the soil skeleton depends on their relative compressibility. Consider the soil element illustrated in Figure 6.4a. An all-round pressure $\Delta\sigma$ is applied to the soil in a situation where no water can drain from the soil. Consider also that the compressibility of the soil and water is given by the following

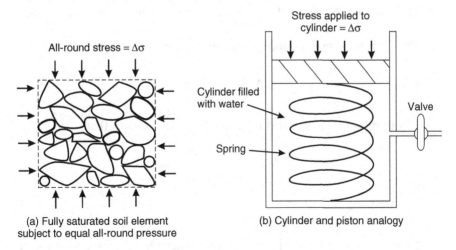

(a) Fully saturated soil element subject to equal all-round pressure

(b) Cylinder and piston analogy

Figure 6.4   Pore pressure response to applied load and cylinder/piston analogy.

expressions:

$$\text{Soil: } \frac{\Delta V}{V} = m_v \Delta \sigma' \tag{6.3}$$

$$\text{Water: } \frac{\Delta V}{V} = m_w \Delta u \tag{6.4}$$

where V is the volume, m_v and m_w are the coefficients of compressibility of the soil and water, respectively, $\Delta \sigma'$ is the change in effective stress, and Δu is the change in pore pressure.

Equation 6.3 applies to three-dimensional compression of the soil. In practice, soil compression takes place predominantly in the vertical direction, and the equation is expressed as

$$\frac{\Delta L}{L} = m_v \Delta \sigma' \tag{6.5}$$

where ΔL is the compression of a soil layer of thickness L subject to a stress increase of $\Delta \sigma'$. The parameter m_v is the coefficient of one-dimensional compressibility. It is a constant of proportionality between stress and strain and therefore similar to Young's modulus for other materials (though expressed in inverse form).

We will assume that the soil element has a volume V and porosity n, and as a result of the stress application, the effective stress and pore pressure rise by $\Delta \sigma'$ and Δu, respectively.

The volume change of the soil skeleton is given by $\Delta V = m_v V \Delta \sigma'$.
The volume change of the pore water is given by $\Delta V = m_w nV\Delta u$.

These must be equal, since no water escapes, and we can write

$$\Delta \sigma' = n \frac{m_w}{m_v} \Delta u$$

and from $\sigma' = \sigma - u$ we also have $\Delta \sigma' = \Delta \sigma - \Delta u$, so that

$$\Delta \sigma - \Delta u = n \frac{m_w}{m_v} \Delta u$$

and rearranging this gives

$$\Delta u = \left(\frac{1}{1 + n(m_w/m_v)} \right) \Delta \sigma = B \Delta \sigma \quad \text{where } B = \left(\frac{1}{1 + n(m_w/m_v)} \right) \tag{6.6}$$

Because the compressibility of the water is negligible compared to the compressibility of the soil skeleton, the value of B is very close to unity for all soils, varying between about 0.995 and 0.9999 (for sands and soft clays, respectively). For practical purposes, for the application of an all-round stress to a fully saturated clay, we can write

$$\Delta u = \Delta \sigma \qquad (6.7)$$

It can be shown Equation 6.7 is also valid for the situation of a vertical stress increase on a fully saturated soil when no lateral deformation of the soil is permitted. This would be the case in the ground beneath the central part of a wide load at the ground surface.

The way soil behaves during undrained or drained loading can be likened to the way a piston supported by a spring in a cylinder full of water behaves when a vertical load is applied to the piston, as shown in Figure 6.4b. Provided the outlet valve from the cylinder is closed, no movement of the piston occurs when the load is applied. The load on the piston is resisted entirely by the water in the cylinder, which is essentially incompressible compared to the spring. This is equivalent to "undrained" behavior. The pressure in the water rises by the value of the applied stress $\Delta \sigma$. If the outlet valve is open at the time the load is applied, the behavior may still be undrained, at least for a brief period, especially if the outlet pipe is very small and the load is applied suddenly. With time, water will drain from the cylinder and the behavior becomes "drained." As this drainage process takes place, the load is being transferred from the water to the spring, analogous to consolidation of soil. If the outlet pipe is large and the load is applied slowly, then water will drain from the cylinder and the load will be taken immediately by the spring. This would be "drained" behavior.

Soil behavior is not necessarily either entirely drained or entirely undrained; it may be intermediate between these states. If there is no pore pressure change as a result of the load application, then the behavior is **fully drained**, and if no movement of soil water occurs during the load application, then the behavior is entirely **undrained**. Any situation intermediate between these states can be termed **partially drained**.

6.4   SOME PRACTICAL IMPLICATIONS OF THE PRINCIPLE OF EFFECTIVE STRESS

6.4.1   Stress State on Soil Element Below Submerged Surface (Bed of Lake or Seabed)

Consider the stress state of a soil element lying beneath a submerged soil surface, as shown in Figure 6.5.

Total vertical stress on the element: $\sigma = \gamma_w H_w + \gamma D$

Pore pressure on the element: $u = \gamma_w (H_w + D)$

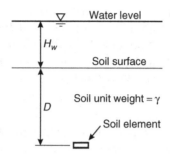

Figure 6.5 Stress state on soil element beneath a submerged surface.

$$\text{Effective stress } \sigma' = \sigma - u$$
$$= \gamma_w H_w + \gamma D - \gamma_w (H_w + D)$$
$$= (\gamma - \gamma_w) D$$

The effective stress is thus totally independent of the depth of water; the sand is in the same state whether submerged beneath 1 cm or 1 km of water. Raising the water level will have no effect on the sand.

6.4.2 Force Resisting Sliding of Concrete Gravity Dam

Consider the stability of the concrete gravity dam shown in Figure 6.6. The reservoir water pressure exerts a force (P) on the dam, so that it tends to slide horizontally on its base. This force is resisted by the friction between dam and the foundation soil or rock it rests on.

The sliding resistance is not simply the weight (W) of the dam multiplied by the coefficient of friction between the dam base and the foundation material. There is likely to be seepage beneath the dam and thus pore pressure acts at the contact of dam and foundation material. There will in effect be an uplift force (U) on the dam base arising from this pore pressure, and the effective weight of the dam will be $W - U$.

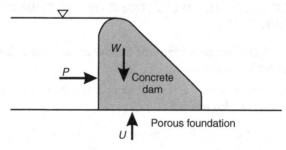

Figure 6.6 Stability with respect to sliding of a gravity concrete dam.

Thus the resisting force is given as $(W - U)\mu$, where μ is the coefficient between concrete and foundation.

The safety factor against sliding is thus given by:

$$F = \frac{(W - U)\mu}{P} = \frac{(W - U)\tan \sigma'}{P}$$

A "friction angle," denoted ϕ', is normally used in soil mechanics to denote the frictional component of soil shear strength; $\tan \phi'$ therefore takes the place of the coefficient of friction μ. This is described fully in Chapter 9.

6.4.3   Influence of Rainfall on Slope Stability

In many parts of the world, slips or landslides are triggered by periods of heavy rainfall. The reason for this is not that the rainfall adds to the weight of the soil or that it lubricates the soil. Many slopes in which rain triggers landslides consist essentially of fully saturated clay all year round, and the rainfall has negligible influence on either the unit weight or the water content of the soil. The real reason for such slides is the influence that rainfall has on the pore pressure state and thus the effective stress in the slope. A typical slope is shown in Figure 6.7. The circular arc shown is a potential failure surface on which slip movement could occur. Experience shows that many slips and landslides commence with movement on surfaces that are approximately circular.

We will consider the influence of changes in water table level on the shear strength of a soil element on this circular arc, namely the soil element at point P. The initial water table shown in the figure is taken as the average value applying when weather conditions are "normal." As a result of prolonged rainfall, the water table rises to the final water level indicated in the figure. We can write the following as reasonable approximations of the stress state at P:

Total stress $= \gamma D$, where γ is the unit weight of the soil

Initial pore pressure $u = \gamma_w H$, where γ_w is the unit weight of water

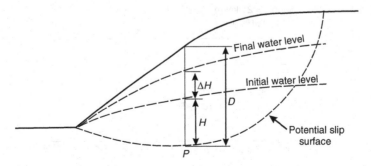

Figure 6.7   Influence of water table on stability of soil slopes.

Initial effective stress $= \gamma D - \gamma_w H$
Final pore pressure $u = \gamma_w(H + \Delta H)$
Final effective stress $= \gamma D - \gamma_w(H + \Delta H)$
Change in effective stress $= -\gamma_w \Delta H$

There is no significant change in the total stress, so the change in effective stress is the same as the change in pore pressure. Because the shear strength of soil is made up partly of a frictional component governed by the effective normal stress acting across the slip plane, this decrease in effective stress reduces the shear strength of the soil. This decrease may result in slip movement.

6.4.4 Ground Settlement Caused By Lowering Water Table

Consider the situation illustrated in Figure 6.8. A layer of sand overlies a layer of relatively soft clay. As a result of drainage activities in the area, the water table is lowered from an initial depth of 1 m to a depth of 3 m. This will cause the ground surface to settle as a result of a decrease in pore pressure and a consequent increase in the effective stress in the clay layer. The sand layer is of very low compressibility and can be considered to be incompressible.

To calculate the settlement of the ground surface we can use Equation 6.3, namely,

$$\frac{\Delta L}{L} = m_v \Delta \sigma'$$

where ΔL is the change in thickness of a layer of thickness L, m_v is a soil property known as the coefficient of one-dimensional compressibility, and $\Delta \sigma'$ is the change in vertical effective stress.

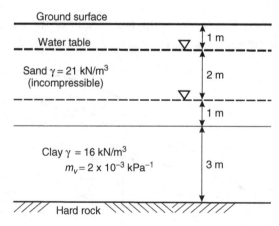

Figure 6.8 Ground settlement from lowering the water table.

To apply this formula to the layer given, we need to calculate the change in effective stress caused by lowering the water table. It is only necessary to do this at the center of the layer, as the change is uniform throughout the layer. The following table shows the calculation:

	Original Situation	**After Water Table Lowering**
Total vertical stress	$4 \times 21 + 1.5 \times 16 = 108$ kPa	108 kPa
Pore water pressure	$4.5 \times 9.81 = 44.1$ kPa	$2.5 \times 9.81 = 24.5$ kPa
Effective stress	$108 - 44.1 = 63.9$	$108 - 24.5 = 83.5$ kPa

Hence the change in effective stress is $83.5 - 63.9 = 19.6$ kPa.

Using the formula gives the value of $\Delta L = 3000 \times (2 \times 10^{-3}) \times 19.6 = 118$ mm.

We should note the following:

- The soil is not subjected to any change in external loading. No surface loads of any sort have been applied to the soil.
- The change in effective stress is the same as the change in pore pressure, though with an opposite sign. The pore pressure has decreased but the effective stress has increased.
- For simplicity, we have used the same unit weight for the sand above and below the water table. In practice the unit weight above the water table is likely to be less than that below because some water will drain out of the sand.

REFERENCES

Bishop, A. W. 1964. Soil properties. Imperial College MSc(Eng) course lecture notes.

CHAPTER 7

PERMEABILITY AND SEEPAGE

7.1 GENERAL

As already described, soils consist of particles with interconnected voids between them. Hence they are **permeable**, that is, water can flow, or seep, through them, even though it may be at a very slow rate in some soils. Soil permeability and seepage are of interest to geotechnical engineers for a variety of reasons, including the following:

1. The rate at which seepage occurs. This is of vital importance in the design of water-retaining structures, especially earth dams, embankments, and canals.
2. The influence that seepage has on stability. The seepage situation governs pore pressures, which in turn govern the effective stress and the strength of the soil. The stability in question may be that of soil slopes, either natural or man made, or it could that of deep excavations below the water table.
3. The rate of transport of contaminants. The growth of environmental concerns over the last several decades means that a great deal of attention is now given to the rate at which contaminants may travel through the ground and the means by which this may be prevented.

7.2 PRESSURE, "HEAD," AND TOTAL HEAD

Consider Figure 7.1, which shows a cross section in natural ground. Imagine that at points A and B instruments (called piezometers) have been installed in the ground to measure the pore pressures at these points. The values obtained are shown. Now think carefully for a moment, will water tend to flow from A toward B or vice versa? Point A is higher than point B so perhaps water should flow from A to B. On the other hand, the pressure at B is higher than at A so perhaps flow should be from B to A.

To answer this question, we need to determine the difference in **total head** (commonly called just **head**) between the two points. The total head can only be expressed in relation to a specific elevation in space, that is, a datum. Any elevation datum is acceptable; we will adopt the line $X-Y$ as our datum. The total head at any point is defined as the sum of the elevation head and the pressure head:

$$h = y + \frac{u}{\gamma_w} = h_e + h_p \qquad (7.1)$$

where y is the elevation head $(= h_e)$, u is the pore pressure, and u/γ_w is the pressure head $(= h_p)$.

In Figure 7.1 (using the datum $X-Y$), the heads at A and B are given by

$$h_A = \text{elevation head} + \text{pressure head} = 3.5 + \frac{20}{9.8} = 3.5 + 2.04 + 5.54\,\text{m}$$

$$h_B = \text{elevation head} + \text{pressure head} = 1.5 + \frac{50}{9.8} = 1.5 + 5.10 = 6.60\,\text{m}$$

The total head at B is thus higher than the head at A, with a head difference of 1.06 m. Water will therefore tend to seep from B to A. Seepage

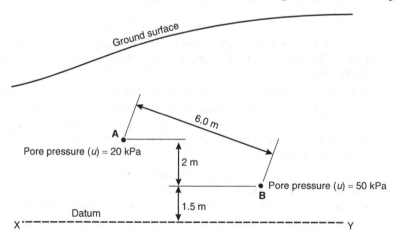

Figure 7.1 Will water seep from point A to point B or vice versa?

occurs as a result of differences in total head, and analysis of seepage behavior is always in terms of total head, or of head difference, and not in terms of pressure. This is the same concept that applies to pipe flow in fluid mechanics.

To better understand the concept of total head, it is helpful to focus on the physical situation it expresses. This is done in Figure 7.2. Instead of piezometers buried within the soil mass as implied in Figure 7.1, we can install vertical pipes or tubes with their tips at points A and B and measure the level to which the water level rises in the pipes. This gives us a direct measure of head, from which we can calculate the pore pressure if we wish to know it. Such piezometers are known as standpipe piezometers and are commonly used in geotechnical engineering.

The water level in the piezometers is a direct measure of the total head, and the head difference between points A and B is immediately apparent. While seepage problems can be solved purely by mathematical manipulation, it is always helpful to keep the physical situation clearly in mind, and one way of doing this is to imagine standpipe piezometers installed in the ground and to focus on the level to which water would rise in them.

We can note in passing that Strack (1989) defines the total head (termed hydraulic head in his book) at a certain point P in a soil body as "*the level to which water rises in an open standpipe with its lower end at point P.*" Although Strack's treatment of groundwater mechanics is highly mathematical, the definition he uses for total head is a simple physical one.

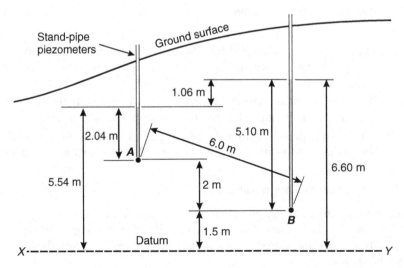

Figure 7.2 Physical portrayal of head and head difference calculated from Figure 7.1.

7.3 DARCY'S LAW

Darcy's law states that the seepage rate is proportional to the hydraulic gradient and is written as follows:

$$q = kiA \qquad (7.2)$$

where

q = flow rate
i = hydraulic gradient
A = cross-sectional area of flow
k = physical property of soil known as **coefficient of permeability** In groundwater studies it is normally called the **hydraulic conductivity**. The terms are synonymous.

The hydraulic gradient is defined as the ratio of change in (total) head to distance; that is,

$$i = \frac{\Delta h}{L}$$

where Δh is the head change and L the distance over which it occurs.

In Figure 7.2 the hydraulic gradient between points B and A is $1.06/6 = 0.18$.

Darcy's Law can also be expressed as

$$v = ki \qquad (7.3)$$

where v is termed the "discharge velocity."

7.3.1 Notes on Darcy's Law

1. Darcy's law is only valid for laminar (streamlined) flow. This is normally the case; only in coarse gravels will turbulent flow possibly arise.
2. The coefficient of permeability is only a constant if the temperature is constant. It is normally assumed to apply to soil at 20°C.
3. Values of k depend primarily on the particle size of the soil, although the actual composition of the particles also has a strong influence.

7.3.2 Note on Seepage Velocity

The velocity in the Darcy expression $v = ki$ is not the true velocity at which the water travels through the soil. It is called a **discharge velocity**—if it is multiplied by the cross-sectional area of the soil, it gives the flow rate, that is, $q = vA$.

The true seepage velocity is different from the Darcy velocity because part of the cross-sectional area is occupied by solid matter. If the effective

Table 7.1   Typical Values of Coefficient of Permeability

Material	Coefficient of Permeability (m/s)	Comment
Gravel	≥ 0.01	Can be drained by pumping, that is, water will flow out of void space under gravity
Coarse sand	$10^{-2}-10^{-3}$	
Medium sand	$10^{-3}-10^{-4}$	
Fine sand	$10^{-5}-10^{-6}$	
Silt	$10^{-6}-10^{-7}$	Water does not generally flow out of void space under gravity
Silty clay	$10^{-7}-10^{-9}$	
Clay	$10^{-8}-10^{-11}$	Almost impermeable (at bottom end of range)

cross-sectional area through which water can seep is given by A_v, then the seepage rate can be expressed as follows:

$$q = v_d A = v_t A_v,$$

where v_d is the Darcy velocity and v_t is the true velocity (or travel time velocity). Hence

$$v_t = \frac{v_d A}{A_v} = \frac{v_d}{A_v/A} = \frac{v_d}{n}$$

where n is the porosity.

The porosity is likely to be between about 0.3 and 0.6 so that the true "travel time" velocity is likely to be 2 or 3 times greater than the Darcy velocity.

The point of the two velocities v_d and v_t is that if we want to calculate seepage rates (flow rates) we must use the Darcy (discharge) velocity, but if we want to determine how long it takes water or contaminants to be transported through the ground by seepage flow we must use the true (travel time) velocity.

Coefficient of permeability: The range of values is very wide, as indicated in Table 7.1. In natural deposits and compacted soils the horizontal permeability is often greater than the vertical permeability.

7.4   MEASUREMENT OF PERMEABILITY

With reasonably coarse-grained soils, measurement of permeability is straightforward and simply involves allowing water to flow through a

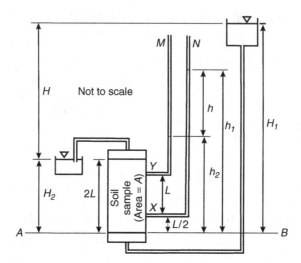

Figure 7.3 Flow through a soil sample in a permeameter.

sample of the soil contained in a permeameter, as shown in Figure 7.3. The soil sample is cylindrical and normally enclosed in a Perspex container. The "header" tank and the receiving tank are set up so that the water level in them remains constant. There are two manometer tubes connected to the side of the cylinder to measure the head at these points. The test is carried out by measuring the flow into the lower container in a known time and recording the water levels in the manometer tubes. We can then write

$$q = \text{flow rate} = V/t,$$

where V is the volume of water collected in time t.
Also

$$q = kiA = k\frac{h}{L}A$$

Hence the coefficient of permeability is given as

$$k = \frac{VL}{thA} \tag{7.4}$$

Note the following with respect to Figure 7.3:

- If $A-B$ is used as datum, the head at $X = h_1$ and the head at $Y = h_2$. In each case this is made up of elevation head plus pressure head. Flow is occurring because of the difference in head between the two points.
- The manometer tubes are used because some head may be lost in the tubes leading from the tanks to the sample.

- If we assume no head is lost in the lead tubes, the head at the start of seepage (at the base of the sample) is H_1 and at the end of the seepage zone (at the top of the sample) is H_2. Head is steadily lost, in a linear fashion, along the length of the seepage zone. Because the flow is the same throughout the sample, the hydraulic gradient must be constant—from Darcy's law. This is analogous to flow through a pipe of constant size.

This procedure is satisfactory for moderate- to high-permeability materials. For low-permeability clays, the flow rate is likely to be so slow that the procedure is impracticable. In this case other procedures are available, such as setting up the sample in a triaxial apparatus and applying a high pressure to ensure a high flow rate or using a different type of permeameter called a falling-head permeameter.

7.5 GENERAL EXPRESSION FOR SEEPAGE IN A SOIL MASS

We can derive a general expression governing the pattern of seepage in the ground by considering the flow into and out of a soil element, as shown in Figure 7.4. We will limit the analysis to two-dimensional (2-D) flow; the element is dx in length and dy in height and of unit width (at right angles to the page). We will consider the general case which includes the possibility of change over time despite the fact that most civil engineering seepage studies do not consider such changes. The reason for doing this will become apparent a little later. Situations where conditions do not change over time are referred to as **steady state**, and those involving change over time as **non–steady state** or **transient**.

Flow entering the element is given as

$$v_x dy + v_y dx.$$

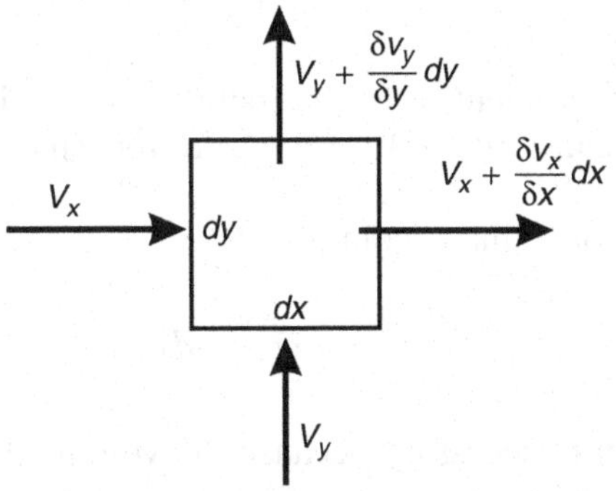

Figure 7.4   Flow through a soil element.

Flow leaving the element is then

$$\left[v_x + \frac{\partial v_x}{\partial x} dx \right] dy \quad + \quad \left[v_y + \frac{\partial v_y}{\partial y} dy \right] dx \qquad (7.5)$$

Net flow out of the element equals flow out minus flow in, or

$$\frac{\partial v_x}{\partial x} dx \, dy + \frac{\partial v_y}{\partial y} dx \, dy = \left[\frac{\partial v_x}{\partial x} + \frac{\partial v_y}{\partial y} \right] dx \, dy$$

From Darcy's law (Equation 7.3), assuming the material is isotropic,

$$v_x = k \frac{\partial h}{\partial x} \quad \text{and} \quad v_y = k \frac{\partial h}{\partial y}$$

Substituting these in the above equation gives the net flow leaving the element:

$$\left[k \frac{\partial^2 h}{\partial x^2} + k \frac{\partial^2 h}{\partial y^2} \right] dx \, dy = \quad k \left[\frac{\partial^2 h}{\partial x^2} + \frac{\partial^2 h}{\partial y^2} \right] dx \, dy$$

For a fully saturated soil, this must equal the rate of volume change of the element, so that we can write

$$\frac{dV}{dt} = \left[k \frac{\partial^2 h}{\partial x^2} + k \frac{\partial^2 h}{\partial y^2} \right] dx \, dy \qquad (7.6)$$

where dV is the volume of the element.

We can derive a second expression for the rate of volume change in terms of the rate of change of effective stress. Volume change is related to effective stress change as follows (from Chapter 6):

$$\frac{\Delta V}{V} = m_v \Delta \sigma'$$

where m_v is the coefficient of compressibility and σ' is the effective stress.

From this we can write $\Delta V = m_v \Delta \sigma' dx \, dy$ (since dx dy is volume of soil element)

In differential form this becomes

$$\frac{dV}{dt} = -m_v \frac{\partial \sigma'}{\partial t} dx \, dy \qquad (7.7)$$

(The negative sign is necessary because the volume decreases as the effective stress increases.).

In most seepage situations of interest to geotechnical engineers, the total vertical stress is a function of the soil depth and remains constant over time at a given depth. In the equation for effective stress change $\Delta\sigma' = \Delta\sigma - \Delta u$, the term $\Delta\sigma$ is therefore zero and $\Delta\sigma' = \Delta u$. Since $u = \gamma_w h$, we have

$$\Delta\sigma' = \gamma_w \Delta h$$

Substituting this is Equation 7.7 gives

$$\frac{dV}{dt} = m_v \gamma_w \frac{\partial h}{\partial t} dx\, dy \tag{7.8}$$

The term $\partial V / \partial t$ is the rate of volume change and must be equal to the expression in Equation 7.6 Hence we have

$$m_v \gamma_w \frac{\partial u}{\partial t} dx\, dy = k \left[\frac{\partial^2 h}{\partial x^2} + \frac{\partial^2 h}{\partial y^2} \right] dx\, dy$$

Rearranging gives

$$\frac{\partial^2 h}{\partial x^2} + \frac{\partial^2 h}{\partial y^2} = \frac{m_v \gamma_w}{k} \frac{\partial h}{\partial t} \tag{7.9}$$

Equation 7.9 is an important equation and we will use it in this book to derive expressions governing steady-state flow, non-steady-state groundwater flow, as well as the consolidation process in soil (the Terzaghi consolidation equation).

7.6 STEADY-STATE FLOW, LAPLACE EQUATION, AND FLOW NETS

With steady-state flow there are no changes in the head with time, and the flow entering and leaving the element is the same. The term on the right-hand side of Equation 7.9 is therefore zero, and the equation becomes

$$\frac{\partial^2 h}{\partial x^2} + \frac{\partial^2 h}{\partial y^2} = 0 \tag{7.10}$$

This is the two-dimensional form of the **Laplace equation**. It also governs flow of heat or flow of electricity through 2-D conductors. Laplace's equation is satisfied by two conjugate harmonic functions ϕ and ψ, the curves ϕ_{xy} = const being the orthogonal trajectories of the curves ψ_{xy} = const. In plainer language, the solution for h in effect consists of two sets of smooth curves that interest at right angles.

To understand the physical significance of these two sets of lines, consider the flow of water beneath a sheet pile wall, as shown in Figure 7.5. A sheet

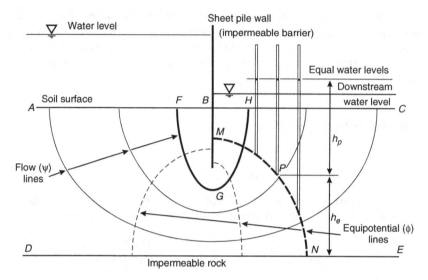

Figure 7.5   Seepage under a sheet pile wall: flow lines and equipotential lines.

pile wall is in effect a wall of driven steel piles that interlock to form an impermeable barrier. They are commonly used around deep excavations. The wall is holding water at a much higher level on one side (the upstream side) than the other (the downstream side). This difference in level causes water to enter the upstream surface AB and flow down and around the wall and come out again along BC. Water entering at some point F will follow a fixed path $F - G - H$. Water entering at any other point will also follow a fixed, but different, path. These paths are called flow lines. There are an infinite number of flow lines, of which only three are shown.

As flow occurs, head is being lost along the flow lines. We can determine the head at any point, for example, at P, by installing a standpipe piezometer and noting the level to which the water level rises. We can use these standpipes to determine lines of equal head, such as the line $M - P - N$. These are called "equipotential" lines, that is, lines of equal potential head (or simply equal head). There are also an infinite number of such lines, of which three are shown.

The head at P (with respect to datum DE) equals the elevation head plus the pressure head:

$$h_e + h_p = h_e + u_p/\gamma_w$$

The elevation head and pressure head are varying all along $M - N$ but the total head is constant. We thus have two sets of lines (which have the form of smooth curves). They intersect at right angles and form a pattern of lines called a **"flow net."** These lines are thus the lines that come from the solution of the Laplace equation, the lines $\phi = $ const are the equipotential lines, and the lines $\psi = $ const are the flow lines. Solution of seepage problems generally involves the establishment of a flow net and then using it to determine what we wish to know, such as flow rates, pore pressures, or uplift pressures below structures.

7.6.1   Flow nets — Conventions Used in Their Construction

A flow net consists of an infinite number of lines — for practical purposes we need to select a limited number of lines only. It is convenient to:

(a) Select flow lines so that the flow between them is always the same.
(b) Select equipotential lines so that the head drop between them is the same.

Consider the flow net shown in Figure 7.6. Assume that by some means we have established the flow lines and equipotential lines such that (a) and (b) above are satisfied. Consider also unit length along the weir.

Consider any two "rectangles" or "elements" and give them the dimensions shown.

Let N_f be the number of flow channels and N_e the number of equipotential drops. The total head loss from start to end of the flow net is h. The head loss between any two equipotentials is given as $\Delta h = h/N_e$.

Applying Darcy's law to the first element gives

$$\text{Hydraulic gradient } i = \frac{\Delta h}{a} = \frac{h}{N_e a}$$

$$q = kiA = k\frac{h}{N_e a}b = \frac{kh}{N_e}\frac{b}{a}$$

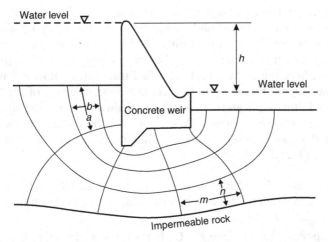

Figure 7.6   A simple flow net.

Similarly, for the second element

$$\text{Hydraulic gradient } i = \frac{\Delta h}{m} = \frac{h}{N_e m} \text{ and}$$

$$q = kiA = k\frac{h}{N_e m}n = \frac{kh}{N_e}\frac{n}{m}$$

Because of the assumptions made, these flows must be the same. Hence to satisfy the intention that flow between the flow lines is equal and head drops between equipotential lines are equal, the ratio of width to length of every "rectangle" must be constant:

$$\frac{a}{b} = \frac{m}{n}$$

It is convenient to take this ratio as 1, so that each "rectangle" then becomes a "square." For this reason it is universal practice to construct flow nets to form "squares," created by the intersection of the two sets of lines. Note that the expression for flow in one channel (channel n) then becomes

$$q_n = \frac{kH}{N_e}$$

and the total flow is the sum of these values for the total number of channels N_f.

$$\text{Total flow } q = kH\frac{N_f}{N_e} \tag{7.11}$$

7.6.2 Boundary Conditions for Flow Nets

In order to understand flow nets and be able to sketch simple flow nets by hand, it is important to understand what the boundary conditions of the seepage zone are. These can be illustrated by considering the homogeneous earth dam shown in Figure 7.7 that has been constructed on an impermeable rock base. Seepage of this kind is termed **unconfined flow** because there is no physical upper boundary to the seepage zone. Seepage of the kind shown earlier in Figures 7.5 and 7.6 is **confined flow** because the upper limit of the seepage zone is confined by the base of the concrete structure. There are basically four different boundary conditions, as illustrated in Figure 7.7: AB, BC, CD, and DA.

1. **Soil/water interface $A-B$:** Along the line AB the head is constant and is defined by the water level in the reservoir. Such lines are clearly equipotential lines.
2. **Impermeable surface $B-C$:** No water can cross this line or originate from this boundary. This line is therefore essentially a flow line.

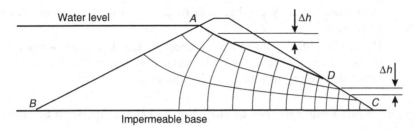

Figure 7.7 Boundary conditions for a flow net (for a coarse-grained soil).

3. **Surface of seepage $C-D$:** On this line, water is seeping out of the soil from within and is crossing $C-D$, so it clearly cannot be a flow line. The elevation head along the surface is varying, but the pressure head is constant. The total head is thus changing and the surface is neither an equipotential nor a flow line.
4. **Line of seepage (phreatic surface or free surface) $D-A$:** The upper surface of flow is known as both a line of seepage and a phreatic surface. As this is the upper limit of seepage, no flow can cross it. It is therefore essentially a flow line.

Note, however, that the way the flow net is drawn in Figure 7.7 is only correct if the material is free draining, such as sand or gravel. For clay, the situation is different and is described later in Section 7.8.

7.6.3   Methods for Solution of Flow Nets

Historically, a range of procedures have been used for establishing flow nets, the main ones of which are listed below. The advent of computers means that these days most methods have been displaced by the computer. However, it is very useful to have sufficient understanding of flow nets to be able to draw simple flow nets by hand sketching.

1. **Mathematical closed form** (that is, an exact analytical expression): These are only practical in cases of very simple boundary conditions and may be obtained using complex variable theory.
2. **Models:** A hydraulic model consisting of sand in a glass-faced tank can be used —a cross section of the actual seepage zone can be set up. By injecting dye the flow lines can be traced, and pressure heads can be obtained from manometer tubes tapped into the side of the tank.
3. **Electrical analogy:** Little used these days since the advent of the computer.
4. **Numerical methods:** These are methods that can be applied to any geometry and to numerous zones of soil of different permeability.

They are extremely cumbersome by hand but ideal for computer programming.

5. **Hand sketching**: This is a very good method for situations involving
one soil type consisting of homogeneous isotropic soil. It is extremely
difficult to apply to multizoned situations.

7.6.4 Basic Requirements of Flow Net and Rules for Hand Sketching Flow Nets

The basic conditions that flow nets must satisfy are the following:

1. Equipotential lines and flow lines must intersect at right angles.
2. The two sets of lines should form a network of "squares." A useful
guide to this condition is to imagine a circle inside each square—the
circle should touch each side of the square.
3. For unconfined flow, there must be equal elevation drops between the
intersection points of equipotential lines with both the phreatic surface and the surface of seepage. This is illustrated in Figure 7.7; the
distance Δh must be equal all along the phreatic surface and surface
of seepage. This is because the pore pressure, and thus the pressure head, is zero on these surfaces, and their elevation thus defines
the head on any equipotential that intersects them.

Flow nets can be sketched with sufficient accuracy for many engineering
situations provided the soil is homogeneous and isotropic. To do this, it is
necessary to understand the basic requirements of flow nets and to adopt
systematic procedures for sketching them. The following "rules" should be
followed in drawing flow nets:

(a) Use a scale such that all lines will start and end within the drawing
(except for boundary lines of infinite extent).
(b) Sort out the boundary conditions so that you are quite clear which
are equipotentials and which are flow lines and which are (or will be)
free surfaces or phreatic lines.
(c) Try to picture how the water will tend to travel through the ground,
that is, what its natural tendency under gravity will be and how this
will be altered by the physical constraints (boundaries) of the seepage
zone.
(d) Sketch two or three trial flow lines to form a whole number of flow
channels; remember that the greater the curvature of flow, the closer
together the flow lines.
(e) Sketch in the equipotential lines, starting from one end of the flow
net and advancing progressively along the net, observing the basic
requirements listed above. The number of equipotential drops may
not come out to a whole number.

(f) Review the general appearance of the flow net and make adjustments as necessary. For unconfined flow nets check that the intersection points of equipotentials with the phreatic surface are at equal elevation changes.

7.6.5 Use of Flow Nets for Practical Purposes

Worked example: Figure 7.8 shows a hand-sketched flow net for seepage beneath a concrete weir (small dam). We wish to determine the following:

(a) Flow rate beneath the dam.
(b) Pore pressure at points A and B.
(c) Time it would take for water to flow from point C at the start of the seepage zone and emerge at the downstream surface.
(d) Exit hydraulic gradient at the downstream end of the weir (that is, at the vertical face). This gradient is of practical concern, as explained in the next Section 7.7.

Using the flow net, each of these is determined as follows:

(a) To determine flow rate we can use Equation 7.11, and extract the values of N_f and N_e from the flow net:

$$q = kH \frac{N_f}{N_e}$$

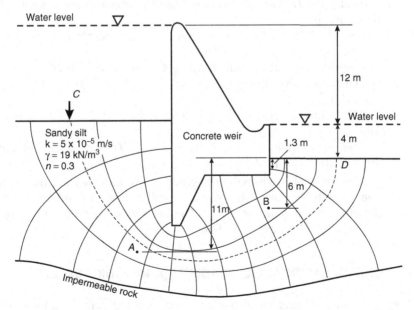

Figure 7.8 Confined flow net for worked example.

where $k = 5 \times 10^{-5}$ m/s, $H = 12$ m, $N_f = 4$, and $N_e = 12$, giving $q = 2 \times 10^{-4}$ m^3/s.

Note that N_f and N_e are the number of flow *channels* and equipotential *drops*, NOT the number of flow lines or equipotential lines.

(b) To determine the pore pressure at any point we first calculate the head loss between each equipotential line and use this to determine the head loss from the start of the flow net to the point concerned. The head loss per equipotential is $H/N_e = 12$ m$/12 = 1$ m. (the fact that this comes to a whole number is purely fortuitous).

Then for point A, the number of equipotential drops from the start of the flow net is 3.6 (approximately) so that the total head loss to that point is 3.6×1 m $= 3.6$ m.

The head at the start of the flow net with respect to point A is $12 + 4 + 11 = 27$ m.

The head at A with respect to A is therefore $27 - 3.6 = 23.4$ m (note that this is the height to which water would rise in a standpipe piezometer installed at point A).

The pore pressure at A is $23.4 \times 9.8 = 229.3$ kPa.

Similarly the pore pressure at point B is $9.8(12 + 4 + 6 - 9.3 \times 1) = 9.8 \times 12.7 = 124.5$ kPa.

(c) To determine travel time for water (or a contaminant in the water) to travel from point C to emerge at the downstream surface, we first establish the path it would follow by sketching the flow line starting at point C. This line follows the pattern of the other flow lines and exits at point D. We can now calculate the true average velocity along this flow line.

Length of line $CD = 49$ m (from scale of diagram).

Head loss along $CD = 12$ m.

Average hydraulic gradient along $CD = 12/49 = 0.24$.

Darcy velocity along $CD = v_d = ki = 5 \times 10^{-5} \times 0.24$ m/s $= 1.2 \times 10^{-5}$ m/s.

True "travel" velocity along $CD = v_d/n = 1.2 \times 10^{-5}/0.3 = 4 \times 10^{-5}$ m/s.

Therefore time to travel from C to $D =$ distance/velocity $= 49/4 \times 10^{-5}$ s $= 14.2$ days.

(d) To determine the exit hydraulic gradient, we can measure the distance over which the last equipotential drop occurs. As indicated on the drawing, this distance is 1.3 m. The head loss over the same distance is 1 m so the hydraulic gradient is $1/1.3 = 0.77$.

7.7 CRITICAL HYDRAULIC GRADIENT (AND "QUICKSAND")

There are some natural situations and many engineering situations associated with hydraulic structures where water seeps upward toward the soil

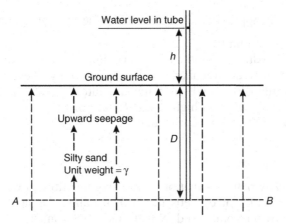

Figure 7.9 Upward seepage and critical hydraulic gradient.

surface, as shown earlier in Figure 7.8. Such seepage can lead to instability. Figure 7.9 shows a situation where water is seeping upward toward the ground surface. Because water is flowing upward, the head at any depth below the ground surface must be greater than at the ground surface. This means that if we install a measuring tube to an arbitrary depth D, the water level will rise to some height above the ground level, which we will call h. We will consider the stability of the block of soil above this depth D. There is in effect an uplift force coming from the seeping water. We can examine this effect by analyzing the stress state on the plane $A - B$.

The total vertical downward stress $= \gamma D (= \text{total stress})$.

The upward or "uplift" pressure coming from the seeping water $= \gamma_w(D + h) = $ pore pressure.

Instability will occur when these become equal, that is, when $\gamma D = \gamma_w(D + h)$.

In this situation the effective stress has become zero, and the sand no longer has any confining stress to keep it stable and the whole block of sand may "lift" slightly, or "boil," and effectively become a liquid. The hydraulic gradient when this occurs is called the critical hydraulic gradient i_c.

Rearranging the expression above, we can obtain

$$\frac{h}{D} = \frac{\gamma}{\gamma_w} - 1$$

Now h/D is the hydraulic gradient, so that

$$i_c = \frac{\gamma}{\gamma_w} - 1 \qquad (7.12)$$

Since $\dfrac{\gamma}{\gamma_w}$ is approximately 2 for sands, the critical hydraulic gradient is normally close to unity.

The exit hydraulic gradient is very important in excavations below groundwater that are retained (i.e., supported) by sheet piles or similar perimeter walls. Water will seep upward into the base of the excavation, and if the gradient is too high, uplift failure of the base of the excavation can occur with disastrous consequences.

7.7.1 Quicksand

There are many natural situations where springs flow upward through sand layers. If the gradient exceeds the critical value, the sand "boils," a phenomenon known as quicksand. Novels and films sometimes portray graphic scenes of people being sucked down into "quicksand" and disappearing out of sight (e.g., *Lorna Doone* and *Jamaica Inn*). As quicksand is of approximately the same density as loose sand and humans are of about half that value, the deepest a person will sink to is approximately waist level, so accounts of people disappearing out of sight in quicksand rightfully belong where they are found—in works of fiction

7.7.2 Worked Example

We will investigate the seepage situation when excavation below the water table is carried out using sheet piles to support the sides of the excavation. Sheet piles are special steel piles that link together to form a seepage barrier. Figure 7.10 illustrates one-half of a symmetrical sheet pile excavation in a 2-m-deep lagoon or river. The soil is uniform sand below which hard rock is found. The 20-m-long sheet piles have been driven to a depth of 17 m below the bed of the lagoon.

We will determine the following:

(a) Hydraulic gradient at base of excavation and safety factor against uplift failure
(b) Seepage rate into excavation
(c) Head and pore pressure at point P

The first step is to draw the flow net. Because the excavation is symmetrical, only one-half of the flow net needs to be drawn. The center line becomes a dividing line between two symmetrical flow nets and can be regarded as an impermeable barrier, or a flow line. For most civil engineering seepage situations, four flow channels are ideal, and we will adopt this number here. Figure 7.10 shows a hand-sketched flow net, which we can now use to determine the above information.

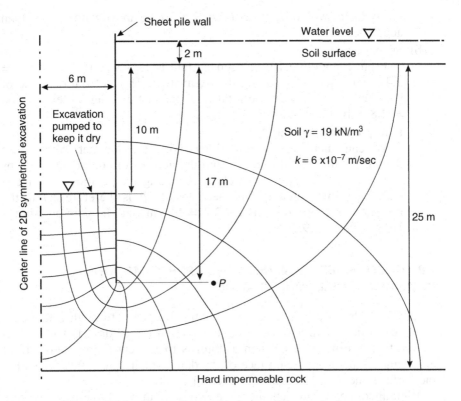

Figure 7.10   Sheet pile excavation used in worked example.

(a) Hydraulic gradient at base of excavation:

Total head loss $= 12\,\text{m}$

Number of equipotential drops $= 12$ (this is pure coincidence)

Head loss per equipotential $= 12/12 = 1.0\,\text{m}$ (the fact that this comes out to a whole number is purely fortuitous).

We must now examine the shortest dimension of any "square" at the base of the excavation. Because the flow entering the base of the excavation is almost vertical, the four squares here are of similar size with side dimensions of $1.5\,\text{m}$.

The exit hydraulic gradient is therefore $1/1.5 = 0.67$.

From Equation 7.12, the critical hydraulic gradient is determined as $19/9.8 - 1 = 0.94$.

The safety factor against uplift failure is given by the ratio of the critical hydraulic gradient to the actual gradient. Hence the safety factor is $0.94/0.67 = 1.4$. In most situations of this sort a desirable safety factor would be at least 2 or 2.5, so the above value is less than adequate

(b) Seepage rate:

From Equation 7.10 the seepage rate is $12 \times 6 \times 10^{-7} \times \ 4/12 = 2.4 \times 10^{-6}\,\mathrm{m^3/sec/m}$ along the excavation. This is from one side of the excavation only so the total flow per metre along the excavation $= 4.8 \times 10^{-6}\,\mathrm{m^3/s}$.

(c) Pore pressure at point P:

We can calculate this from either end of the flow net. We will use the "upstream" end. From this end the number of equipotential drops to point P is 2.7 approximately (by judging the position of point P within the third square). The head loss to P is therefore 2.7 m, the pressure head at P is $17 + 2 - 2.7 = 16.3$ m, and the pore pressure is $16.3 \times 9.8 = 159.7\,\mathrm{kPa}$.

7.8   UNCONFINED FLOW NETS AND APPROXIMATIONS IN CONVENTIONAL FORMULATION

Figure 7.11 shows a flow net for unconfined seepage through a homogeneous earth dam with a seepage collector drain ("toe" drain) at its downstream toe. The purpose of such a drain is to prevent seepage flow from emerging on the downstream face of the dam, which may soften the soil at the surface and lead to erosion or even instability.

With unconfined flow nets there are two ways in which the pore pressure can be determined:

(a) Using the procedure already described above for confined flow nets.

(b) Using the intersection point of the equipotential through P with the phreatic surface. This intersection point directly gives (defines) the head along that line. Thus the pore pressure at P is given by $u_p = \gamma_w h$.

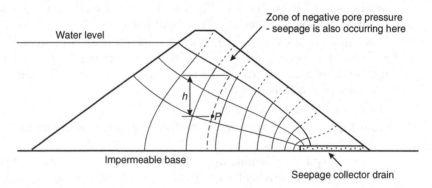

Figure 7.11   Unconfined flow through a homogeneous earth dam with a seepage collector drain.

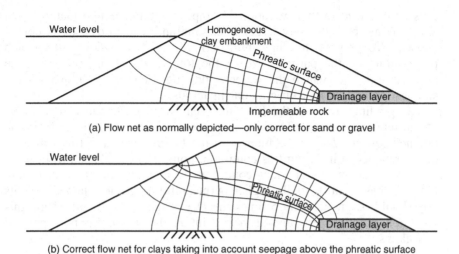

(a) Flow net as normally depicted—only correct for sand or gravel

(b) Correct flow net for clays taking into account seepage above the phreatic surface

Figure 7.12   Influence of boundary condition assumptions on seepage pattern and phreatic surface.

The treatment of unconfined flow in Figure 7.11, and earlier in Figure 7.7, follows normal practice in soil mechanics. However, this treatment is not strictly correct and violates the point made in Chapter 4, namely that in fine-grained soils the phreatic surface is not the upper boundary of the seepage zone; it is simply a line of zero (atmospheric) pore pressure. Homogeneous earth dams, such as that in Figure 7.11, can only be built of fine-grained soil (normally clay), so it is inevitable that seepage will occur above the phreatic surface. The flow nets in Figure 7.12 illustrate this point. In the top diagram, for a coarse grained material, the phreatic surface is the upper limit of the seepage zone, while in the bottom diagram, for a clay, the surface of the embankment is the upper limit of the seepage zone.

It is clear that the flow net is significantly different in each case, but there is little change in the position of the phreatic surface. Thus, despite its limitations, the assumption that the phreatic surface is the upper limit of the seepage zone in an earth dam will not normally involve serious errors in the estimation of pore pressures in the embankment. It may alter somewhat estimates of the seepage rate.

7.9   USE OF FILTERS IN DESIGNED STRUCTURES

Where seepage passes from a fine-grained soil to a coarser material, as, for example, in the collector drainage layers in Figures 7.11 and 7.12, there is a danger that soil particles will be eroded from the finer material at the interface and be carried away through the coarse material. This **internal**

erosion can lead to the formation of "pipes" within the fine material that may become progressively larger over time and lead to disastrous consequences. To prevent this happening, criteria have been developed which place limits on the particle size of the filter material in relation to the retained soil from which the water is flowing. Figure 7.13 illustrates the simplest criteria, used with granular materials.

The grading curve of the filter material is related to the D_{15} and D_{85} of the retained material. In order to prevent erosion of particles from the retained soil, the D_{15} of the filter must not be greater than 5 times the D_{85} of the retained soil. Using this condition, point d is established from point b. This is the most important condition. In addition, in order to ensure the filter has a substantially higher permeability than the retained soil, its D_{15} should be greater than 5 times the D_{15} of the retained soil. Using this condition, point c is established from point a. The criteria can be expressed as follows:

$$5D_{15(r)} \leq D_{15(f)} \leq 5D_{85(r)}$$

where $D_{15(r)}$ and $D_{85(r)}$ are the 15% and 85% sizes from the retained soil and $D_{15(f)}$ is the 15% size from the filter material. The permissible range of grading for the filter is then established by drawing curves through points c and d with similar shape to the grading curve of the retained soil.

The above criteria have been found to be very satisfactory for essentially granular materials of reasonably uniform grading. They cannot be applied to

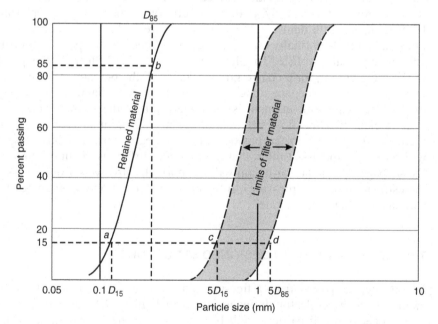

Figure 7.13 Filter criteria applicable to granular materials.

fine-grained materials, especially clays, because they result in unrealistically fine grained filters that would be very difficult to produce. However, it is found that for moderate- to high-plasticity clays, the tendency for the soil to erode is restricted by its cohesive nature, and filters can be much coarser than the above criteria would suggest. For some very broadly graded soils and soils that are gap graded, the above criteria have been found to be inadequate, and modified criteria have been developed. The work of Sherard and Dunnigan (1989) and the more recent work of Foster and Fell (2001) give recommended criteria for these soils.

7.10   VERTICAL FLOW THROUGH SINGLE LAYERS AND MULTILAYERS

Figure 7.14 shows a single clay layer through which water is seeping due to the head difference between the top and bottom of the layer. We wish to know the pore pressure at point P. The important point in this situation is to recognize that the hydraulic gradient is given by

$$i = \frac{h_1 - h_2}{T} = \frac{h}{T}$$

It is the difference in head divided by distance, not the difference in pore pressure.

Head is being lost linearly as seepage travels through the layer. The head loss Δh from the top of the layer to point P is given by

$$\frac{\Delta h}{T - d} = i = \frac{h}{T}$$

so that

$$\Delta h = \left(\frac{T - d}{T}\right) h$$

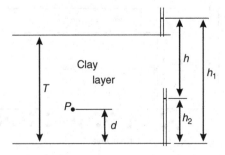

Figure 7.14   Seepage through a single clay layer.

The head at the top of the clay layer with respect to $P = h_1 - d$ so that the head at P will be this value minus the head loss between the top of the layer and point P, that is, Δh.

Thus

$$\text{Head at point } P = (h_1 - d) - \left(\frac{T - d}{T}\right) h$$

$$= (h_2 - d) + \frac{d}{T} h$$

Figure 7.15 shows seepage through a series of clay layers of different thickness and permeability. The ground surface of layer 1 marks the start of the seepage zone and is submerged by a depth of water h_w. The bottom of layer 3 marks the end of the seepage zone, where the flow exits into a highly permeable gravel layer, in which the head is h_1 above the top of the layer. We may be interested in the overall seepage rate, or the head and pore pressure at some point in the layers.

The total head lost is h and it occurs over a distance of $T_1 + T_2 + T_3$, but we do not know the head loss or hydraulic gradient in each layer. However, we do know that the flow rate through each layer must be the same, since the flow coming out of one layer must be the same as the flow entering the next layer. We can therefore calculate the head loss in each layer in terms of the flow rate and put the sum of these equal to h.

Let the head loss in layers 1, 2, and 3 be Δh_1, Δh_2, and Δh_3, respectively. Then the hydraulic gradient is given by

$$i_n = \Delta h_n / T_n$$

where n is the nth layer.

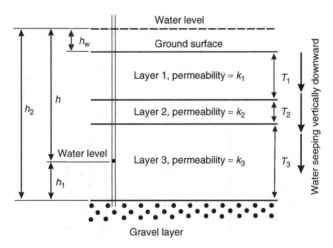

Figure 7.15   Vertical seepage through a series of clay layers.

The flow rate is given as

$$q = kiA = k_n \left(\frac{\Delta h_n}{T_n} \right) = \left(\frac{k_n}{T_n} \right) \Delta h_n$$

where q is the flow rate per unit area

Therefore each head is given by

$$\Delta h_n = q \left(\frac{T_n}{k_n} \right)$$

The sum of these must equal the total head loss so that:

$$h = q \Sigma \left(\frac{T_n}{k_n} \right) \text{ and } q = \frac{h}{\Sigma (T_n / k_n)} \qquad (7.13)$$

and the head loss through each layer is given by

$$\Delta h_n = \frac{T_n}{k_n} \frac{h}{\Sigma (T_n / k_n)} \qquad (7.14)$$

The parameter T_n / k_n is a measure of the resistance to flow of the layer, and Equation 7.14 simply states that the head loss through a particular layer is proportional to the resistance of that layer

7.11 NOTE ON GROUNDWATER STUDIES AND GROUNDWATER MECHANICS

In recent years geotechnical engineers have become more involved in groundwater studies because of environmental concerns, especially contamination of groundwater. Groundwater studies are normally carried out to investigate the use of groundwater as a resource, or the effect of some external disturbance on the groundwater state, and also the possible impact of contaminants on the groundwater. The theoretical side of groundwater studies is known as **groundwater mechanics** and involves the same basic laws as those used in soil mechanics as described above. However, groundwater studies are different in several important aspects, including the following:

1. The assumption is invariably made that the phreatic surface is the upper boundary of the seepage zone. This implies that the material is coarse grained, which may well be the case if the focus of the studies is the exploitation of groundwater as a resource.

2. It is also assumed that the phreatic surface is relatively flat, so that equipotential lines can be assumed to be vertical. The hydraulic gradient is then given by the slope of the phreatic surface. This is known as the Dupuit assumption.
3. Flow is often transient, and the governing equation is expressed using different parameters from those used by geotechnical engineers.

The purpose of including comments on groundwater studies here is to make clear the connections between the parameters used in groundwater mechanics and those used in geotechnical engineering. Groundwater mechanics normally uses the following parameters:

S_s = specific storage = volume of water released per unit volume of soil per unit change in head

S = storativity = volume of water released per unit area per unit change in head

= $S_s b$, where b = depth of seepage zone

T = permitivity = kb, where k is the coefficient of permeability (or hydraulic conductivity, as it is called in groundwater mechanics) and b is the depth of the seepage zone

For a fully saturated soil the volume of water released per unit volume of soil must equal the volume change of the soil, so we can write

$$\frac{\Delta V}{V} = S_s \, \Delta h = \frac{S_s}{\gamma_w} \, \Delta u, \text{ since } u = \gamma_w h,$$

We know also that, $\dfrac{\Delta V}{V} = m_v \Delta \sigma' = -m_v \Delta u$ (since in this situation $\Delta u = -\Delta \sigma'$) so that (ignoring the sign difference)

$$m_v = \frac{S_s}{\gamma_w} = \frac{S}{b\gamma_w} \qquad (7.15)$$

There is thus a direct relationship between the soil mechanics compressibility parameter m_v and the groundwater mechanics storativity parameters S_s and S. This is to be expected since in fully saturated soils the volume of water to flow out of a soil due to a change in pore pressure is governed directly by its compressibility.

Also by definition

$$k = \frac{T}{b} \qquad (7.16)$$

We can now recall the equation derived earlier for transient flow, namely, Equation 7.9, and substitute into it these relationships, giving

$$\frac{\partial^2 h}{\partial x^2} + \frac{\partial^2 h}{\partial z^2} = \frac{m_v \gamma_w}{k} \frac{\partial h}{\partial t} = \frac{S}{T} \frac{\partial h}{\partial t}$$

That is,

$$\frac{\partial^2 h}{\partial x^2} + \frac{\partial^2 h}{\partial z^2} = \frac{S}{T}\frac{\partial h}{\partial t} \tag{7.17}$$

This is the **general equation** used in groundwater studies, and computer programs are available for its solution. The relationship in Equation 7.15 is valid provided the soil is fully saturated. In partially saturated soils air can enter the soil and the volume change will be less than the volume of water flowing out of the element. In this case Equation 7.15 no longer applies. Also, if the zone of interest is very deep, the high stresses involved may mean that the compressibility of water needs to be included in the analysis.

7.12 FLOW INTO EXCAVATIONS, DRAINS, AND WELLS

Equations governing steady-state seepage flow into a drain (or an excavation with a vertical side) and into a circular well can be developed using the Dupuit assumption mentioned above, namely, that equipotential lines are vertical, and the hydraulic gradient is the slope of the phreatic surface. The assumption is valid for flow that is predominantly horizontal above a level impervious base, as shown in Figure 7.16. This impervious base is normally taken as the reference level for head measurements.

For two-dimensional flow as in Figure 7.16a, the seepage rate per unit width at any point with head h can be written as

$$q = kiA = k\frac{dh}{dx}h$$

where k is the coefficient of permeability and x is the horizontal ordinate.

We can rearrange and integrate this between any two points such as those at distances d_0 and d_1 from the edge of the drain. Then

$$\frac{q}{k}dx = hdh$$

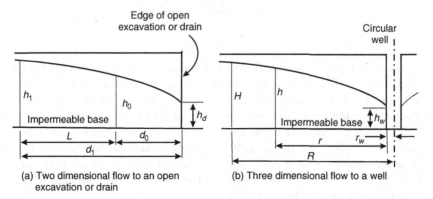

(a) Two dimensional flow to an open excavation or drain

(b) Three dimensional flow to a well

Figure 7.16 Seepage flow into a drain (or excavation) and a circular well.

Therefore

$$\frac{q}{k} \int_{do}^{d1} dx = \int_{ho}^{h1} h\, dh$$

and

$$\frac{q}{k}(d_1 - d_o) = \frac{h_1^2 - h_2^2}{2}$$

giving

$$q = \frac{k\left(h_1^2 - h_2^2\right)}{2L} \tag{7.18}$$

For flow into the drain (from one side only) or excavation this becomes

$$q = \frac{k\left(h_1^2 - h_d^2\right)}{2d_1} \tag{7.19}$$

For 3-D flow, as in Figure 7.16, the seepage rate toward the well is given by

$$q = kiA = k\frac{dh}{dr}2\pi rh$$

We can again rearrange and integrate this between the radius of the well r_w and the radius R:

$$\frac{q}{2\pi k}\frac{dr}{r} = h\, dh,$$

Therefore

$$\frac{q}{2\pi k} \int_{rw}^{R} \frac{dr}{r} = \int_{hw}^{H} h\, dh$$

giving

$$q = \frac{\pi k\left(H^2 - h_w^2\right)}{\ln \dfrac{R}{r_w}} \tag{7.20}$$

The question we may well ask at this stage is: of what use are Equations 7.18 to 7.20? The answer is that they are of rather limited use for several reasons. First, the geometry on which they are based is very simplistic. The drain, excavation, or well must extend all the way to the impermeable base, which is assumed to be level. Such drains or wells are termed "fully penetrating." Second, the assumption is made that the source of water (the recharge source) is an isolated source some distance from the drain or well. In practice, surface rainwater may be the main recharge source. Third, the formulas are for steady-state flow only.

Despite these limitations, the equations can be used to calculate the seepage rate if the coefficient of permeability and the values of head at two locations are known. Alternatively, the flow rate can be measured and the coefficient of permeability calculated. It is not unusual for wells to be installed specifically for this purpose. However, the performance of wells and the aquifers in which they are installed often requires evaluation under transient conditions, in which case more sophisticated analysis is required.

REFERENCES

Foster, M. and R. Fell. 2001. Assessing embankment dam filters that do not satisfy design criteria. *ASCE Journal of Geotechnical and Geoenvironmental Engineering*, Vol. 127, No 5, 398–407.

Sherard, J. L. and L. P. Dunningan. 1989. Critical filters for impervious soils. *ASCE Journal of Geotechnical Engineering*, Vol. 115, No 7, 927–947.

Strack, O.D.L. 1989, *Groundwater mechanics*. Englewood Cliffs, NJ: Prentice Hall.

EXERCISES

1. Figure 7.17 shows a concrete gravity dam constructed on a layer of sandy clay. Sketch a flow net and use it to:

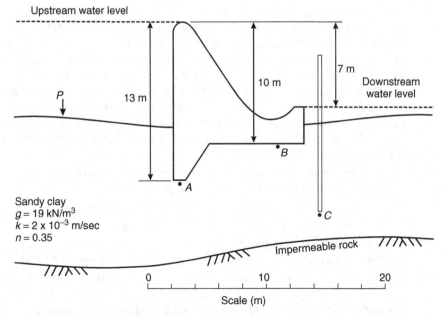

Figure 7.17   Seepage beneath a concrete weir (for exercise 1).

(a) Derive the formula for seepage flow rate, explaining carefully the steps in your derivation.

(b) Determine the pore pressure at points A and B. Also determine the height, above the downstream water level, to which water will rise in the standpipe piezometer at point C. **(103.5 kPa, 47.8 kPa, 1.69 m)**

(c) Determine the safety factor with respect to uplift ("piping") failure at the downstream toe. **(1.54)**

(d) Determine the time it would take for a contaminant to travel from point P to exit at the downstream surface. **(8.4 hours)**

2. At a level site water is seeping upward through a deep sandy clay layer to the surface. The sandy clay has a unit weight of 17.6kN/m^3 and a coefficient of permeability of $5 \times 10^{-6} \text{m/s}$. A standpipe piezometer installed in the sand to a depth or 7.0 m shows a rise in water level to 3.5 m above the ground surface. Estimate the following:

(a) Hydraulic gradient in sandy clay layer **(0.5)**

(b) Total and effective vertical stress at depth of 5 m in sandy clay layer **(88.0 kPa, 14.5 kPa)**

(c) Maximum height to which water could rise in standpipe before uplift failure would occur **(5.57 m)**

(d) Seepage rate per square meter **($2.5 \times 10^{-6} \text{m}^3/\text{s/m}^2$)**

3. A former sand quarry has a wide level base and is to be used as a rubbish tip (a landfill). The sand has been excavated to just reach the water table, which is fixed at a constant level by its connection with a nearby lake. A clay lining layer is to be placed immediately on top of the sand to form a low-permeability barrier to minimize seepage of leachate (contaminated water from the landfill) into the groundwater. A limited quantity of low-permeability clay with a coefficient of permeability of 10^{-8}m/s is available but would only make possible a layer 0.5 m thick. Also available is a different clay with a coefficient of permeability of $0.8 \times 10^{-7} \text{m/s}$ that would make possible an additional layer 1.0 m thick. Assuming a depth of leachate above the clay liner of 0.1 m, determine the following:

(a) Seepage rate with only a 0.5-m-thick layer consisting of low-permeability material **($1.2 \times 10^{-8} \text{m}^3/\text{s/m}^2$)**

(b) Seepage rate if a composite liner (double layer) is used consisting of 0.5 m of low-permeability and 1.0 m of higher permeability clay **($2.56 \times 10^{-8} \text{m}^3/\text{sec/m}^2$).**

(c) Pore pressure at boundary of two (composite) layers **(7.6 kPa)**

4. Figure 7.18 shows sheet piling driven to form a cofferdam so that an excavation can be carried out inside it. Excavation has been carried out under water until the level of the sand within the cofferdam is 7 m

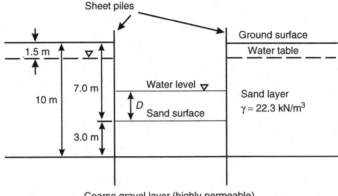

Figure 7.18 Sheet pile excavation (for exercise 4).

below ground level. Pumping is then commenced to reduce the water level within the cofferdam.

Find the level of the water within the cofferdam when "uplift" instability of the base of the excavation is expected. The unit weight of the sand is $22.3\,\text{kN/m}^3$. Assume that the pumping does not affect the groundwater level outside the cofferdam or the head in the gravel layer. **(1.67m)**

Note: this is a simple stability situation—flow net construction is not needed.

5. Figure 7.19 shows a shallow pond from which water is seeping vertically through the two clay layers to the underlying sand layer. The sand layer is of very high permeability and a standpipe piezometer (measuring tube) installed in the sand layer shows the head in the sand is 1.0 m above the top of the sand.

Determine the seepage rate from the pond and also the total stress, pore pressure, and effective stress at the boundary of the two clay layers. **($Q = 3.76\,\text{m}^3/\text{s/m}^2$, $\sigma = 47.3\,\text{kPa}$, $u = 22.7\,\text{kPa}$, $\sigma' = 24.6\,\text{kPa}$)**

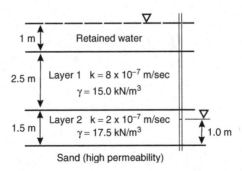

Figure 7.19 Vertical seepage situation for exercise 5.

CHAPTER 8

COMPRESSIBILITY, CONSOLIDATION, AND SETTLEMENT

8.1 GENERAL CONCEPTS

When loads are applied to a soil, as, for example, when a building or an embankment is built, as shown in Figure 8.1, the soil deforms and compresses and the structure settles. A similar process happens if the water table is lowered by drainage measures or by pumping from wells.

Most of the deformation that occurs is due to water being squeezed out of the soil, with the result that the soil particles move closer together. Compression of the soil particles themselves is negligible compared to the reduction in void space between them. This is indicated in Figure 8.2.

The process by which an increase in stress causes water to flow out of the soil accompanied by volume reduction is referred to as **consolidation**. This term should not be confused with **compaction**, which is a mechanical process in which dynamic energy is used to make the soil more compact. Compaction does not involve any removal of water from the soil; it is described fully in Chapter 15. Consolidation is a process that occurs in clays and silts. With sands and gravels, pore water drainage from the voids occurs almost instantaneously as load is applied and is not normally referred to as consolidation. There are a variety of methods that can be used to estimate the settlement resulting from building loads. These are described in the following sections.

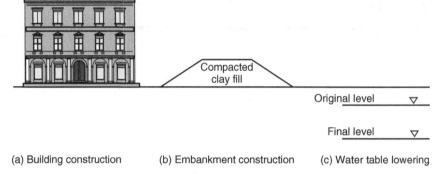

(a) Building construction   (b) Embankment construction   (c) Water table lowering

Figure 8.1   Situations where stress changes cause settlement of the ground surface.

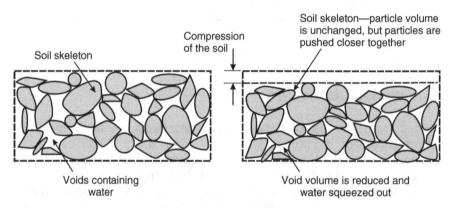

Figure 8.2   Compression of soil due to an increase in vertical stress acting on the soil.

8.2   ESTIMATION OF SETTLEMENT USING ELASTICITY THEORY

Elasticity theory solutions for determining stresses within a soil mass resulting from applied external loads are described in Chapter 5. Solutions are also available for estimating the settlement of foundations on the surface of an elastic medium, as illustrated in Figure 8.3. As with the earlier solutions, these are based on the assumption that the "soil" consists of a uniform elastic material of infinite depth.

For a uniformly loaded circular foundation the maximum settlement is at the center and is given by

$$\delta = qD\frac{1 - \nu^2}{E} \tag{8.1}$$

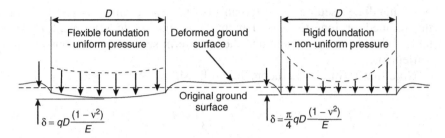

Figure 8.3 Elasticity theory solutions for flexible and rigid circular foundations.

For a rigid circular foundation

$$\delta = \frac{\pi}{4} q D \frac{1 - v^2}{E} \tag{8.2}$$

where D = diameter
$\quad\quad E$ = Young's modulus of the soil
$\quad\quad v$ = Poisson's ratio of the soil

The uniformly loaded foundation is flexible; otherwise the pressure would not remain uniform. The resulting settlement is variable with the maximum value at the center and the minimum at the edges. With the rigid foundation, the pressure across the foundation is not uniform. The minimum value occurs at the center and the maximum value at the edge. Large storage tanks for water or oil will normally perform as flexible foundations, while reinforced-concrete foundations for buildings will perform as rigid foundations.

In general, soils are not elastic, though in some situations the assumption of elastic behavior is reasonable. If we are to use the above formula, both the Young's modulus and Poisson's ratio of the soil must be known.

8.2.1 Drained and Undrained Behavior

In using elastic theory solutions, both drained and undrained behavior must be considered, as the stiffness of the soil is not the same in each case. There are two sets of elastic parameters, one for undrained loading (E and v) and one for drained loading (E', v'). For fully saturated soils, undrained loading takes place under conditions of zero volume change, which means that $v = 0.5$.

For soils of very low permeability (clays of high plasticity), there may be no significant consolidation during construction and application of load to the foundation. The load application is thus undrained, and the use of undrained values of E and v in the formula above will give the undrained or "immediate" settlement. We can also use the drained values to obtain the drained or long-term settlement. The immediate settlement is often termed the elastic settlement and the long-term settlement the consolidation settlement.

Note that from elastic theory it can be shown that

$$E' = \frac{2(1 + v')}{3} E \qquad (8.3)$$

Since $v' < 0.5$, E' will be less than E; hence the deformations under drained loading will be larger than under undrained loading.

8.2.2 Limitations of Elasticity Theory

The above elastic theory solutions are not very widely used in practice, for the following reasons:

(a) Soils are not elastic; their stress–strain curves are not linear, and deformations occurring during loading are not normally recoverable on unloading.
(b) Measurement of the values of Young's modulus and Poisson's ratio is difficult, especially the drained values.
(c) Soils are not normally homogeneous; the soil profile may consist of several distinct layers of quite different stiffness.

8.3 ESTIMATION OF SETTLEMENT ASSUMING 1-D BEHAVIOR

To overcome the problems associated with the use of elastic theory, settlement estimates are commonly made on the basis of one-dimensional behavior. It is assumed that no lateral deformation occurs as a result of the vertical loading. In this situation the soil stiffness (or compressibility) is expressed in terms of one-dimensional parameters, rather than in terms of the elastic theory parameters E and v. We have already described the simplest one-dimensional parameter m_v in Chapter 6, defined by Equation 6.5 as follows:

$$\frac{\Delta L}{L} = m_v \Delta \sigma'$$

Using elastic theory it can be shown that m_v is related to the drained elastic parameters E' and v' as

$$m_v = \frac{1}{E'} \left(1 - \frac{2v'^2}{1 - v'} \right) \qquad (8.4)$$

Because both soil properties and the stress increase induced by the foundation load normally change with depth, it is necessary to divide the soil profile into a number of sublayers and individually calculate the compression of each layer. The procedure is demonstrated in the example shown in Figure 8.4. The soil is assumed to be homogeneous with an m_v value of $2 \times 10^{-4} \, \text{kPa}^{-1}$.

A circular tank applies a pressure of 100 kPa at the surface. To calculate the settlement, the soil layer is divided into seven sublayers. Equation 5.2 or Figure 5.5 are used to calculate the stress increase in each layer. Equation 6.5 is then used to calculate the compression of each layer. The sum of the values gives the settlement of the center of the tank. The calculation is set out in Table 8.1.

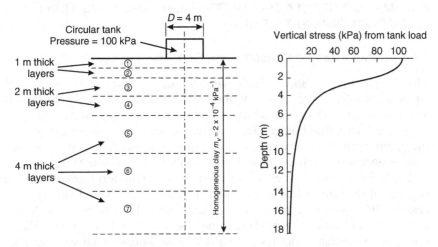

Figure 8.4   Settlement calculation using sub-layers and assuming one-dimensional compression.

Table 8.1   Settlement Calculation

Sublayer	Layer Thickness L (m)	Depth to Layer Center z (m)	R/z	I_σ	$\Delta\sigma'$ (kPa)	ΔL (mm)
1	1	0.5	4	0.986	98.6	19.7
2	1	1.5	1.33	0.839	83.9	16.7
3	2	3.0	0.67	0.427	42.7	17.1
4	2	5.0	0.40	0.200	20.0	8.0
5	4	8.0	0.25	0.087	8.7	7.2
6	4	12	0.167	0.040	4.0	3.2
7	4	12	0.125	0.023	2.3	1.8

Total: 73.7 mm

We can also calculate the settlement using elasticity theory. Assuming a midrange value for Poisson's ratio of 0.25, Equation 8.4 gives $E' = 4167\,\text{kPa}$.

From Equation 8.1 we obtain $\Delta L = 90\,\text{mm}$.

Elasticity theory thus gives an estimate about 20 percent higher than the one-dimensional analysis. This is because the one-dimensional method implies complete lateral restraint, while elastic theory predicts some horizontal deformation. These estimates are for the drained or consolidation (or long-term) settlement. We will now give more detailed consideration to the question of immediate settlement and its relationship to the long-term settlement.

8.4 IMMEDIATE ("ELASTIC") SETTLEMENT AND LONG-TERM (CONSOLIDATION) SETTLEMENT

8.4.1 Immediate and Consolidation Settlement in Sands

The permeability of sands is such that any pore pressures generated will dissipate as rapidly as the load is applied, and the behavior is always drained. The exception to drained loading is earthquake loading, which will not be considered here. The immediate settlement is the main component of settlement, although there may still be some slow, long-term "creep" settlement. This is normally small compared to the immediate settlement.

The relative magnitudes of immediate and long-term settlement will depend on the rate of load application. If the load is applied slowly over a considerable time, then most of the settlement will be induced as the load is applied, and long-term settlement will be small, and vice versa. We can note in passing that buildings normally take months or even years to build, so the rate of load application is relatively slow. In contrast, storage tanks can be built quickly and can be filled in a matter of hours or days, so that the rate of loading is high.

8.4.2 Immediate and Consolidation Settlement in Clays

In clays, the situation is different from sands and more complex. The load application and the response of the soil occur in one of the following three ways:

Case 1: Completely undrained when the load is applied followed by slow consolidation. This will be the case if the soil is of low permeability and the load is applied over a short period of time.

Case 2: Completely drained at all times. This will be the case if the soil is of relatively high permeability and the load is applied slowly. The consolidation settlement will all occur as the load is applied.

Case 3: Partially drained. This situation is intermediate between the above two and is probably the most common case. Some drainage (consolidation) occurs as the load is applied, and further consolidation continues with time under constant load.

We can use elastic theory, and the undrained elastic parameters of the soil to estimate the immediate settlement that occurs in case 1 above. We can also use the effective stress (drained) parameters, or the one-dimensional method described Section 8.3, to estimate the consolidation settlement that occurs in case 2. This drained settlement will naturally be greater than the undrained value.

Figure 8.5a illustrates conceptually the settlement and deformations that will occur during cases 1 and 2 above. The full line shows only the undrained deformation that forms part of case 1. It is seen that in addition to settlement there is ground heave away from the foundation, and considerable lateral deformation beneath the edge of the foundation, which is to be expected since no volume change occurs. The first dashed line shows the fully drained deformations of case 2. There is now less lateral deformation and no significant ground heave, which is also to be expected since volume change occurs as the load is applied.

If soils were elastic, the total settlement from case 1 after consolidation is complete would be the same as that from case 2, since the final stress state is the same. However, because soils are not elastic, the final settlement from case 1 is normally greater than case 2, as indicated by the second dashed line in the figure.

In practice, the foundation loading is likely to be between fully undrained and fully drained (case 3 above), so that the total settlement will be greater than that arising from fully drained loading (case 2) but less than that arising from fully undrained loading followed by consolidation (case 1). Figure 8.5 (b) illustrates the settlement versus load curves corresponding to the above cases. Curve a-b_1-c_1 is for undrained loading followed by consolidation. Curve a-c_3 is for fully drained loading. Curve a-b_2-c_2 is for partially drained loading followed by consolidation.

What we are saying here is that the final deformations are dependent on the sequence in which the stresses are applied, and not just on the final stress state. This sequence is referred to as the **stress path** and is explained in detail in Section 9.6. The Mohr's circle of stress, also explained in Chapter 9, and the effective stress paths for a typical soil element representing the loading conditions in the above cases are shown in Figure 8.5c. For drained loading the soil follows stress path $a-c$, and for undrained loading it follows stress path $a-b$. During consolidation it follows path $b-c$. As a general rule, the closer the stress path is to the failure line, the greater will be the deformations. This is because the soil stiffness is least when it is close to failure. Thus the deformations are greater for stress path $a-b-c$ than for $a-c$.

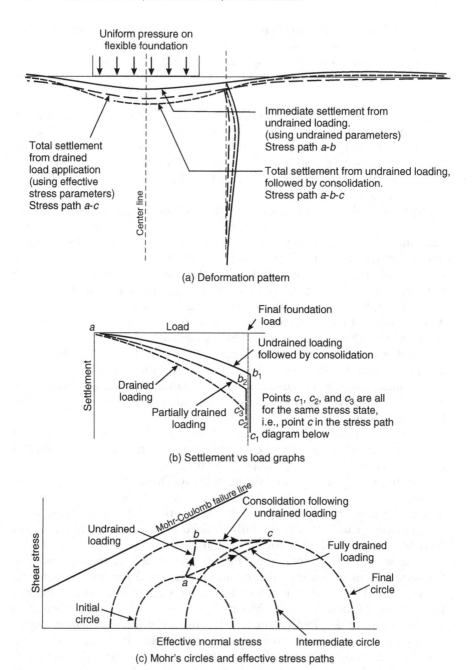

Figure 8.5 Immediate and long-term settlement and deformation in clays.

Most settlement estimates for foundations on clay assume one-dimensional behavior, despite the fact that in theory this underestimates the settlement. In practice, it is still a reasonable approach as other effects, such as sample disturbance, are likely to result in overestimates of settlement. Skempton and Bjerrum (1957) proposed a method for estimating the total settlement by summing the immediate settlement and the consolidation settlement after the latter has been corrected to account for the actual pore pressure induced during the undrained loading. The method does not take into account the construction time and any consolidation occurring during construction. It may be a reasonable approach for sedimentary soils but of doubtful applicability to residual soils. For most situations, it is generally sufficiently accurate to calculate the settlement using one-dimensional effective stress parameters without attempting to divide it into immediate and consolidation components. In addition to the linear parameter m_v, there are alternative one-dimensional parameters based on a log scale for pressure; we will examine these in the following sections.

8.5   CONSOLIDATION BEHAVIOR OF CLAYS (AND SILTS)

Settlement of buildings or embankments on low-permeability clays may take several years, or even decades, to complete. Both the magnitude of compression and the rate at which it occurs are of interest to geotechnical engineers. There is a wide variation in the compression characteristics of clays, depending on their composition and structure. As a starting point to understanding these characteristics, we will examine the results of laboratory compression tests on a range of soil types.

8.5.1   Odometer Test

To measure both magnitude and rate, consolidation tests are carried out in a device known as an odometer, a diagrammatic view of which is shown in Figure 8.6. An undisturbed sample of the soil is trimmed into a circular ring having a diameter of about 80 mm and a thickness of 20 mm. The porous stones allow water to drain from top and bottom of the sample when vertical pressure is applied. The sample is then placed between porous stones in a holding cell and set up in a loading frame.

A series of known vertical pressures is then applied to the sample using a weight and lever system that forms part of the loading frame of the odometer. As each pressure increment is applied, readings of vertical compression are taken at regular time intervals until movement ceases; the next load is then applied. The measurements made provide a picture of both compression versus stress behavior and compression versus time behavior. Once the applied pressure reaches the vertical effective stress that acted on the sample in the field, water is added to the odometer cell to ensure no water is lost by evaporation.

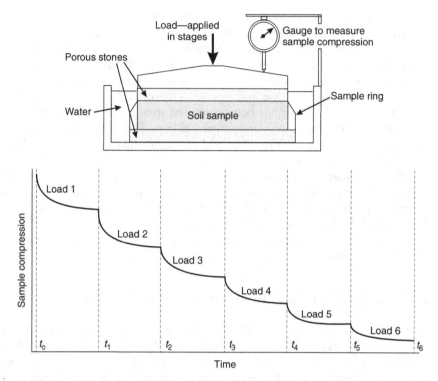

Figure 8.6 Odometer test for measuring consolidation behavior.

8.5.2 Consolidation Characteristics — Magnitude

Sedimentary Soils We can investigate the compression behavior of a sedimentary soil by simulating its formation process in an odometer test. The natural formation process, described in Chapter 1, involves sedimentation in a lake or marine environment followed by compression as further soil is deposited above it. To simulate this, we can prepare a sample at a very high water content, close to or above its liquid limit, and then apply a series of loads in sequence, leaving each load in place until movement ceases. A graph obtained in this way is shown in Figure 8.7.

During the test, the behavior during unloading and reloading has also been investigated. The load was raised in steps to point b and then reduced to point c, then reloaded to point d and continued on to point e. The load was then reduced to point f and reloaded again to point g. It is evident that the line $a-b-e$ is the line representing its behavior when first compressed. This line is known as the **virgin consolidation line**. When the pressure is reduced, as, for example, from point b to point c, and then reapplied to point d, the soil almost returns to its former void ratio and rejoins the virgin consolidation line.

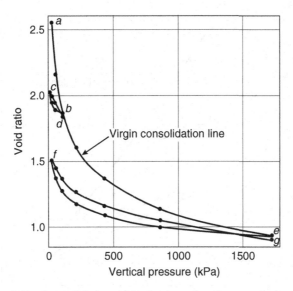

Figure 8.7 Consolidation of a clay, starting from a slurry state.

It is very apparent from Figure 8.7 that the soil does not behave in an elastic manner. Its void ratio–pressure graph is not a straight line and in general is irreversible. The soil clearly becomes stiffer as the pressure increases. Even the unloading and reloading curves are not linear—they form a closed loop and involve a form of hysteresis. It is not possible with this type of behavior to define simple parameters to express the compressibility characteristics of the soil. The linear parameter m_v varies with both stress level and the sequence in which the stress is applied.

The data in Figure 8.7 is replotted in Figure 8.8 using a logarithmic scale for pressure. When this is done, a different picture emerges. The virgin consolidation line now becomes almost a straight line, and the unloading and reloading curves also become much more linear. There is no theoretical reason why this should be the case, and not all soils produce a virgin consolidation line as straight as that in Figure 8.8. However, it is certainly the case that the logarithmic scale generally produces graphs similar to those in Figure 8.8, and for this reason the logarithmic plot is normally used for presenting the results of odometer tests. We will see later that this is not necessarily good practice, especially with residual soils.

As a result of the behavior illustrated in Figure 8.8, it has become standard practice to portray the compression behavior of sedimentary clays in the idealized fashion shown in Figure 8.9 and to define this behavior in terms of two parameters, C_c and C_s. The parameter C_c is the slope of the virgin consolidation line, which is assumed to be linear on the

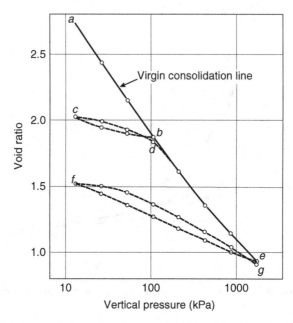

Figure 8.8 Consolidation of a clay, using a logarithmic scale for pressure.

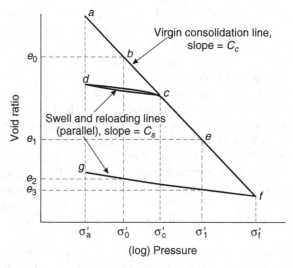

Figure 8.9 Idealized representation of compression behavior of sedimentary soils.

logarithmic plot. It is termed the compression index and is defined by

$$C_c = \frac{e_o - e_1}{\log \frac{\sigma'_1}{\sigma'_o}} = \frac{\Delta e}{\log \frac{\sigma'_1}{\sigma'_0}} \tag{8.5}$$

where log denotes the logarithm to the base 10.

The parameter C_s is the slope of the unloading (or swell) and reloading lines, which are assumed to be linear and parallel regardless of stress level. It is termed the swell index and is defined by

$$C_s = \frac{e_3 - e_4}{\log \frac{\sigma'_1}{\sigma'_0}} \tag{8.6}$$

The values of C_c and C_s are most easily determined by taking the change in void ratio over one log cycle, since the denominator in Equations 8.5 and 8.6 then becomes unity.

It is assumed that there is a sharp change of slope where the reloading lines, such as $d-c$ or $g-f$, meet the virgin consolidation line $a-f$. If a natural soil has been consolidated to a stress σ'_c and then undergone stress reduction (from overburden removal) to $\sigma'_d, (= \sigma'_a)$ then it is assumed to follow path $d-c-f$ when reloaded as a result of foundation construction or the placing of fill on it.

This representation (or "model") is an idealization only, based on the behavior of remolded soils starting from a slurry state. Many soils do not conform to this picture, as we shall see shortly. The following shortcomings in this model should be noted:

(a) Natural soils harden with age due to the development of bonds between particles or other chemical effects. This means that the consolidation graph for the natural soil may no longer relate to the virgin graph produced when the soil was originally deposited.

(b) The virgin consolidation line may not be linear, in which case C_c is not constant

(c) The swell and reloading lines are not coincident, and often not parallel to each other, so that C_s is not a constant. This is seen to be the case with the soil portrayed in Figure 8.8.

(d) There is not necessarily a sharp change of slope when the reloading line meets the virgin consolidation line. Figure 8.8 illustrates this point; when the reloading line $f-g$ meets the virgin consolidation line, its slope is nearly the same as the virgin line.

We thus have two different coefficients (or sets of coefficients) for representing the one-dimensional compression behavior of clay. The first and simplest is the linear parameter m_v and the second is the log parameter C_c and its companion parameter C_s. The choice of which parameter to use in practice will become clearer when we consider settlement estimates later.

Normally Consolidated Soils, Overconsolidated Soils, and Preconsolidation Pressure

If a natural sedimentary soil, after formation, has been subject to an overburden stress equal to that at point $c(\sigma_c')$ in Figure 8.9 and no reduction in stress has occurred, it is termed a **normally consolidated soil**. If, however, some of the soil above it has been eroded by the action of water or glaciation and the stress reduced to that at point $d(\sigma_a')$, then the soil is termed an **overconsolidated soil**. A soil existing at point g would also be overconsolidated. The stress at points c and f is termed the **preconsolidation pressure**. The ratio of the preconsolidation pressure to the existing effective overburden pressure (σ_c'/σ_a') is termed the **overconsolidation ratio (OCR)**. The soil existing at point g clearly has a higher OCR than the soil existing at point d.

We will now examine some consolidation test results from undisturbed sedimentary soils to see how they behave and to what extent the "model" portrayed in Figure 8.9 is a valid representation of real soil behavior. Figure 8.10 shows typical odometer test results from a "normally consolidated" clay plotted using both logarithmic and linear scales.

The soil is normally consolidated from a geological point of view, meaning that its geological history since deposition indicates it has not been subject to a vertical effective stress greater than its current in situ value, which is about 95 kPa. Figure 8.10a shows a very sharp change in gradient of the consolidation graph, or preconsolidation pressure, at about 200 kPa, which is over twice the in situ stress. The soil thus appears to have an OCR of about 2.

The linear plot in Figure 8.10b confirms the picture given in the log plot. The soil initially behaves as a stiff material, but a sharp increase in compressibility occurs between about 100 and 200 kPa. This behavior is common; there are in fact very few soils that behave as strictly normally consolidated materials. As mentioned in Chapter 1, sedimentary soils, after deposition, undergo further processes that alter their chemical composition, and their properties change over time. They are likely to develop weak bonds between particles and become stiffer of harder. They may also undergo leaching processes, which can dramatically alter their properties. These hardening or leaching processes can have a more significant influence than stress history.

For the above reasons, the term **vertical yield stress** is a better term than preconsolidation pressure to denote the pressure at which the soil softens and the stiffness decreases and is being increasingly used in this context. This vertical yield stress can be caused by a number of factors, only one of which is stress history, and stress history influence is limited to sedimentary soils. A further point which should be noted in Figure 8.10a is that the swell

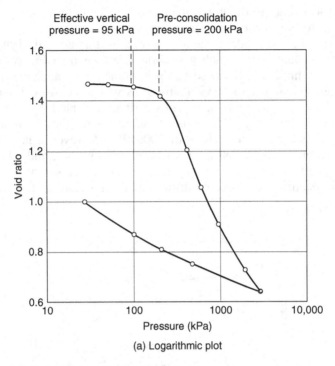

(a) Logarithmic plot

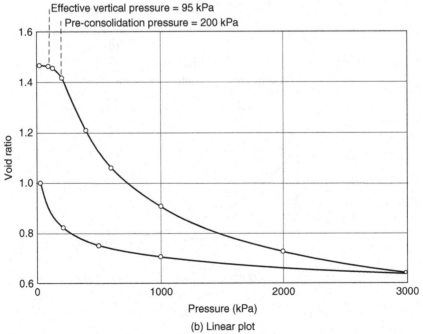

(b) Linear plot

Figure 8.10   Odometer test results from a normally consolidated clay.

line is not parallel to the initial loading line. The latter is almost horizontal while the former has quite a steep gradient.

We will now consider a heavily overconsolidated soil. Typical results of an odometer test for such a material are shown in Figure 8.11, again using logarithmic and linear scales. It is immediately apparent that there is no longer a clearly defined yield stress, as was the case with the normally consolidated soil in Figure 8.10. Considering the behavior illustrated earlier in Figure 8.8, the log plot seems to indicate a preconsolidation pressure, or yield pressure, in the vicinity of 1000 kPa. However, a large proportion of the loading curve consists of a smooth circular arc, and when the data are replotted using a linear scale, there is no indication of a yield stress. The soil steadily increases in stiffness as the pressure increases. It should

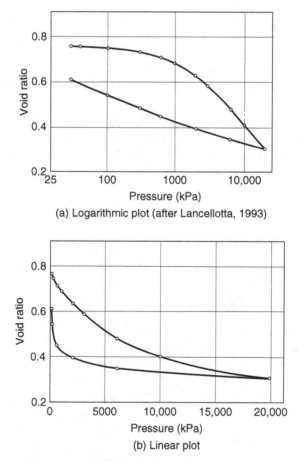

Figure 8.11 Odometer test results from an overconsolidated clay. (Adapted from Lancellotta, 1995)

be noted again from Figure 8.11a that the unloading line is not parallel to the initial loading section of the curve.

Important points to emerge from these odometer tests on real soils are the following:

(a) The behavior is only "moderately" similar to that portrayed in Figures 8.8 and 8.9 for the remolded slurry soil and the idealized "model." The behavior of the normally consolidated soil is similar to curve $c-d-e$ of Figure 8.8, and that of the overconsolidated soil is similar to curve $f-g$.

(b) The behavior of the normally consolidated soil, however, differs in that the "preconsolidation" pressure evident in the graphs does not coincide with the true value known from its geological history.

(c) The idealized representation of sedimentary soil behavior shown in Figure 8.9 is very much an approximation and is clearly defective on several points. In particular, the unloading and reloading curves are not parallel, and with heavily overconsolidated soils there is no clear preconsolidation pressure (or vertical yield stress).

It must be emphasized that the behavior in Figures 8.10 and 8.11 is not representative of all natural sedimentary soils. A wide range of behavior is found with these soils, which cannot be covered here. It is believed the behavior in Figures 8.10 and 8.11 is valid for a large proportion of such soils and is adequate to illustrate the most significant trends in real soil behavior in comparison to the idealization in Figure 8.9.

Residual Soils

As described in Chapter 1, the formation of residual soils does not involve the processes of sedimentation and consolidation that are essential components of sedimentary soil formation. There is therefore no reason to expect their compression behavior to follow that of sedimentary soils. Residual soils do not have a virgin consolidation line and there is no logical reason at all to plot odometer test results using a log scale for pressure. The use of this scale arises from the behavior of a slurry soil illustrated in Figure 8.8, which has no relevance to residual soils. Despite this, the log scale continues to be widely used (rather blindly) for residual soils and the assumption is made that residual soil behavior is the same as that of sedimentary soil. This is not necessarily the case, as the following sections illustrate.

Figure 8.12 shows the results of odometer tests on four undisturbed samples of residual soils. In Figure 8.12a they are plotted in the conventional manner, and in Figure 8.12b they are plotted using a linear scale for pressure and percentage compression in place of void ratio. The use of percentage compression makes it easier to compare the compressibility of the four soils.

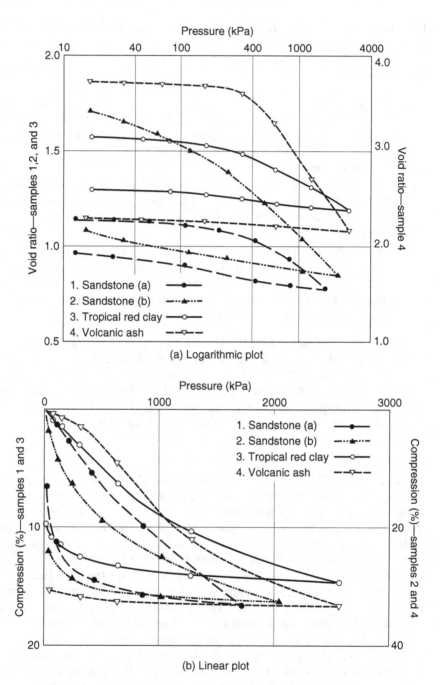

Figure 8.12 Odometer test results from four residual soils.

The four samples are from fairly common types of residual soils. Samples 1 and 2 are clays weathered from sandstones, sample 3 is a tropical red clay of volcanic origin, and the last is a clay derived from volcanic ash. They are all true clays, the weathering process having progressed to a state where virtually no trace remains of their parent material. Two of the samples are considerably more compressible than the others, and for this reason the vertical scales used in the figure are not the same for all the samples. The prime purpose of the figures is to examine the relative shapes of the compression curves, rather than to compare magnitudes.

The graphs in Figure 8.12a with the log scale convey the impression that the behavior of these soils is similar to that given earlier for sedimentary soils. The graphs are all concave from below, and their shape possibly suggests yield stresses at points of maximum curvature. However, the linear plot in Figure 8.12b shows that only one sample (sample 4) has a yield stress, which is between 300 and 400 kPa. The other three samples show no evidence of a yield stress; instead they show a steady increase in stiffness as the pressure increases. At high stress levels all the samples show steadily decreasing compressibility. Figure 8.12 emphasizes the importance of presenting compression behavior using linear plots, especially with residual soils, to gain a true impression of their behavior.

To illustrate more clearly the behavior at low stress levels likely to be relevant to engineering situations, the initial part of Figure 8.12b, up to a stress of 750 kPa, is reproduced in Figure 8.13. This shows the range of

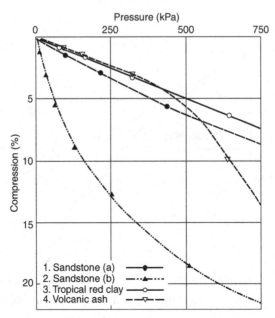

Figure 8.13 Compression behavior at stresses below 750 kPa.

behavior to be expected in residual soils. Many residual soils, when plotted using a linear scale for pressure, exhibit behavior that is close to linear, at least over the lower stress range. This is the case with samples 1 and 3. Other residual soils show a clear yield stress, as is the case with sample 4, while others become progressively stiffer over the complete stress range, as is the case with sample 2.

All three types of behavior are common with residual soils and can sometimes be found within the same geological soil type. There is thus no clear link between soil type and compression behavior and it should not be assumed that the behavior in Figures 8.12 and 8.13 will necessarily be found with these particular soil types. Weathered sandstones and volcanic ash clays can exhibit all three types of behavior in samples taken at different depths from the same site. In the author's experience, tropical red clay shows more uniform behavior without a yield stress.

The significance of Figure 8.13 is not limited to residual soils. With sedimentary soils, it is often the case that normally consolidated soils show behavior similar to the top curve, while lightly or moderately overconsolidated soils approximate to the two middle curves and heavily overconsolidated soils show behavior similar to the bottom curve. Figure 8.14 shows a general representation of soil compressibility valid for all natural soils, sedimentary or residual. This presents a more realistic picture of soil compressibility, especially in the stress range of interest to geotechnical engineers, than the traditional picture shown earlier in Figure 8.9.

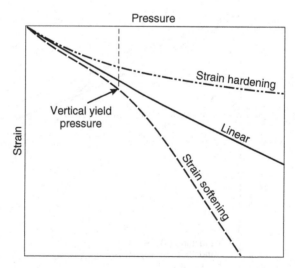

Figure 8.14 General representation of soil compressibility for residual and sedimentary soils.

It is important to recognize that applying stress to a natural soil has two effects:

1. The soil particles are pressed closer together. This will tend to make the soil harder and increase its "stiffness."
2. The structure of the soil will be damaged or destroyed. This will normally tend to make the soil softer, resulting in reduced stiffness.

The shape of the compression curve depends on which effect is the greater. Compact unstructured soils will display strain hardening, and highly structured "noncompact" soils will display strain softening.

The advantages of the linear plot over the log plot are the following:

(a) It gives a true picture of the compressibility of the soil. In particular, it avoids giving the impression that the compression behavior of all soils is similar.
(b) It establishes the presence or not of a yield (or preconsolidation) stress.
(c) The compressibility parameter m_v can be calculated directly from the curve.
(d) It allows direct comparison of the compressibility of different soils.

In the author's view, for natural (undisturbed) soils, the linear plot is clearly preferable to the logarithmic plot, except for very soft normally consolidated soils.

The use of odometer test results for settlement estimates is described in Section 8.4. At this stage we will only derive the relevant formulas used in settlement calculations.

(a) Using the linear parameter m_v:
   The formula is self-evident from the way m_v is defined, as in Equation 6.5, so we can write

$$\delta = m_v H \, \Delta\sigma' \qquad (8.7)$$

where H is the layer thickness, δ the compression, and $\Delta\sigma'$ the stress increase.

(b) Using the log parameter C_c:
   This parameter is defined by the expression

$$C_c = \frac{\Delta e}{\log(\sigma'_1/\sigma'_o)} \quad \text{(Equation 8.5)}.$$

So we can write

$$\frac{\delta}{H} = \frac{\Delta e}{1 + e_o} = \frac{C_c}{1 + e_o} \log \frac{\sigma'_1}{\sigma'_o}$$

and hence

$$\delta = \frac{C_c}{1 + e_o} H \log \frac{\sigma_1'}{\sigma_0'} \tag{8.8}$$

8.5.3 Consolidation Behavior–Time Rate

In addition to magnitude, geotechnical engineers are also interested in whether compression and settlement will occur quickly, or will continue for many years. Unlike sands, where compression is almost instantaneous on the application of load, compression of clay occurs slowly, with the result that settlement of building foundations may continue for a long time after the completion of the building. In extreme cases, significant settlement may still be occurring decades after completion of the building. There are two reasons for this time-dependent compression behavior:

1. The first and most important reason is that compression is governed primarily by the rate at which pore water can drain out of the soil. When stress is applied to a soil, as in an odometer test, or from construction of a building foundation, pore pressures are induced in the soil, which will cause water to flow out of the soil. With soils of low permeability, this consolidation process can be very slow. Compression whose rate is governed by pore water dissipation is called **primary consolidation.**

2. A secondary reason is that regardless of the rate at which water can drain from a soil, all soils show time-dependent compression behavior to some extent. Even if water could drain from soils instantaneously, there would still be some delay in the occurrence of compression. Compression may thus continue to occur even when pore pressures are fully dissipated. Such compression is referred to as **secondary consolidation** or secondary compression.

Terzaghi's Theory of One-Dimensional Consolidation An equation governing the rate of consolidation has been derived by Terzaghi based on the following principal assumptions:

(a) Compression of the soil and flow of pore water occur only in the vertical direction.

(b) The soil is homogeneous; in particular, the value of the coefficient of compressibility m_v is uniform and constant.

(c) No volume change occurs to the soil particles or the water.

(d) The soil is fully saturated.

(e) Darcy's law applies.

(f) The total and effective stress is always uniform on any horizontal plane.

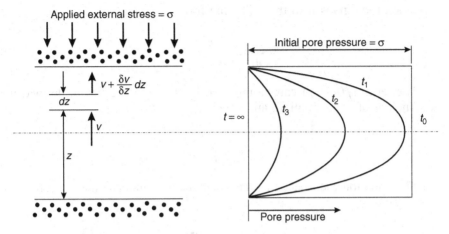

Figure 8.15   One-dimensional consolidation of a soil layer.

The equation can be derived by considering a soil element undergoing consolidation, as illustrated in Figure 8.15. The element is at an elevation z above the base of the layer with thickness dz and unit area. The soil layer is bounded at the top and bottom by sand layers, which act as free-draining surfaces. An external stress σ has been applied to the soil, causing a uniform increase in pore pressure of the same amount. Water will immediately start to drain toward the top and bottom boundaries, resulting in a decrease in pore pressure over time, with a distribution as shown in the right-hand diagram of Figure 8.15. Because consolidation is occurring, the velocity of water leaving the element is not the same as the velocity of water entering it.

This situation is a special case of the general seepage case considered in Section 7.5, which led to Equation 7.9:

$$\frac{\partial^2 h}{\partial x^2} + \frac{\partial^2 h}{\partial y^2} = \frac{m_v \gamma_w}{k} \frac{\partial h}{\partial t}$$

Although not explicitly stated, the Terzaghi assumptions also applied in the derivation of that equation, except for (a) and (f) above. In this case the term involving the horizontal ordinate (x) disappears and we are left with the expression

$$\frac{\partial^2 h}{\partial y^2} = \frac{m_v \gamma_w}{k} \frac{\partial h}{\partial t} \tag{8.9}$$

Since $u = \gamma_w h$, we can replace h with u in this equation. It is normal practice to use z rather than y as the vertical ordinate in one-dimensional consolidation, and we will follow this convention here. Rearranging the equation then gives it in the following form:

$$\frac{\partial u}{\partial t} = \frac{k}{m_v \gamma_w} \frac{\partial^2 u}{\partial z^2} \tag{8.10}$$

The term $k/(m_v \gamma_w)$ can be replaced by a single coefficient, termed the coefficient of consolidation, and designated c_v, that is,

$$c_v = \frac{k}{m_v \gamma_w} \tag{8.11}$$

The equation then becomes the Terzaghi one-dimensional consolidation equation:

$$\frac{\partial u}{\partial t} = c_v \frac{\partial^2 u}{\partial z^2} \tag{8.12}$$

The pore pressure u during consolidation is, as expected, a function of the soil properties, the time, and the elevation within the soil layer. The rate at which consolidation occurs clearly depends upon:

(a) The speed at which water can flow out of the soil, governed by the permeability k
(b) The volume of water that has to flow out of the soil, governed by the compressibility of the soil, and thus the parameter m_v

Hence c_v is a function of both k and m_v. The value of c_v decreases with decreasing k or increasing m_v; low values of c_v indicate slow rates of consolidation. During consolidation of a particular soil, both k and m_v normally decrease, so that c_v may remain approximately constant. As a general trend, increasing plasticity of a soil means a decrease in permeability and an increase in compressibility, so that the value of c_v decreases rapidly with increasing plasticity index.

The solution of the Terzaghi differential equation depends on the initial pore pressure condition and on the boundary conditions. We will consider only the simplest initial condition, where the initial *excess* pore water pressure is uniform ($u = u_i$) throughout the layer. The boundary condition may be either one-way or two-way drainage, as shown in Figure 8.16.

By expressing the solution to the Terzaghi equation in terms of d, the maximum length of the drainage path, the same solution can be applied to both cases. The solution can be written as

$$u = f \left\{ u_i, z, \frac{c_v t}{d^2} \right\}$$

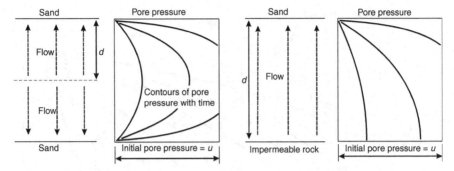

Figure 8.16   Boundary conditions and drainage path length.

and has the form of a series,

$$\text{or } u = f(u_i, z, T),$$

where the time factor T is given as

$$T = \frac{c_v t}{d^2} \tag{8.13}$$

The maximum drainage path length is therefore half the layer thickness when drainage can occur to both boundaries but is the layer thickness when drainage can only occur to one boundary. Note that the pore pressure u in the above equations is the **excess pore pressure**, that is, the pressure caused by the application of load, and is in excess of the equilibrium pore pressure in the soil prior to the load application, to which it returns when consolidation is complete. This equilibrium pressure is normally the hydrostatic pressure.

The solution to the Terzaghi equation is available in the form of graphs. The graph for pore pressure is given in Figure 8.17. The pore pressure is expressed in dimensionless form as u/u_0, where u_0 is the initial excess pore pressure; the depth is similarly expressed in dimensionless form as z/d where z is the distance from a drainage boundary measured toward the undrained boundary and d is the distance from the drained to the undrained boundary (the maximum drainage path length). The time t is expressed using the dimensionless time factor T.

For routine settlement calculations in engineering practice, we are generally interested only in the rate of settlement over time, not with actual values of the pore pressure u at any point or time. For this reason the solution is also available in terms of the average degree of consolidation U, which is defined as

$$U = \frac{\text{settlement at time } t}{\text{total eventual settlement}}$$

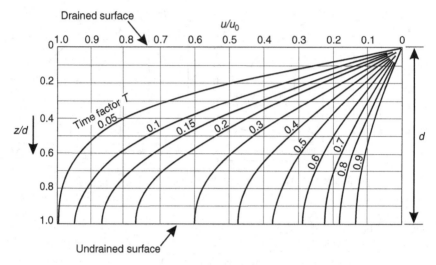

Figure 8.17   Pore pressure related to time factor T and elevation within the soil layer.

The solution in terms of U and T is given approximately by the following expressions:

$$U \leq 5\%: \qquad U^2 = \frac{4}{\pi}T \tag{8.14}$$

$$U \geq 50\%: \qquad U = 1 - \frac{8}{\pi^2}e^{-(\pi^2/4)T} \tag{8.15}$$

The relationship between U and T is presented in Figure 8.18. Note that for a given soil and boundary conditions graphs of actual time t will be identical in shape to those of the time factor T.

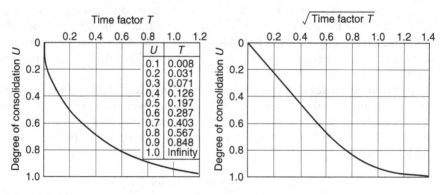

Figure 8.18   Relationship between degree of consolidation U and time factor T.

It is seen that the rate at which consolidation takes place decreases steadily over time and becomes asymptotic to the line $U = 1.0$. The time for consolidation to be 100% complete is theoretically infinite. When U is plotted against the square root of the time factor T, the graph becomes linear up to a value of U of about 0.6. This property is made use of when determining the value of the coefficient of consolidation c_v from results of odometer tests.

Comparison of Actual Soil Behavior with Consolidation Theory For many soils, it is found that the behavior revealed in odometer tests is close to that predicted by consolidation theory, apart from one important deviation. This is illustrated in Figure 8.19.

This shows that the laboratory curve follows the theoretical curve until consolidation is approaching completion, at which point it departs from the theoretical curve and continues to diverge further from it. According to the theory, consolidation will level off at a $U = 1.0$, but in practice the soil continues to consolidate indefinitely, though at a steadily decreasing rate, as shown. The consolidation behavior that obeys the theoretical concept is governed by the rate of pore pressure dissipation and is the primary consolidation mentioned earlier, while the consolidation which continues on for an indefinite time is not governed by pore pressure dissipation and is termed secondary consolidation. It occurs under constant effective stress.

Determination of Coefficient of Consolidation (c_v) from Odometer Test Results

The method most commonly used for determining c_v from odometer tests is that of Taylor (1948) and is known as the root time method. The readings

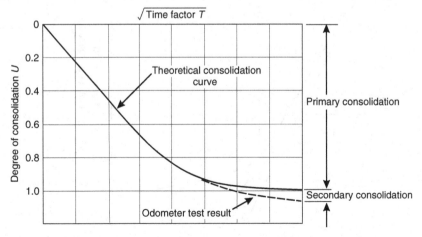

Figure 8.19 Comparison of odometer test result with consolidation theory prediction.

of compression taken at regular intervals when each load is added in the odometer test are plotted as a graph of compression versus the square root of time. The purpose of this is twofold:

1. To provide a picture of whether the soil is behaving in the manner expected from consolidation theory
2. To provide a means by which the coefficient of consolidation can be determined

Convenient time intervals for taking readings and the root time values they correspond to are shown in Table 8.2. Readings can be continued to higher values of time as needed, depending on the soil type.

Figure 8.20 shows a typical graph obtained from an odometer test and illustrates the root time method. Before using the method it is important to first examine whether the sample is behaving according to consolidation

Table 8.2   Time Intervals for Deflection Readings in Odometer Tests

Time	2.4 s	9.6 s	21.6 s	38.4 s	1 min	2.25 min	4 min	9 min	16 min	25 min	36 min
Root time ($\sqrt{}$ min)	0.2	0.4	0.6	0.8 m	1	1.5	2	3	4	5	6

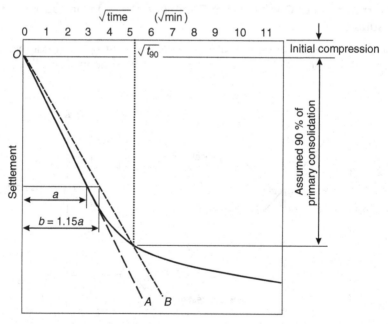

Figure 8.20   Root time method for determining t_{90} and coefficient of consolidation (c_v) from odometer tests.

theory. If the plot clearly shows a significant number of readings on a straight line, then it is reasonable to assume the soil is behaving as theory predicts. The rate of compression is being governed by the rate of pore pressure dissipation and thus by the parameter c_v. To determine c_v we must use that part of the graph that is in accordance with theory. This is generally assumed to be up to 90 percent of primary consolidation, and the value of t_{90} (the time to reach 90 percent of primary consolidation) is therefore determined from the graph and used to calculate the coefficient of consolidation.

A simple construction enables the value of t_{90} to be determined. The line OA is drawn through the points lying on a straight line. This line may not necessarily intersect the vertical axis at its origin because of yield of the apparatus and "bedding" in of the porous stones. This initial compression is not considered part of the soil behavior and is ignored. A second straight line, OB, is then drawn in such a way that the horizontal ordinate $b = 1.15a$. The point of intersection of OB with the laboratory curve defines the value of $\sqrt{t_{90}}$ from which t_{90} is easily calculated. The theoretical value of the time factor T for a degree of consolidation of 90 percent is 0.848 (see Figure 8.18), so that, using Equation 8.13, we can write

$$T_{90} = \frac{c_v t_{90}}{d^2} \tag{8.16}$$

and

$$c_v = \frac{T_{90}d^2}{t_{90}} = \frac{0.848d^2}{t_{90}} \tag{8.17}$$

where d in the odometer test is half the sample thickness.

In this way, a value of c_v can be determined from each loading stage in the test. In general, for most soils, the values do not vary greatly. However, with some soils, the effect of increasing stress is to destroy important structural characteristics of the soil, in which case the c_v value may decrease dramatically with increasing stress. Units for c_v have traditionally been cm^2/s, but $m^2/year$ is a more practical unit and is gaining in popularity. Values of c_v cover a very wide range, reflecting primarily the wide range of coefficients of permeability found in natural soils (see Section 7.3). A rough guide to the values expected for various soil types is shown in Table 8.3.

Note that Equation 8.16 can also be written as

$$t = \frac{Td^2}{c_v} \tag{8.18}$$

This means that for a given soil type and boundary conditions the time to reach a particular degree of consolidation is proportional to the square of the drainage path length (since the value of T has a fixed relationship to the

Table 8.3 Values of Coefficient of Consolidation

Soil Type	Coefficient of Consolidation (m^2/year)
Silts and clays of low plasticity	1–500
Clays of medium plasticity	$10^{-3}-10$
Clays of high plasticity	$10^{-4}-10^{-2}$

average degree of consolidation, U). From intuitive reasoning this is not surprising—as the thickness of the soil layer increases, not only does the drainage path length increase in direct proportion to the thickness but the volume of water to drain from the soil also increases in direct proportion, with the net result that the time taken increases in proportion to the square of the thickness.

With some soils, the odometer test will not produce graphs having a linear portion in the form shown in Figure 8.20. In this case a reliable value of c_v cannot be determined. The reason for this is explained in the next section on residual soil behavior, although the absence of a linear portion is not restricted to residual soils; it may also occur with sedimentary soils.

Behavior of Residual Soils and Limitations of the Odometer Test for Determining c_v Residual soils generally tend to be of higher permeability than sedimentary soils and pore water can therefore drain from them more rapidly. It is not uncommon for the rate of drainage to be so high that the coefficient of consolidation can no longer be accurately measured in conventional odometer tests. Figure 8.21 shows graphs of root time versus compression for three typical residual soils. For convenience in comparing the graphs, the vertical scale is compression in percent, rather than direct deflection; the shape of the graph remains the same.

None of these graphs show the straight-line section predicted by consolidation theory. The reason is that with these soils pore pressures dissipate so rapidly that the shape of the curves is no longer governed by the mechanics of pore pressure dissipation. The shape of the graphs suggests pore pressures are fully dissipated even before the first reading is made. With manual recording, it is not possible to take readings at closer intervals than those in Figure 8.21, although it is apparent from the shape of the curves that continuous recording of deflection would not significantly alter their appearance.

There is thus an upper limit to the value of the coefficient of consolidation that can be measured in conventional consolidation tests. It is easy to show that with a 20-mm-thick soil sample this limit is approximately 0.1 m^2/day ($= 36.5\,m^2$/year $= 0.012\,cm^2$/s.). Readings taken in the first minute will only lie on a straight line if the c_v value is less than this. If reliable values of c_v are required for soils with higher values, it is necessary to use a different method of measurement, such as a pore pressure dissipation test in a triaxial

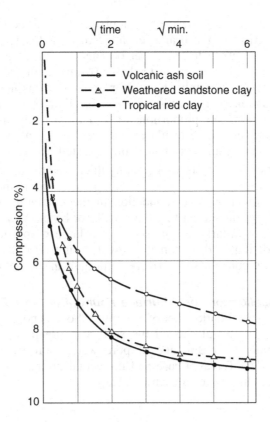

Figure 8.21 Typical root time graphs from three residual soils.

cell. It can be seen from Table 8.3 that many soils have c_v values greater than $0.1\,\mathrm{m^2/day}$. While the majority of such soils are likely to be residual soils, it is true that a significant number of sedimentary soils also fall into this category.

Because graphs of the shape shown in Figure 8.21 are common with residual soils and not uncommon with sedimentary soils, their physical significance should be clearly understood. The important points are the following:

(a) If the odometer test does not produce a linear deformation–root time plot, it means that the deformation rate is not governed by pore pressure dissipation, and the coefficient of consolidation is greater than the limiting value mentioned above.

(b) In terms of the concepts in Figure 8.19, nearly all the compression is secondary compression, as it is occurring under constant effective stress. The curves therefore reflect the creep behavior of the clay when tested in this particular way.

(c) The shape of the curves is not primarily a property of the soil. It is a function of the drainage path length in the test. If the soils in Figure 8.21 were tested in an odometer with a much greater sample thickness (say 1 m instead of 20 mm), then the root time plots would almost certainly show clear linear sections in accordance with consolidation theory followed by a section representing secondary consolidation. The proportions of settlement made up of primary and secondary settlement would then be very different and would be governed by the thickness of the sample tested.

(d) In practice, soil layers are generally thick, and the drainage path length may be several orders of magnitude greater than in the odometer. The absence of primary consolidation in the laboratory test is therefore not a direct indication that there will not be primary consolidation in the field. Similarly, the presence of a large proportion of secondary consolidation in the odometer test does not necessarily indicate that secondary consolidation will be important in the field.

Rate of Consolidation for a Surface Foundation on a Deep Soil Layer

The Terzaghi consolidation theory presented above is not applicable to foundations of finite size placed on the surface of deep soil layers, because neither settlement nor drainage of pore water is restricted to the vertical direction. The Terzaghi case and the actual drainage pattern beneath a foundation are illustrated in Figure 8.22.

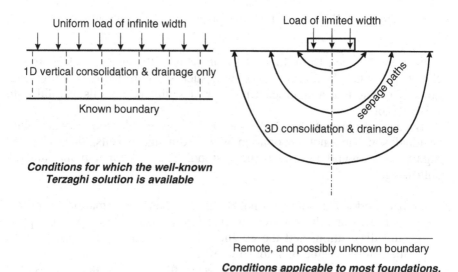

Figure 8.22 One-dimensional Terzaghi consolidation and the situation beneath a surface foundation.

For circular or rectangular foundations, drainage will be three-dimensional, while for a strip foundation (very long compared to its width), it will be two-dimensional. The analysis of these foundation situations is much more complex than the one-dimensional case, and analytical solutions are not available. Davis and Poulos (1972) have obtained solutions using numerical methods and computer programming. An abbreviated version of their solutions is given in Figure 8.23. Davis and Poulos presented their solution using the same time factor as the Terzaghi solution, which involves the thickness of the soil layer. With deep layers

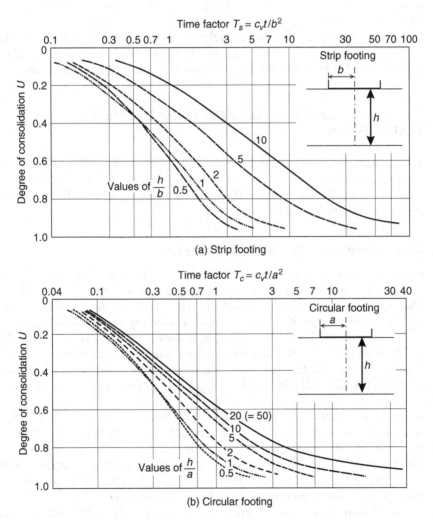

Figure 8.23 Degree of consolidation beneath surface foundations. (Based on analysis of Davis and Poulos, 1972.)

of firm to stiff clay (on which building foundations are likely to be built), the thickness of the layer is often deep and poorly defined. For this reason the solutions are presented here in a slightly different form, defining the time factor T in terms of the foundation dimensions, as follows:

Strip foundation: $T = \dfrac{c_v t}{b^2}$ where b is half the width of the foundation.

Circular foundation: $T = \dfrac{c_v t}{a^2}$ where a is the radius of the foundation.

In the two cases given in Figure 8.23, the lower boundary of the layer undergoing consolidation is considered to be rigid and impermeable. The base of the foundation is also considered to be impermeable. These situations are considered the most likely to be encountered in practice. The complete Davis and Poulos solutions also cover the cases of a permeable boundary at the base of the clay layer and the base of the foundation.

8.6 ESTIMATION OF SETTLEMENT FROM ODOMETER TEST RESULTS

To illustrate the way in which the settlement is estimated from the results of odometer tests, we will consider two typical examples. The first is for a building foundation on a stiff soil, and the second is a fill on a soft, normally consolidated clay.

8.6.1 Settlement of a Building Foundation

Figure 8.24 shows a foundation 3 m wide by 4 m long founded at a depth of 1 m on a stiff clay layer 13 m thick. The stress from the foundation is 150 kPa. The water table is 6 m deep. The clay is a fine-grained residual soil of moderate plasticity derived from weathering of the sandstone. It has a unit weight of $17.5\,\text{kN/m}^3$ and it is assumed to be fully saturated. The result of an odometer test carried out on a sample of the clay is shown in Figure 8.25 using both a log and linear scale. The c_v value was measured as $0.12\,\text{m}^2/\text{day}$. We will consider the magnitude and rate of settlement in turn.

Magnitude Following the procedure described earlier in Section 8.3 we will divide the soil into three sublayers only (for simplicity), as shown in Figure 8.24. This will give a reasonable estimate, although in practice a larger number would normally be used to obtain greater accuracy. As the stress increase is greatest near the surface, it is desirable that the sublayers close to the surface be thinner than those deeper down.

Calculation of the initial effective stress and the stress increase are shown in Table 8.4. The stress increase in each sublayer is calculated using Figure 5.6. The estimate is for settlement at the center of the foundation, so

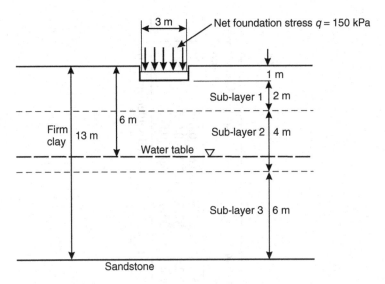

Figure 8.24 Soil conditions, foundation details, and sublayers used in settlement estimate.

that to use Figure 5.6 the foundation is divided into four identical quarters. All stress values (in kPa) are calculated at the midheight of each sublayer (to represent average values within each layer). The depth z is from the base of the foundation to the midheight of each layer. In this table:

σ'_o = initial effective stress
$\Delta\sigma'$ = increase in effective stress
σ'_1 = final effective stress
Δe = change in void ratio as a result of stress increase

There are several ways to proceed to determine the compression of each layer. We can use the graphs directly as follows:

$$\text{Compression per unit thickness} = \frac{\Delta L}{L} = \frac{\Delta e}{1+e} \qquad (8.19)$$

$$\text{Total compression of sublayer} = \frac{\Delta e}{1+e}h \qquad (8.20)$$

where h is the thickness of the sublayer.

With the initial and final effective stresses from Table 8.4 we can go to the graphs in Figure 8.25 and read off the corresponding values of Δe and use Equation 8.20 to obtain the compression. This is illustrated for layer 1, using the log plot; it could equally well be done using the linear plot. The values of Δe and the compression of each layer are shown

Table 8.4 Settlement Estimate

Sublayer No.	Thickness h (m)	Depth z (m)	$m = B/z$	$n = L/z$	I_σ	$4I_\sigma$	$\Delta\sigma = 4I_\sigma q$	σ'_0	σ'_1	Δe	Compression (mm)
1	2	1	1.5	2.0	0.226	0.904	135.6	35.0	170.6	0.039	36.4
2	4	4	0.38	0.5	0.067	0.268	40.2	87.5	127.7	0.012	22.4
3	6	9	0.17	0.22	0.017	0.068	10.2	135.8	138.4	0.003	8.4

Total settlement = 67.2 mm

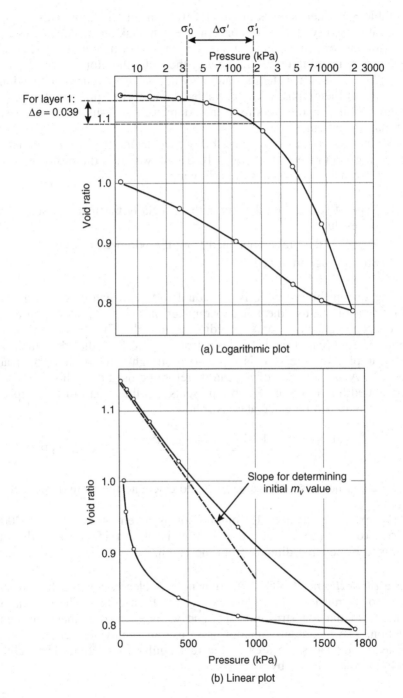

Figure 8.25 Graphs of void ratio versus pressure from an odometer test on the clay in Figure 8.24.

in Table 8.4. Their sum is the total settlement of 67.2 mm. This procedure illustrates clearly the basic concept used in making settlement estimates from odometer test results—a stress-versus-deformation curve is obtained which is assumed to represent the way the soil will deform in the field. This procedure, however, is tedious and may not be very accurate, depending on the scale of the graphs. It is preferable, and is normal practice, to determine compressibility parameters from the odometer test results and use these for estimating settlement.

We have a choice of compressibility parameters: the linear parameter m_v or the log parameters C_c and C_s. To decide which is the more appropriate to use, we need to consider the following:

1. Type of soil we are dealing with, especially whether it is sedimentary or residual
2. Shape of the graphs obtained from the odometer test
3. Stress range involved

The soil in this example is a residual soil, in which case it is generally more sensible to use the linear parameter m_v, since C_c and C_s are based on concepts applicable only to sedimentary soils. The stress range is from about 35 to 170 kPa, and it is seen from Figure 8.25 that the graph using a linear plot for pressure is close to a straight line from zero to about 400 kPa. A single value of m_v can therefore be used over this stress range. The dotted line in Figure 8.26b represents this linear part of the graph, and the m_v value can be calculated as follows:

$$m_v = \frac{\Delta e/(1 + e_0)}{\Delta \sigma'} = \frac{(1.145 - 1.00)/2.145}{500} = 1.35 \times 10^{-4} \text{ kPa}^{-1}$$

and we can now use this value of m_v to calculate the compression of each layer.

For example, in layer 1, compression is given as $hm_v \, \Delta \sigma' = 2000 \times 1.35 \times 10^{-4} \times 135.6 = 36.6$ mm, which is almost identical to the value obtained by reading directly from the graph.

Rate of Settlement   The situation here is clearly not the 1-D Terzaghi case, so we must use the solutions given in Figure 8.23. The foundation is neither circular nor strip in shape, and we are forced to make an approximation. The best approximation is to use an "equivalent circle" that is a circle having the same area as the rectangular foundation. The radius of such a circle is given by

$$a = \sqrt{\frac{3 \times 4}{\pi}} = 1.95 \text{ m}$$

Then

$$\frac{h}{a} = \frac{12}{1.95} = 6.15 \, \text{m}$$

We could determine the time for any degree of consolidation, but we are normally interested in the time for consolidation to be almost complete. Therefore we will determine the time for 90 percent consolidation. The value of U is 0.9 and the corresponding value of the time factor T_{90} from Figure 8.24b is 5 (approximately).

Therefore

$$t_{90} = \frac{Ta^2}{c_v} = \frac{5 \times 1.95^2}{0.12} = 158 \text{ days} \approx 5 \text{ months}$$

This is a relatively short time compared with that expected with many soil conditions. It is not uncommon for t_{90} to be several years or even decades. The construction period for a reasonably large building would generally be several months, so that a large proportion of the primary consolidation would take place during the construction period.

8.6.2 Settlement of Fill on Soft Clay

It is not uncommon for areas of soft clay to be made usable by placing a reasonably thick layer of good-quality fill over them and allowing the soil to consolidate. In this way the soft clay becomes firmer, and the layer of fill provides a firm foundation for the construction of roads, services, and light buildings. Figure 8.26 shows an example of such a situation. The clay layer is 12 m thick and overlies dense sand. According to its geological history, the clay is normally consolidated, except for the top 1 m, which is quite hard due to the influence of surface drying. The water table is at a depth of 1 m. A 4-m-thick layer of well-compacted sand (unit weight $21 \, \text{kN/m}^3$) is to be placed on top of the clay layer. The results of odometer tests on two undisturbed samples taken at 1.8 and 8.2 m deep are shown in Figure 8.27. The unit weight and void ratio of the clay vary with depth; reasonable average values are unit weight $13.8 \, \text{kN/m}^3$ and void ratio 3.2. The measured coefficient of consolidation has an average value of $0.0047 \, \text{m}^2/\text{day}$.

Magnitude   Unlike the foundation example considered above, the stress increase will now be uniform throughout the clay layer because the fill extends over a wide area (edge effects are ignored). However, because the soil is normally consolidated, there will still be more compression near the surface than deeper down, and it is therefore desirable to divide the clay layer into sublayers, with thin layers close to the surface and thicker layers deeper down. We will adopt five sublayers as indicated in Figure 8.26. The

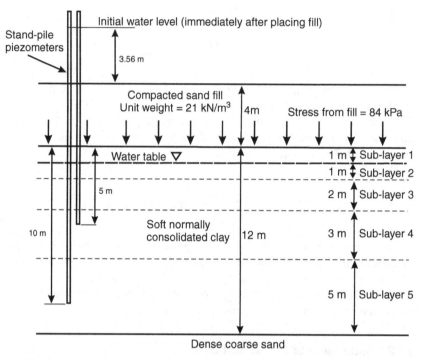

Figure 8.26 Settlement of fill on a normally consolidated clay layer.

initial effective stresses, calculated at the midpoint of each sublayer, are easily determined and are given in Table 8.5. From this point onward we need to examine the odometer curves in Figure 8.27 and decide how best to use them for the settlement estimate.

The most significant features of the curves are the following:

1. Both the log and linear plots show an initial section of low compressibility followed by a section of much greater compressibility separated by a distinct "yield" or preconsolidation pressure. This is typical of young normally consolidated clays

2. Estimates of the in situ vertical effective stress (σ_0') and the preconsolidation pressure (σ_c'), shown on the graphs, indicate that the soil is not strictly normally consolidated. Rather, it is lightly overconsolidated. Note that the in situ vertical effective stress is calculated directly in the usual manner, but the preconsolidation pressure is a matter of judgment. It is normally at or a little greater than the point of maximum curvature on the graphs. (Several empirical constructions have been proposed for determining the preconsolidation pressure; they are described later but are only marginally more reliable than direct

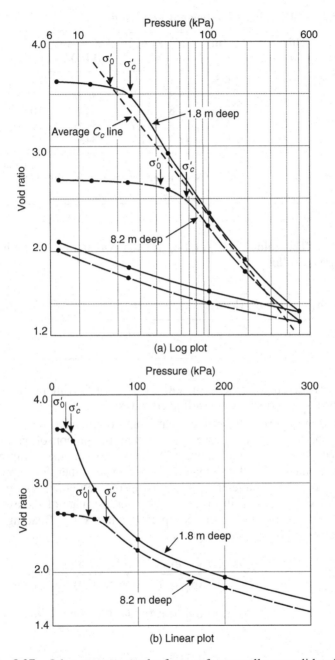

Figure 8.27   Odometer test results from soft, normally consolidated clay.

Table 8.5   Calculation of settlement magnitude

Sublayer	Thickness h (m)	z (m)	σ'	σ'_c	$\Delta\sigma' = \Delta\sigma$ (kPa)	σ'_1	Settlement (m)
1	1	0.5					
2	1	1.5	15.8	23.7	84.0	99.8	0.247
3	2	3.0	21.8	32.7	84.0	105.8	0.402
4	3	5.5	31.8	47.7	84.0	115.8	0.456
5	5	9.5	47.8	71.7	84.0	131.8	0.522

Total = 1.627

Allowing for a small amount of secondary settlement we will adopt a value of 1.65m

judgment.) The values of σ'_0 and σ'_c for the two samples, together with the OCR, are as follows:

Depth (m)	σ'_0 (kPa)	σ'_c (kPa)	OCR Ratio (σ'_c/σ'_0)
1.8	17.0	25	1.47
8.2	42.6	66	1.55

The average value of the OCR is about 1.5, which is not uncommon for recent normally consolidated clays.

3. The only part of the curves that is approximately linear is the section of the log plots at stresses greater than the preconsolidation pressure. With a pressure increment of 84 kPa, most of the compression will take place over this linear section, and it is therefore appropriate to use the log parameter C_c for making our settlement estimate.

The line drawn in Figure 8.27a is an estimate of the average slope for determination of the compression index C_c. Reading off values between 20 and 80 kPa gives:

$$C_c = \frac{e_0 - e_1}{\log(\sigma'_1/\sigma'_0)} = \frac{3.5 - 2.5}{\log(80/20)} = 1.66$$

We can now use Equation 8.8 to estimate the compression of each sublayer:

$$\delta = \frac{C_c}{1 + e_0} h \log \frac{\sigma'_0 + \Delta\sigma'}{\sigma'_0}$$

We can see from Figure 8.27 that compression governed by the parameter C_c (along the virgin consolidation line) only occurs when the pressure exceeds the preconsolidation pressure. Below this pressure, between the in situ pressure (σ'_0) and the preconsolidation pressure (σ'_c), some minor compression will occur, governed by the parameter C_s. For a theoretically rigorous estimate, this compression should be included in our estimate. However, because the stress range involved is small, and C_s is small compared to C_c, the resulting compression will be small and will be ignored. We will therefore only estimate the compression occurring between the preconsolidation pressure and the final effective stress. To obtain the preconsolidation pressure for each sublayer, we will assume that the overconsolidation ratio of 1.5 applies throughout the clay layer. The values estimated in this way and the complete settlement estimate are shown in Table 8.5. Because the top 1 m is hard compared to the rest of the clay, compression of this layer will be ignored. The total settlement of 1.65 m may seem surprisingly large, but it is not unusual in soft, normally consolidated clay layers. The author's experience is that in these clays the settlement can be expected to be between 30 percent and 50 percent of the applied fill thickness.

Rate of Settlement   Estimation of the rate of settlement is straightforward in this situation because it conforms to the 1-D situation and the Terzaghi solution can be applied directly. We will estimate a complete settlement-versus-time graph. The simplest way to do this is to determine the time t corresponding to progressive degrees of consolidation, as is done in Table 8.6. The maximum drainage path length (d) in this case is 6 m, as water can drain to both the upper and lower boundaries.

The settlement-versus-time curve is illustrated in Figure 8.28. Curves such as this are very useful in situations involving fill on soft clays, as it is common practice to monitor the behavior of the clay by installing instruments to measure settlement (and possibly pore pressures); the records so obtained can then be compared with the theoretical predictions.

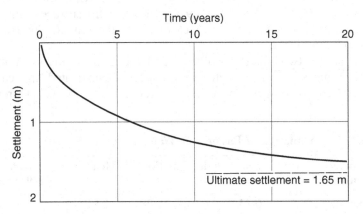

Figure 8.28   Settlement rate of fill on a soft clay layer.

Table 8.6   Determination of Settlement-Versus-Time Curve

Degree of consolidation U	0.20	0.40	0.60	0.80	0.90	1.0
Settlement (m) = $U \times 1.65$	0.33	0.66	0.99	1.32	1.485	1.65
Time Factor T	0.031	0.126	0.287	0.567	0.848	∞
Time (years) = Td^2/c_v	0.65	2.6	6.0	11.0	17.8	∞

Pore Pressure During Consolidation   We may wish to monitor the progress of consolidation by installing standpipe piezometers in the clay. Two such piezometers are shown in Figure 8.26 at depths of 5 and 10 m. We can estimate the expected water levels in these piezometers at any time we choose; we will choose 5 years. Firstl we can estimate the water level immediately after the fill is placed. The applied total vertical stress is 84 kPa. Since the clay is fully saturated and the loading is undrained, the pore pressure throughout the clay will rise by the same value. The head will therefore rise by 84/9.81 = 8.56 m. This means that in any standpipe the water level will rise by 8.56 m. Thus the height above the fill surface will be 8.56 m − (4 + 1) m = 3.56 m. To determine the water levels at 5 years we need to note the following:

1. The coefficient of consolidation is 0.0047 m²/day = 1.72 m²/year.
2. We can calculate both the time factor T and the values of z/d for each piezometer. We can then use Figure 8.14 to obtain values of u/u_0 from which to calculate the required values of pore pressure. For this situation $T = c_v t/d^2 = 1.72 \times 5/36 = 0.24$.
3. In this case $d = 6$ m and the values of z are measured from the drained boundary toward the undrained boundary (i.e., the center of the layer).

The calculation is set out in Table 8.7. It shows that with the piezometer at 5 m the water level is 0.73 m above the fill surface, while for the 10-m-deep piezometer the water level is 2.09 m below the fill surface. This is to be expected as this piezometer is much closer to the drainage boundary than the 5-m-deep piezometer.

The above two examples illustrate the principles involved in settlement estimates. Such estimates may appear sophisticated and can easily be accorded an unwarranted degree of reliability. It is well to remember

Table 8.7   Calculation of Piezometer Water Levels at 5 Years

Piezometer Depth	z (m)	z/d	u/u_0	u (kPa)	Head (m)	Height Above Fill Surface (m)
5 m	5	0.83	0.67	56.28	5.73	0.73
10 m	2	0.33	0.34	28.56	2.91	−2.09

that they contain many assumptions and other sources of error and are at best little more than crude approximations. Sources of error are considered in the following section.

8.7 APPROXIMATIONS AND UNCERTAINTIES IN SETTLEMENT ESTIMATES BASED ON ODOMETER TESTS

8.7.1 Interpretation of Void Ratio–Stress Curves and Sample Disturbance

One of the principal sources of error in estimating settlement magnitude arises from sample disturbance and the interpretation of the curves. Sampling and trimming almost invariably soften the soil to some extent and the compressibility indicated by the odometer test will generally be greater than the true value.

In addition to sample disturbance, the void ratio–pressure curve can easily be interpreted wrongly because of preconceived ideas of how soil should behave. The conventional concept that soil compressibility can be represented by two straight lines on a plot using a log scale for pressure (Figure 8.9) is only valid for a limited range of soils, primarily normally consolidated and lightly overconsolidated clays, and even for these soils the concept may be only a crude approximation. It is essential therefore that results of odometer tests be plotted on linear scales as well as log scales in order to gain an accurate picture of the compression behavior. By examining the shape of these graphs and comparing them with the relevant stress range, an appropriate choice can then be made between the linear parameter m_v and the log parameters C_c and C_s. In some cases, for example, the soil in Figures 8.11 and 8.12, there is no linear section on either the linear or log plot, and different values of m_v will need to be determined for each stress increment involved in the settlement calculation.

Schmertman's Method of Correcting for Sample Disturbance There are no sure ways of making corrections to take account of sample disturbance, although a method proposed by Schmertman (1953) for sedimentary soils has some merit, even if only as a concept to aid understanding of the problem of sample disturbance. To use the method, it is necessary to first estimate the preconsolidation pressure of the soil. A well-known method for doing this, proposed by Casagrande (1936), is illustrated in Figure 8.29.

The line $A-B-C$ is the odometer curve. The point of maximum curvature, B, is identified (by judgment) and a horizontal line $B-E$ drawn through it. A second line $B-F$ is drawn tangential to the odometer curve, and the angle EBF is bisected to give the line $B-G$. The straight part of the odometer curve is extended toward point D. The intersection of this line with $B-G$ gives the preconsolidation pressure.

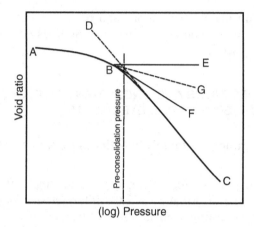

Figure 8.29   Determination of the preconsolidation pressure. (After Casagrande, 1936.)

Schmertman's construction, based on the results of considerable experimental work, is illustrated in Figure 8.30. The curve $A-B-C$ is the odometer curve. The pressure σ'_0 is the in situ vertical effective stress, and σ'_c is the preconsolidation pressure obtained using the construction in Figure 8.29. The in situ void ratio of the soil is e_0. The point O therefore represents the state of the soil in the ground. We can see immediately that

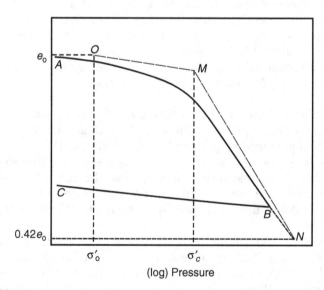

Figure 8.30   Construction of the true field consolidation curve. (After Schmertman, 1953.)

the odometer curve cannot be the true field curve, because it does not pass through the point O. Schmertman's construction is a way of correcting for this anomaly, which inevitably arises from sample disturbance.

The line $O-M$ is drawn parallel to the rebound (or swell) line BC to intersect the preconsolidation pressure and establish point M. Point N is established by extending the odometer curve beyond B in a straight line to intersect the line corresponding to a void ratio of $0.42e_0$. The line $O-M-N$ is then deemed to be the true field curve. This construction appears reasonable in view of the behavior illustrated earlier in Figures 8.8 and 8.9. However, it is still very much an idealization and makes assumptions that are valid only for certain soils.

The following limitations of the concept should be noted:

(a) The slope of the rebound line $B-C$ is not constant and can actually be considerably steeper than the initial part of the odometer curve $O-M$. This is clearly the case for the soils in Figure 8.10 and 8.12, and the procedure cannot be applied to these soils.

(b) The use of the log plot often leads to the erroneous identification of a preconsolidation pressure followed by the (equally erroneous) use of the Schmertman construction.

(c) The method has no relevance to residual soils. There is no reason at all for the rebound lines from such soils to be related to the initial compressibility of the soil.

8.7.2 Assumptions Regarding Pore Pressure State

The methods of estimating settlement described above assume that the pore pressure state in the ground is the same after the completion of consolidation as it was prior to the application of the load. This assumption is true for soils with shallow water tables, in particular, normally consolidated soils in low-lying areas. However, with deep water tables, the assumption may not be valid. In Chapter 4, an account is given of possible pore pressure states above the water table, as illustrated in Figures 4.4 and 4.5.

The effect of constructing a foundation and covering the site with a building is to close off the avenues for drying or wetting, and the pore pressure state will then tend toward the static equilibrium situation. The implication of this is that we know approximately what the long-term pore pressure state will be after the foundation is built, but we do not know the pore pressure state prior to construction. The foundation may be built at the end of a long dry summer or at the end of a very wet winter. Hence, there will be unknown changes in the pore pressure, and thus in the effective stress state in the ground, as a result of construction of the foundation. These are normally ignored in settlement estimates, simply because there is no reliable way of estimating them. The error arising from this can be large. If the foundation is built at the end of a long dry summer, the soil may swell rather than settle, as the effect of covering the surface may outweigh the influence of the foundation load.

8.7.3 Lateral Deformation

We have already covered this issue in Section 8.3, which suggested that using one-dimensional analysis and ignoring lateral deformation result in an underestimate of settlement by about 20 percent.

8.7.4 Submergence of Fill Loads

With large fill loads on soft clays, such as in the second example above, it is not unusual for the consolidation to be so great that the original ground surface ends up below the equilibrium water table. This means that some of the fill load also ends up below the water table, and the effective applied stress is less than the actual fill weight. In the example above the water table was 1 m deep, but the settlement magnitude was 1.63 m so that 0.63 m of the fill was below the water table. This was not allowed for in the settlement estimate, and a correction should therefore be made to the above calculation to take this into account.

8.7.5 Use of Terzaghi Theory of Consolidation for Nonlinear Soils

We should recognize also that the calculation of time rate is normally made using Terzaghi's consolidation theory, which assumes the soil has uniform compressibility and permeability. This means that calculations such as that given in Section 8.6.2 for a soft, normally consolidated soil are mixing two different "models" of soil compressibility. The magnitude of settlement uses a log expression of compressibility, while the time rate uses a linear expression.

With some soils, such as that in Figure 8.25, the assumption of linear behavior is valid, at least over the stress range involved in the settlement estimate. However, with other soils, such as that in Figure 8.27, this assumption is not valid. Figure 8.31 illustrates the values of coefficient of compressibility (m_v) associated with soils that are normally consolidated or close to normally consolidated.

One graph is for the soil described in Section 8.6.2 and Figure 8.27, and the other is for an "ideal" normally consolidated clay with the water table at the ground surface. With the ideal normally consolidated soil the value of m_v decreases with depth, the decrease being most pronounced close to the surface. With the other soil, the assumption was made that the compressibility of the top 1 m of soil was very low and could be ignored in estimating settlements. This results in the discontinuity in the second graph.

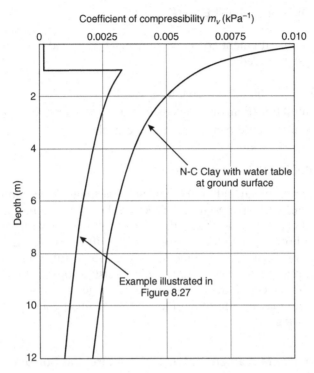

Figure 8.31 Coefficient of compressibility for an N–C soil and the soil in Figure 8.27.

In both cases the compressibility of the soil is far from uniform and clearly not in keeping with the Terzaghi assumptions.

8.7.6 Influence of Inadequate Data on Actual Soil Conditions

Determination of the coefficient of consolidation (c_v) from odometer or triaxial dissipation tests is not difficult and is routinely carried out. However, the extent to which c_v values determined from laboratory tests are representative of the soil mass in situ is problematical. Field consolidation rates often turn out to be much faster than predictions based on laboratory values. The reason is believed to be the presence of discontinuities in the clay, particularly in the form of thin silt or sand layers (lenses). Such layers provide drainage paths that allow water to escape rapidly and thus greatly accelerate the rate of consolidation. These layers can easily be missed during conventional site investigations. Ideally, long undisturbed samples from continuous coring are necessary in order to identify the presence or otherwise of such layers.

8.8   ALLOWABLE SETTLEMENT

Constraints on settlement arise in a number of ways and the following issues need to be considered:

- Total (or absolute) settlement
- Differential settlement between a structure and surrounding ground
- Differential settlement of a structure itself

8.8.1   Total (or Absolute) Settlement

If a filled area, or an individual fill, or a building settles uniformly, then the settlement will have no direct effect on the fill or the building itself and will not be of concern as far as the integrity of the structure is concerned. There may however be other factors which will require a limitation on the settlement; for example, in some situations where fill is placed over wide low-lying areas for development as light industrial or housing subdivisions, there may be effects on the performance of services (storm water and sewer lines) or there may be a minimum level to be maintained to provide security against flooding.

8.8.2   Relative Movement between Structure and Surrounding Ground

Relative movement in this case will normally involve a building settling relative to the surrounding ground, although the reverse may sometimes be true. Occasionally, fill is placed over a wide area of soft ground, and roads and services are put in place and building sites made available. Some settlement of the fill may still be occurring. The buildings constructed in such a situation may be placed on piled foundations, either because the bearing capacity of the fill and underlying soft soil may be inadequate or because the settlement from the building loads would be unacceptable. Hence the buildings will not settle at all and the surrounding ground may settle relative to the buildings. This may adversely affect access levels and service connections.

8.8.3   Differential Settlement of Buildings

With large multistory buildings there is a tendency for differential settlement to occur because the internal columns generally carry more load than the outside columns. Some buildings have central "cores" for lifts and services, for example, that involve heavy shear walls to provide lateral stiffness against earthquake forces. This means there is a concentration of load, which tends to accentuate the likelihood of differential settlement. Differential settlement can affect the building in two ways. First it may damage the building and threaten its structural integrity, and second it may affect architectural and functional features of the building. Unsightly cracks may appear in finishing elements or decorative features, and windows or doors may jam and become difficult to operate.

Various methods have been proposed to determine acceptable limits of differential settlement with such buildings, as illustrated in Figure 8.32. The criterion most commonly used as an indicator of possible damage is the ratio of the differential settlement δ to the distance L over which it occurs, after eliminating the effect of any tilt of the building. This is illustrated in Figure 8.32. This ratio is termed the "angular distortion" or the "relative rotation" and is designated by the angle β. Other terms used in describing differential settlement are the tilt of the structure (or part thereof) $\omega = (S_L - S_P)/B$ and the angular strain α, as indicated in the figure. The angular strain is similar to the angular distortion, but it is simpler to use for internal joints because taking account of the tilt of the structure is not straightforward in this case.

With conventional buildings having uniform distances between columns, the maximum differential settlement is most likely to occur between outside footings and the adjacent internal footings. This is because the loads on internal footings are likely to be very similar, while the outside footings will carry substantially less load.

The most commonly used limit for this angular distortion, expressed as δ/L, is 1/500, although some published papers suggest a value as high as 1/300 may be acceptable. These values relate to "nuisance damage," for example, cracking of walls and partitions or jamming of doors. For structural damage to occur, considerably greater distortion would have to take place; published papers suggest the value of δ/L would need to exceed a figure in the vicinity of 1/150 before structural damage became likely.

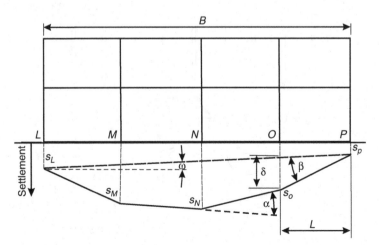

Figure 8.32 Differential settlement and angular distortion.

The limit to be adopted in practice will therefore depend on the nature of the building (since some types of architecture are more susceptible to damage than others) and the rate of settlement. If the structure is built on a soil that consolidates rapidly, most of the settlement will occur during construction of the frame of the building (since this is when the major loads are applied), and postconstruction settlement affecting the architectural features will be relatively small. A larger value of δ/L would be permissible in this case (possibly as much as 1/150) than in the case of a soil which consolidates very slowly.

8.9   RADIAL FLOW AND SAND (OR "WICK") DRAINS

Consolidation rates in soft, normally consolidated clays can be very slow, and various measures are possible to accelerate the process. The best known is probably the use of sand or "wick" drains. Sand drains are simply vertical sand columns closely spaced within the soil mass intended to act as drainage paths so that water can escape more rapidly from the soil. A conceptual sketch of the principle is shown in Figure 8.33. Compacted fill has been

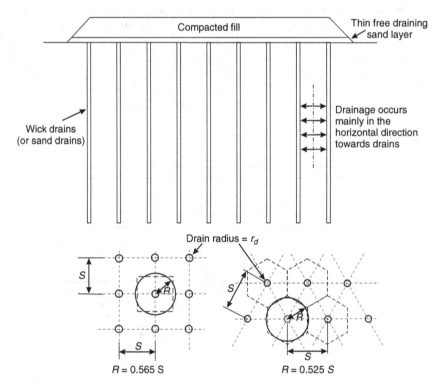

Figure 8.33   Wick (or sand) drains to accelerate consolidation of soft clays.

placed on the surface of a soft clay to provide a building platform or possibly a highway embankment. A thin layer of sand is first placed on the surface to act as a drainage blanket through which water coming from the drains can escape. The drains are normally installed after the drainage blanket is in place, as the sand provides a firm surface on which machinery can operate. The drain spacing is normally about 1–4 m. Drains may be installed on a square or triangular grid, though the latter is generally preferred as it is more efficient. The consolidation process is thus one in which pore water flow occurs horizontally in a radial pattern, while compression occurs in the vertical direction. It is clear that the drainage path length is greatly reduced (at least in theory) by this method as the distance to the nearest drain is likely to be only 1 or 2 m whereas the distance to the surface may be 5 or 10 m.

Sand drains have been used for many years, but in recent years various types of "prefabricated" drains have become much more popular. The best known of these are termed wick drains and consist of a flat plastic core with drainage channels around which is wrapped a layer of filter cloth. The filter cloth retains the soil and allows the center of the wick to act as a drainage channel. These wick drains are installed by the use of a mandrel—a long steel channel—which is pushed into the ground using a drilling rig or a modified machine digger. The wick drains are supplied in the form of large coils the end of which is attached to the mandrel so that it is pushed into the ground along with the mandrel. When the mandrel is withdrawn, the wick drain becomes detached and remains in the ground. Wick drains are typically about 100 mm wide and 5–10 mm thick. Their equivalent diameter is generally assumed to be that of a circle of equal circumference.

There are numerous cases in the literature where sand or wick drains appear to have functioned as intended. However, there are also cases where they have not been effective. There are undoubtedly also many cases where they have been installed and appear to have functioned satisfactorily, but with no proof that the same result would not have been achieved without the drains. The method of installation is important. If the soil at the edge of the hole is disturbed and remolded, then its permeability may be greatly reduced, rendering the drains much less effective. Thus the holes should be formed by a process that cuts the soil cleanly without remolding or "smearing" it.

8.9.1 Theory for Design of Sand and Wick Drains

The equation governing radial drainage, derived from an analysis similar to that for the Terzaghi consolidation equation for 1-D consolidation, is as follows:

$$\frac{\partial u}{\partial t} = c_h \left(\frac{\partial^2 u}{\partial r^2} + \frac{1}{r} \frac{\partial u}{\partial r} \right) \tag{8.21}$$

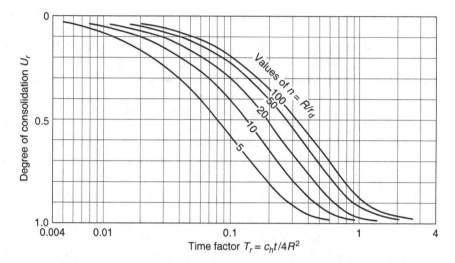

Figure 8.34 Solution for the consolidation process involving radial drainage.

where r is the radius and c_h is the coefficient of consolidation for this situation.

It is assumed that flow occurs only in the horizontal direction. The form is clearly not too different from the Terzaghi equation. Barron (1948) obtained a solution for Equation 8.21 for the simple boundary condition of a circle which is available in the form of charts, not dissimilar to the Terzaghi charts. They are expressed in terms of the degree of consolidation U_r and time factor T_r. The time factor T_r in this case is defined as

$$T_r = \frac{c_h t}{4R^2} \tag{8.22}$$

where R is the radius of influence of the drain.

Charts of U_r versus T_r are available for a range of values of n where $n = R/r_d$ and r_d is the radius of the drain. Figure 8.34 shows the solution.

In Figure 8.33 each drain takes the flow from the square or hexagon of which it forms the center. The solution is in terms of a circular area, so an equivalent radius must be assigned to the area covered by each drain. These areas are square or hexagonal for a square or triangular pattern of drains, respectively, as shown in Figure 8.33. The equivalent circle is normally calculated assuming equal areas; the values obtained in this way are also shown in the figure.

8.10 SETTLEMENT OF FOUNDATIONS ON SAND

Laboratory tests are not normally used to measure the compressibility of sands in order to undertake settlement estimates, because of the extreme difficulty of obtaining undisturbed samples of sand. In place of laboratory tests field tests are used, for which empirical correlations with compressibility and foundation settlement are available. The most common tests are the standard penetration test (SPT) and the static Dutch cone penetrometer test (CPT). A full description of these tests is given in Chapter 10.

The simplest way to estimate the settlement of a foundation on sand is to use the elastic theory relationships given at the beginning of this chapter (Equations 8.1 and 8.2). Provided SPTs or CPTs have been carried out, the value of the Young's modulus E can be estimated from empirical correlations such as those given in Chapter 10 (Figure 10.13). This procedure will give a reasonable, though approximate estimate of the likely settlement. However, the method does not take account of the nonelastic behavior of soils and is not suitable for situations where the sand is not of uniform properties. To help take account of these factors, various additional empirical methods have been proposed, as described in the following sections.

8.10.1   Schmertman Method Using Static Cone Penetrometer Results

One of the more commonly used methods is that proposed by Schmertman and colleagues (Schmertman, 1970; Schmertman et al, 1978). Schmertman pointed out that theoretical studies and model tests show that the vertical strains beneath a loaded foundation are greatest not immediately beneath the surface, where the stress is highest, but at a significant depth below the surface. To take account of this, he introduced a strain influence factor I_z and expressed the strain at a particular depth as

$$e = \Delta p \frac{I_z}{E'} \tag{8.23}$$

where e is the vertical srain, Δp is the net applied pressure, and E' is the Young's modulus.

The value of the influence factor I_z is shown in Figure 8.35. These linear graphs are a simplification of the actual curves produced by theoretical and experimental studies and are different depending on whether the foundation is axisymmetric (circular or square) or a long strip. For axisymmetric foundations the peak value of the factor I_z occurs at a depth of $0.5B$, while for a strip foundation it occurs at about $1.0B$ (B = width or diameter of footing). Schmertman recommends that E' be obtained from the CPT cone resistance (described in Chapter 10) using the relationships given in Figure 8.35. The expression for the total settlement therefore has the form:

$$s = \Sigma e \Delta z$$

where s is the total settlement and Δz is the thickness of the sublayers used in making the estimate; I_z and E' are the appropriate values for each layer.

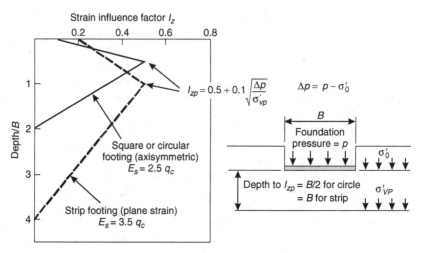

Figure 8.35 Schmertman's strain influence factors for estimation of settlement on sand.

Schmertman introduces two correction factors C_1 and C_2 to take account of the depth of the footing and the additional settlement due to creep. The final expression for settlement is therefore as follows:

$$s = C_1 C_2 \Delta p \Sigma \frac{I_z}{E'} \Delta z \qquad (8.24)$$

The values of C_1 and C_2 are given by the following expressions:

$$C_1 = 1 - 0.5 \frac{\sigma_0'}{\Delta p} \qquad (8.25)$$

$$C_2 = 1 + 0.2 \log \frac{t}{0.1} \text{ (where } t \text{ is the time in years)} \qquad (8.26)$$

8.10.2 Burland and Burbidge Method

An equally well known and widely used method is that proposed by Burland and Burbidge (1985). This is an empirical method based on a detailed, systematic, statistical analysis of a very large amount of field data. The principal variables focused on are the foundation pressure, the foundation width, and the compressibility of the sand as indicated by SPTs. The analysis leads to the following simple expression for the settlement of a strip footing on a normally consolidated sand or gravel:

$$s = \left(\frac{1.71 B^{0.7}}{\overline{N}_{60}^{1.4}} \right) p \qquad (8.27)$$

where s = settlement (mm)

$\quad B$ = foundation width (m)

$\quad \overline{N}_{60}$ = average N_{60} over depth of influence of foundation

$\quad p$ = foundation pressure (kPa)

The term $1.71B^{0.7}/N_{60}^{1.4}$ is designated a_f, the foundation subgrade compressibility (mm/kPa), so that Equation 8.26 can be written as

$$s = a_f p \qquad (8.28)$$

The depth of influence (Z_i) over which the average value of N_{60} is to be taken is given with sufficient accuracy by $Z_i = B^{0.75}$ provided N is constant or increases with depth. If N shows a consistent decrease with depth, then Z_i should be taken as $2B$. Burland and Burbidge recommend the following correction factors to the above expression:

Shape factor f_s: This is small and can be taken as

$$f_s = \left(\frac{1.25(L/B)}{(L/B) + 0.25} \right)^2 \qquad (8.29)$$

where L and B are the length and width of the foundation, respectively.

Thickness of compressible layer f_1: If the thickness of the sand or gravel layer (H_s) is less than the depth of influence Z_i, then the following correction factor should be used:

$$f_1 = \frac{H_s}{Z_i} \left(2 - \frac{H_s}{Z_i} \right) \qquad (8.30)$$

Time-dependent settlement f_t:

To take account of creep (or secondary settlement) the following factor is recommended (for static loads):

$$f_t = 1.3 + 0.2 \log \left(\frac{t}{3} \right) \qquad (8.31)$$

where t is in years and must be equal to or greater than 3 years.

Equation 8.27 is for normally consolidated sands. For overconsolidated sands, it is recommended that the compressibility at pressures less than the preconsolidation pressure be reduced to one-third of the value given by Equation 8.27. Therefore, if the applied pressure p is less than the preconsolidation pressure σ_{vo}', the settlement is given by

$$s = \frac{1}{3} a_f \, p \qquad (8.32)$$

while if p exceeds σ_{vo}', the settlement is given by

$$s = \frac{1}{3}a_f\sigma_{vo}' + a_f\left(p - \sigma_{vo}'\right) = a_f\left(p - \frac{2}{3}\sigma_{vo}'\right) \qquad (8.33)$$

Taking account of all the above factors we can thus express the settlement as follows:

For values of p greater than σ_{vo}':

$$s = f_s f_l f_t \left(\frac{1.71B^{0.7}}{\overline{N}_{60}^{1.4}}\right)\left(p - \frac{2}{3}\sigma_{vo}'\right) \qquad (8.34)$$

For values of p less than σ_{vo}':

$$s = f_s f_l f_t \frac{1}{3}\left(\frac{1.71B^{0.7}}{\overline{N}_{60}^{1.4}}\right)p \qquad (8.35)$$

Other points that should be noted in using the Burland and Burbidge method are the following:

(a) The N values from the SPT tests are not corrected for overburden pressure. This is because the overburden pressure is believed to influence both the N value and the compressibility in a similar manner.
(b) The correction for fine sands or silty sands described in Chapter 10 (Equation 10.4) should still be applied.
(c) The method can be used when only the results of static CPTs are available by using the correlation given in Chapter 10 (Figure 10.11).
(d) The level of the water table beneath the foundation does not appear to have a significant influence on the settlement. This is believed to be because any influence from the water table is reflected in the SPT blow count.
(e) The data used in the analysis come from hard-grained sands and may not be applicable to soft-grained sands or gravels.
(f) The method is based on situations where the safety factor with respect to bearing capacity is greater than 3. If the method is applied to situations where this is not the case, it may give unreliable results.

8.10.3 Worked Example

A square foundation 4 m by 4 m is to be constructed at a depth 1 m below the ground level and will apply a pressure of 200 kPa on the soil. The site consists of a deep layer of sand, believed to be a recent deposit and assumed to be normally consolidated. Figure 8.36 shows the results of a CPT and SPTs carried out in the sand down to a depth of 15 m. The unit weight

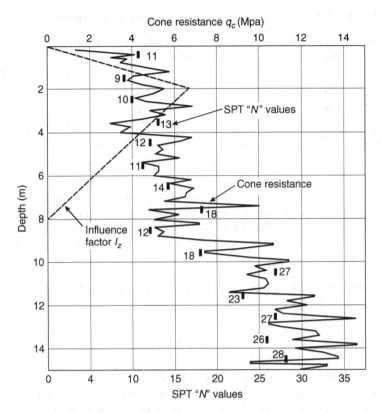

Figure 8.36   SPTs and CPTs in sand for settlement estimate.

of the sand is estimated to be $22\,kN/m^3$. We will estimate the expected settlement of the foundation using elastic theory, Schmertman's method, and the Burland and Burbidge method.

(a) Elastic Theory   The settlement of a circular foundation is given by

$$\delta = qD\frac{1-v^2}{E} \quad \text{(Equation 8.1)}$$

where q is the applied pressure, D is the diameter, and E and v are the effective stress elastic parameters. Because we do not have a solution for a square foundation, we will convert the square foundation to a circle of "equivalent" diameter on the basis of equal area. This diameter is given by

$$D = \sqrt{\frac{4 \times 4^2}{\pi}} = 4.51\,m$$

Table 8.8 Settlement Calculation Using Schmertman Method

Layer	Depth (m)	Thickness (m)	q_c (MPa)	E' (MPa)	I_z	Strain (%)	s (mm)
1	0–2	2	4	10	0.33	5.9	11.8
2	2–4	2	4	10	0.55	9.8	19.6
3	4–8	4	5	12.5	0.22	3.1	12.5
							Total: 44 mm

We can estimate the value of E from the correlations given in Figure 10.13 of Chapter 10. The average value of the N number is about 10, corresponding to an E of about 20 MPa. We do not have any information on the value of Poisson's ratio so we will adopt a midrange value of $v = 0.35$. Inserting these values into the above formula gives $\delta = 32$ mm.

(b) Schmertman Method   The basic equation for the Schmertman method (Equation 8.24) is

$$s = C_1 C_2 \Delta p \Sigma \frac{I_z}{E'} \Delta z$$

The values of I_z and E' vary with depth. The empirical construction proposed by Schmertman for I_z is shown in Figure 8.34 along with the penetrometer test record. To take account of this, it is necessary to divide the soil into several layers within which we can estimate average values of these two parameters. In this case we will select three layers as indicated in Table 8.8. The values of q_c and I_z estimated for each layer are shown in the table. Applying the above formula we can determine the strain and compression (Δz) in each layer and then sum these to get the total settlement. Note that Δp is $200 - 22 = 178$ kPa, as the foundation is 1m deep.

Schmertman's corrections for embedment depth and long-term creep are as follows:

Embedment depth $C_1 = 1 - 0.5 \dfrac{\sigma'_0}{\Delta p} = 0.94$

Long-term creep $C_2 = 1 + 0.2 \log \dfrac{t}{0.1} = 1.2$ for 10-year period

This gives the final settlement as $44 \times 0.94 \times 1.2 = 50$ mm.

(c) Burland and Burbidge Method   The equation is

$$s = \left(\frac{1.71 B^{0.7}}{\overline{N}_{60}^{1.4}} \right) p$$

Table 8.9 Summary of Settlement Estimates

Analysis Method	Settlement (mm)	
	Immediate	After 10 Years
Elastic theory	32	Not allowed for
Schmertman	44	50
Burland and Burbidge	32	45

where $B = 4\,\text{m}$, $p = 178\,\text{kPa}$ (the net foundation pressure), and N_{60} is the average N value immediately beneath the foundation, which we will take as 10. This gives $s = 32\,\text{mm}$. The correction factors (given above) are $f_s = 1$, $f_1 = 1$, and $f_t = 1.4$ for a 10-year period. Thus the corrected settlement is $1.4 \times 32 = 45\,\text{mm}$.

The results are summarized in Table 8.9.

These values are in reasonable agreement, especially those from elastic theory and the Burland and Burbidge method. The Schmertman method generally tends to give somewhat higher values.

REFERENCES

Barron, R. A. (1948). Consolidation of fine-grained soils by drain wells. *Trans. ASCE*, Vol. 113, 718–742.

Burland, J. B., and M. C. Burbidge. 1985. Settlement of foundations on sand and gravel. *Proc. Institution of Civil Engineers*, Part 1, 78, Dec., 1325–1381.

Casagrande, A. 1936. The determination of the pre-consolidation load and its practical significance. In *Proceedings of the First International Conference on Soil Mechanics and Foundation Engineering*, Harvard University, Vol. 3, pp. 60–64.

Davis E. H. and H. G. Poulus. 1972. Rate of settlement under two and three-dimensional conditions. *Geotechnique*, Vol. 22, No. 1, 95–114.

Lancellotta, R. 1995. *Geotechnical Engineering*. Rotterdam: A. A. Balkema, pp. 108–109.

Schmertman, J. H. 1953. Estimating the true consolidation behavior of clay from laboratory test results. *Proceedings ASCE*, Vol. 79, 1–26.

Schmertman, J .H. 1970. Static cone to compute static settlement over sand. *J. Soil Mechanics and Foundation Division, ASCE*, Vol. SM3, 1011–1043.

Schmertman, J. H., J. D. Hartman., and P. R. Brown. 1978. Improved strain influence factor diagrams. *Journal of the Geotechnical Engineering Division*, Vol. GT8, 1131–1135.

Skempton, A. W. and L. Bjerrum. 1957. A contribution to the settlement analysis of foundations on clay. *Geotechnique*, Vol. 7, 168–178.

Taylor, D. W. 1948. *Fundamentals of Soil Mechanics*. New York: John Wiley and Sons.

EXERCISES

1. A 5-m-thick clay layer has an m_v value of $1.5 \times 10^{-4}\,kPa^{-1}$. Construction of an embankment is expected to increase the effective vertical stress throughout the layer by 30 kPa. Estimate the expected compression that will occur in this layer. (**22.5 mm**)

2. If the clay layer in exercise 1 has a c_v value of $0.8\,m^2/year$, estimate the time for consolidation to be 90 percent complete assuming (a) there are sand layers above and below the clay layer and (b) there is impermeable rock below the clay and a sand layer above it. (**6.6 years, 26.5 years**)

3. A layer of residual soil (fully saturated) extends from the ground surface to a depth of 45 m below which solid rock is found. An odometer test is carried out on a sample of the clay taken from near the surface, giving the following deformation-versus-pressure results:

Pressure (kPa)	0	10	20	40	80	160	320	640	1280	2560
Sample thickness (mm)	20.00	19.995	19.965	19.858	19.630	19.228	18.576	17.598	16.443	15.090

The following properties are also measured: unit weight $= 18.4\,kN/m^3$, water content $= 35.4\%$, specific gravity $= 2.72$. The water table is at a depth of 2 m.

A storage tank 40 m in diameter is to be built at the surface and will apply a uniform pressure of 100 kPa to the soil. Assuming that the clay is uniform and the single test result is representative of the whole layer, determine the settlement to be expected at the center of the tank. (**914 mm**)

4. An odometer test is carried out on a sample of clay 20 mm thick assumed to be fully saturated. The time to reach 90 percent consolidation is 10.5 min. Estimate the following:
   (a) Time to reach 50 percent consolidation in odometer test
   (b) Time for a 9-m-thick layer of this clay to reach 90 percent consolidation assuming the same drainage conditions apply and it is subject to the same load increment as in the odometer test (**2.44 min, 1,477 days**)

5. A layer of soft, normally consolidated clay is 14 m thick and overlies dense sand. The water table is just below the ground surface. A sample taken from the clay shows it to have the following properties:

$$\text{Unit weight} = 13.92\,\text{kN/m}^3, \quad \text{Water content} = 112.0\%,$$
$$e = 3.00, \quad C_c = 1.40$$

Determine the value of the parameter m_v that will apply in the top 1 m and the bottom 1 m of the clay layer if the clay is loaded at the surface over a wide area with a uniform pressure of 30 kPa. **(0.014, 0.0022 kPa^{-1})**

6. Odometer tests have been carried out on undisturbed samples of two different clays. The deformation readings for the pressure increment from 40–80 kPa are given in the following table. Determine the value of the coefficient of consolidation (c_v) or the two clays for this pressure increment. The readings are in mm $\times$ 10^{-2}. **(0.66 m^2/year, NA)**

Time	0	2.4 s	9.6 s	21.6 s	38.2 s	1.0 min	2.25 min	4 min	9 min
Soil A	1810	1792	1784	1773	1764	1755	1732	1707	1660
Soil B	1810	1710	1665	1640	1625	1610	1588	1574	1560

Time (min)	16	25	36	49	64	81	100
Soil A	1615	1572	1543	1524	1513	1504	1496
Soil B	1542	1528	1525	1519	1513	1509	1505

7. A soft clay layer is 12 m thick below which a layer of dense sand is found. The water table is virtually at the ground surface. To make the site acceptable for the construction of buildings, a layer of granular fill 3 m thick is to be placed at the surface. The fill unit weight is 21/0 kN/m^2. Two standpipe piezometers are installed in the clay layer, an upper one 5 m deep and a lower one 10 m deep. After 3 years the water level in the upper standpipe is 1.5 m above the surface of the fill. Estimate:

(a) Coefficient of consolidation (c_v) of clay
(b) Water level to be expected in lower standpipe at same time (above or below surface of fill)

Repeat the above estimates when a thin highly permeable sand layer exists at a depth of 9 m and the water level in the upper standpipe is again 1.5 m above the fill surface. Assume the clay properties are uniform above and below the thin sand layer and that the sand layer acts as a free-draining permeable boundary.

[(*a*) $c_v = 2.76 \, \text{m}^2/\text{year}$, **0.62 m below fill surface**

(*b*) $c_v = 0.61 \, \text{m}^2/\text{year}$, **1.91 m below fill surface**]

8. A clay layer is 12 m deep below which a layer of hard sandstone is found. The coefficient of compressibility m_v of the clay is 1.6×10^{-4}. The water table is at a depth 4 m below the ground level. Drainage works carried out near the site result in lowering of the water table by 2.5 m. Determine the expected settlement of the ground surface. **(47 mm)**

CHAPTER 9

SHEAR STRENGTH OF SOILS

9.1 BASIC CONCEPTS AND PRINCIPLES

Soil strength differs from that of other engineering materials (steel, concrete, wood, etc.) in two important aspects:

1. Only the shear strength is of interest. The principal design situations addressed by engineers are the bearing (or load-carrying) capacity of foundations, earth pressures on retaining walls, and stability of slopes. All of these are directly dependent on the shear strength of the soil. The failure modes that govern these situations are shown in idealized form in Figure 9.1: they are discussed comprehensively in Chapters 11–14. In each case the soil tends to fail by shear movement on specific failure surfaces within the soil mass. Compressive or tensile strength is not relevant to the soil behavior in these situations.
2. Shear strength is not constant for a particular soil. Deep in the ground, soil is stronger than it is near the surface. In an embankment the lower layers are stronger than those at the top. This is because the strength is dependent on the confining pressure coming from the layers above and must be expressed in a manner that takes this into account. The soil strength may also undergo changes with time, possibly as a result of natural effects such as rainfall, or the influence of human activity on the slope.

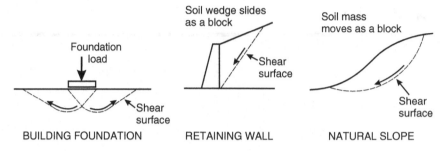

Figure 9.1   Modes of failure for foundations, retaining walls, and soil slopes.

9.1.1   General Expression for Shear Strength

The general expression for soil shear strength, used almost universally in the world today, is

$$s = c' + (\sigma - u) \tan \phi' \qquad (9.1)$$

$$\text{or } s = c' + \sigma' \tan \phi'$$

where s = shear strength or shearing resistance.
 σ = total normal stress on shear plane
 u = pore pressure on shear plane
 σ' = effective normal stress on shear plane
 c' = cohesion intercept in terms of effective stress
 ϕ' = angle of shearing resistance (often called "friction angle") in terms of effective stress

The parameters c' and ϕ' are commonly called the shear strength parameters in terms of effective stress. Equation 9.1 is known as the Mohr–Coulomb failure criterion. If the shear stress on any plane within the soil mass exceeds the value given by Equation 9.1, then movement (or yield) will occur on that plane.

The parameters c' and ϕ' are almost independent of the method used for measuring them and can be regarded as constants for a given soil in a given state (e.g., its undisturbed state). The strength can be considered to consist of two components:

(a) **A cohesive component (c')**, resulting from some form of bonding between particles, or a dense state of packing of the particles. This has a constant value.

(b) **A "frictional" component ($\sigma' \tan \phi'$)**, dependent on the effective normal stress on the shear plane. The true composition of this component is open to debate, as it is not all strictly friction. However, because this component is proportional to the normal stress, it is reasonable to think of it as a frictional component.

9.1.2   Undrained Shear Strength (s_u)

This is a special case and, as discussed in Chapter 6, refers to situations where no change in water content occurs, either in field situations or in laboratory measurements of shear strength. If the soil is undrained and fully saturated, no volume change can occur, which in turn means that no change in effective stress occurs. In accordance with Equation 9.1 above, there is no change in the frictional component of strength, and the strength therefore remains constant, regardless of any changes in the total stresses acting on the soil. In terms of total stress, the soil behaves as though its friction angle is zero. For this reason, the undrained situation is often referred to as the $\phi = 0$ case. It is important to recognize that this $\phi = 0$ case arises from two essential and important factors:

(a) The soil is fully saturated.
(b) The conditions are undrained.

The $\phi = 0$ case is not dependent on the soil type; the same behavior applies equally to clays and sands provided the above two conditions apply. However, as we shall see later, the $\phi = 0$ case has practical relevance only for clays and silts. It has little or no practical relevance for coarse-grained soils. Analysis in terms of the undrained strength of the soil implies analysis in terms of total stress only, since undrained soil behavior is directly related only to the total stress.

9.1.3   Relationship between Strength in Terms of Effective Stress and Undrained Strength

The shear strength in terms of effective stress and of total stress is illustrated graphically in Figure 9.2. The upper diagram (a) shows the stresses acting on a shear (or failure) plane within a soil mass. The total normal stress (σ) is made up of its two components—the pore pressure (u) and the effective stress (σ'). The associated shear strength (s) acts along the plane. The lower diagram (b) shows the Mohr–Coulomb failure criterion (Equation 9.1) in graphical form. To obtain the shear strength on the plane in terms of effective stress, we simply plot the effective stress σ' and read off the strength at point A. This strength will also be the drained strength on this plane, provided the soil is sheared sufficiently slowly that any pore pressures generated by the shearing process will dissipate as shearing occurs.

The situation with respect to the undrained shear strength on the same plane is more complicated. The undrained strength is by definition the strength available on the plane when it is sheared to failure under undrained conditions. Such loading could be done in the field (with great difficulty) or in a laboratory. During this undrained shearing the pore pressure will change; this change is designated Δu, and results in a different effective

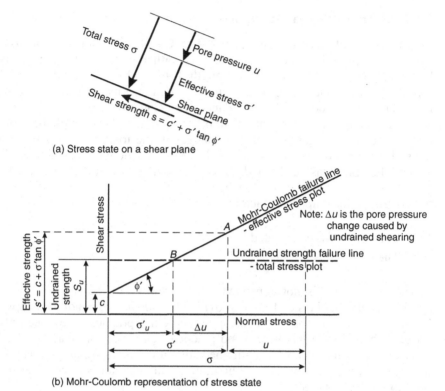

(a) Stress state on a shear plane

(b) Mohr-Coulomb representation of stress state

Figure 9.2 Graphical representation of soil shear strength.

stress at failure designated σ'_u in Figure 9.2b. The undrained shear strength is then given by point B on the Mohr–Coulomb failure line.

The pore pressure change during shearing is not necessarily positive; it could be negative, in which case the effective stress would increase and the undrained strength would be further up on the Mohr–Coulomb failure line and greater than the strength at point A. There are two situations in which the strength in terms of effective stress and the undrained strength will be the same:

1. When the soil is on the point of failure, that is, the shear stress on the plane is already equal to the available strength. In this case no additional stress is needed to cause failure and no change in pore pressure occurs. Bishop and Bjerrum (1960) make this same point when analyzing the failure state of a vertical soil cutting.

2. When the pore pressure neither rises nor falls during the shearing process, which will only be the case in exceptional circumstances. These concepts will become clearer in the subsequent sections on measurement of shear strength.

To summarize, the following important points should be noted:

(a) The undrained strength of a soil element is not normally the same as the strength determined by considering the effective stresses acting on the soil. The value of undrained strength, and strength in terms of effective stress, will only be the same if the soil is on the point of failure.

(b) Despite this difference, the strength is nevertheless always governed by the effective stress on the shear plane, regardless of whether the strength is expressed in terms of effective stress or total stress.

(c) The effective strength parameters (of a particular soil in its undisturbed state) can be regarded as soil constants. For most soils, in most practical situations, their value is almost independent of the method used to measure them.

(d) The undrained shear strength (s_u), however, cannot be regarded as a soil constant in the same way as c' and ϕ'. Its value depends on the method of measurement, whether in the field or the laboratory. Each method generates a different value of pore pressure and thus a different value of undrained shear strength. This will be explained more fully in Section 9.8.2. The value of s_u also depends on the water content of the soil; consolidation of the soil to a lower water content results in a higher value of s_u.

(e) The Mohr–Coulomb failure criterion normally refers to the expression in Equation 9.1 However, a similar expression is also possible in terms of total stress, having the form

$$s = c + \sigma \tan \phi \qquad (9.2)$$

where c and ϕ are now the shear strength parameters in terms of total stress.

Although this form of the equation was widely used in the early years of soil mechanics, it has no general validity as the values of c and ϕ are dependent on the test method used to measure them and are not of practical relevance. The only exception is the case of undrained strength. In this case the values of c and ϕ become

$$c = s_u \text{ (undrained shear strength of soil)}$$

$$\phi = 0$$

In an undrained situation the soil behaves *as though* it has no frictional component of shear strength. This reflects the test conditions under which the strength is measured; in reality it still has a frictional component of shear strength dependent on its ϕ' value.

9.2   MEASUREMENT OF SHEAR STRENGTH

The test method should be such that we know the following information during the test:

1. Principal stresses σ_1 and σ_3 or normal and shear stresses on shear plane
2. Pore water pressure so that effective stresses can be calculated
3. Strain

Measurement of stress and strain involves little difficulty. Measurement of pore pressure requires somewhat greater care but is not a problem as long as we pay attention to the following factors:

(a) The drainage condition in the test, that is, whether it is drained or undrained.
(b) The rate of strain. The rate must be slow enough to ensure that the pore pressure throughout the sample is essentially uniform, so that the "known" or measured value at the end of the sample is representative of the value throughout the sample.

Strength tests are usually carried out in two stages:

Stage 1: Application of normal (or confining stress)—consolidation stage
Stage 2: Application of shear stress until failure occurs—loading stage

9.2.1   Direct Shear Test (or Shear Box Test)

This is the earliest form of shear test and was first used by Coulomb in 1776. It is illustrated conceptually in Figure 9.3. The apparatus consists of a split box into which the sample is placed. Porous plates are placed above and below the sample to allow water to drain into or out of the sample during the test. A hanger and weight system is then used to apply a normal (vertical) stress to the sample. A mechanical jack then applies a horizontal force to the lower half of the box while the upper half remains stationary. The horizontal force is applied at a constant deformation rate and measurements made of both displacement and force as deformation progresses.

A series of tests is done at varying normal stresses. Each test result is plotted in graphical form, first as stress versus displacement and second as peak shear stress (failure value) versus normal stress, as indicated in Figure 9.3. A line through these points defines the c' and ϕ' values of the material.

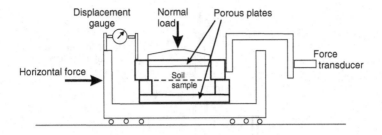

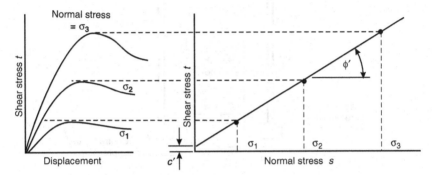

Figure 9.3   Direct shear (or shear box) test.

Advantages and limitations of the direct shear test include the following:

- It is simple and easy to perform.
- Undistorted samples are difficult to prepare because of the square cross section.
- Drainage cannot be controlled, so that undrained tests are not possible.
- Principal stresses are not known
- Area of sample changes as the test proceeds, and area corrections are indeterminate.
- A stress–displacement curve is obtained, not a stress–strain curve

9.2.2 Triaxial Test

Triaxial tests are undoubtedly the most popular and common method used today for measuring shear strength. They are preferred for theoretical reasons and because of their versatility. All types of strength tests can be carried out in the triaxial cell as well as other types of tests for measuring permeability or consolidation characteristics. The apparatus is shown diagrammatically in Figure 9.4.

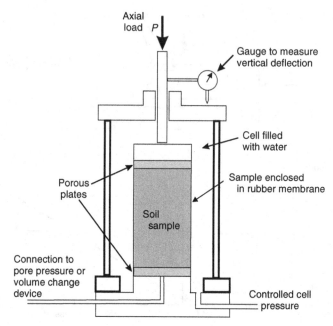

Figure 9.4 Triaxial test.

Three types of triaxial tests are commonly carried out: undrained, consolidated undrained, and drained tests.

Conditions during the consolidation stage (stage 1) and the loading stage (stage 2) of each test type are as follows:

(a) Undrained Tests (Also Called Unconsolidated Undrained Tests) No drainage is permitted during either stage. Pore pressure is not normally measured.

(b) Consolidated Undrained Tests Drainage is permitted during the consolidation stage until the sample is fully consolidated, that is, until all pore pressure has dissipated to zero. During the loading stage no drainage is permitted, and pore pressure is normally measured.

(c) Drained Tests Full drainage is allowed during both stages. Pore pressure is therefore zero. Volume change is normally measured during stage 2.

During the tests the following are measured:

1. Vertical deflection—to determine strain and to correct the area of the sample
2. Vertical load (force P)
3. Pore pressure—during the loading stage in consolidated undrained tests

4. Volume change—during the consolidation stage of both consolidated undrained tests and drained tests and during the loading stage of drained tests

Undrained tests are carried out to determine the undrained shear strength (s_u) of the soil. Consolidated undrained tests and drained tests are carried out to determine the effective stress strength parameters c' and ϕ'. The choice between consolidated undrained tests and drained tests is governed primarily by the permeability of the soil. Drained tests are the easiest to perform and normally used on sand because of its high permeability. With clays, the low permeability presents problems in conducting drained tests. Even if drainage is permitted at the ends of the sample, this does not ensure that the pore pressure is zero throughout the sample. The pore pressure in the central part of the sample may be substantially higher than that at the ends, in which case the assumption of zero pore pressure is not valid. To ensure the pore pressure is zero throughout the sample, drained tests on low-permeability clays must be carried out very slowly, sometimes requiring many days for completion.

For this reason, consolidated undrained tests are normally preferred for clays. In these tests, water does not have to drain out of the sample and the testing rate can be much faster than with drained tests. However, it is still necessary to use a strain rate sufficiently slow to ensure that pore pressure is uniform throughout the sample, and the value measured at the top or bottom of the sample is representative of the remainder of the sample.

9.2.3 Mohr's Circle of Stress

To obtain the Mohr–Coulomb failure line (and thus the values of the shear strength parameters c' and ϕ'), it is normal practice to make use of a graphical construction known as the Mohr circle of stress, and this will now be explained. Mohr's circles are widely used in stress analysis, and the explanation given here is limited to their use in soil mechanics. We will consider the state of stress on an element within a body where the major principal stress and minor principal stress act in the vertical and horizontal directions, respectively. This is illustrated in Figure 9.5 and corresponds to the stress state in a triaxial test.

We wish to know the stress state on any other plane at an angle α to the horizontal since failure cannot occur on either the vertical or horizontal plane. There will be a normal stress σ_n and a shear stress τ on this plane. We can determine the values of these by considering the equilibrium of the element.

Resolving parallel to the inclined plane yields

$$\tau a + \sigma_3 a \sin\alpha\cos\alpha = \sigma_1 a \cos\alpha\sin\alpha$$

$$\tau = (\sigma_1 - \sigma_3)\sin\alpha\cos\alpha$$

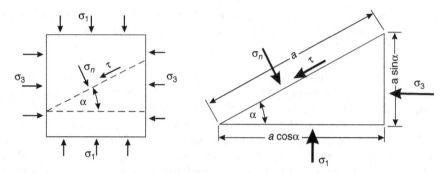

Figure 9.5   Stresses on a triangular element used to derive Mohr circle equations.

and as $\sin 2\alpha = 2 \sin \alpha \cos \alpha$, we can write

$$\tau = \frac{\sigma_1 - \sigma_3}{2} \sin 2\alpha \qquad (9.3)$$

Resolving normal to the plane gives

$$\sigma_n a = \sigma_1 a \cos \alpha \cos \alpha + \sigma_3 a \sin \alpha \sin \alpha$$

That is,

$$\sigma_n = \sigma_1 \cos^2 \alpha + \sigma_3 \sin^2 \alpha$$

Now

$$\cos 2\alpha = 2 \cos^2 \alpha - 1 \quad \text{and} \quad \cos 2\alpha = 1 - 2 \sin^2 \alpha$$

Rearranging gives

$$\sigma_n = \frac{\sigma_1 + \sigma_3}{2} + \left(\frac{\sigma_1 - \sigma_3}{2} \right) \cos 2\alpha \qquad (9.4)$$

Now consider the Mohr's circle shown in Figure 9.6. A line at angle α has been drawn from point A to intersect the circle at point P. We will examine the magnitude of OD and DP.

The horizontal axis represents normal stress and the vertical axis represents shear stress. By plotting the values of principal stress on the x axis, it is possible to immediately determine the values of normal and shear stress on any other plane. In the above diagram we can write for the dimensions of *OD* and *PD*

$$PD = CP \sin 2\alpha = \frac{OB - OA}{2} \sin 2\alpha$$

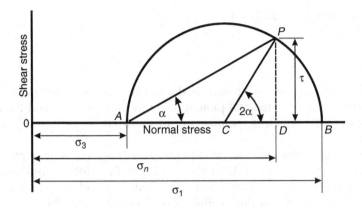

Figure 9.6 Mohr's circle of stress

That is,

$$\tau = \left(\frac{\sigma_1 - \sigma_3}{2}\right) \sin 2\alpha \tag{9.5}$$

$$OD = OC + CD = \frac{OA + OB}{2} + CP \cos 2\alpha$$

and

$$\sigma_n = \frac{\sigma_1 + \sigma_3}{2} + \left(\frac{\sigma_1 - \sigma_3}{2}\right) \cos 2\alpha \tag{9.6}$$

These equations are identical to those above (9.3 and 9.4) for the stresses σ_n and τ derived from the equilibrium analysis. Thus the distance OD gives the value of σ_n and DP gives the value of τ on a plane inclined at angle α to the horizontal. This gives us a simple graphical method for plotting the results of triaxial tests.

9.2.4 Use of Mohr's Circle for Plotting Triaxial Test Results

The stress state after application of the cell pressure is an equal all-round stress. This stress remains constant throughout the test and is therefore the minor principal stress σ_3. The vertical stress is increased by the load applied to the loading ram, force P in Figure 9.4. The increase in vertical stress is then equal to P/A, where A is the area of the sample. Hence the total vertical stress, which is the major principal stress σ_1, is given by

$$\sigma_1 = \sigma_3 + \frac{P}{A}$$

which can be rearranged to give

$$\sigma_1 - \sigma_3 = \frac{P}{A} \qquad (9.7)$$

The term $\sigma_1 - \sigma_3$ is known as the deviator stress. Triaxial test results are normally calculated and plotted using the values of σ_3 and the deviator stress $\sigma_1 - \sigma_3$, which is the diameter of the Mohr's circle, as illustrated in Figure 9.7.

By doing a series of tests at different cell pressures we can plot a series of circles. The tangent to these circles defines the **Mohr–Coulomb failure envelope**. No state of stress can exist which would be represented by a circle that crosses this line. Failure would occur before this could happen, at the stress state at which the circle touches the line.

Triaxial tests may be used to determine either the **undrained shear strength** or the **effective stress parameters c' and ϕ'**. To determine the former, the results are plotted using total stresses, while for the latter they are plotted in terms of effective stresses.

Undrained tests—Normally only the total stresses are known, and hence only a total stress plot is possible. The value of c is thus the undrained shear strength s_u and the ϕ value is zero, as indicated in Figure 9.8. In an undrained situation each increase in confining stress (σ_3) is accompanied by an identical increase in pore pressure, so no change in effective stress occurs (see Section 6.3).

Hence the strength remains constant. If we were to measure the pore pressure in each test and then calculate the effective stress and plot the effective stress circles, we would find that these circles coincide, as indicated in the figure. We are still unable to determine the effective stress parameters because a single circle is insufficient to determine the failure line.

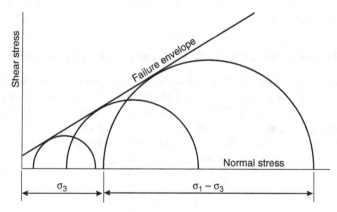

Figure 9.7 Mohr's circles plotted to establish the Mohr–Coulomb failure envelope.

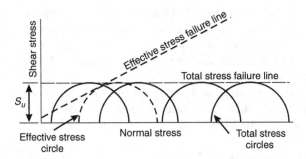

Figure 9.8   Result of an undrained triaxial test on a fully saturated soil.

Consolidated undrained tests—The pore pressure is measured and hence effective stresses are calculated and results plotted in terms of effective stresses, as is illustrated in Figure 9.9

$$\text{Note that } \sigma_1' = \sigma_1 - u \text{ and } \sigma_3' = \sigma_3 - u$$

$$\text{so that } \sigma_1' - \sigma_3' = (\sigma_1 - u) - (\sigma_3 - u) = \sigma_1 - \sigma_3$$

The value of deviator stress is thus the same whether expressed as total stress or effective stress.

Drained tests—Pore pressures are known (equal to zero) and hence effective stress is the same as the total stress. The results thus give plots of Mohr's circles in terms of effective stresses similar to those in Figure 9.9.

9.2.5   Soil Behavior in Consolidated Undrained and Drained Tests

Consolidated undrained tests and drained tests are both used to measure the effective stress parameters c' and ϕ'. The choice of test type is dependent

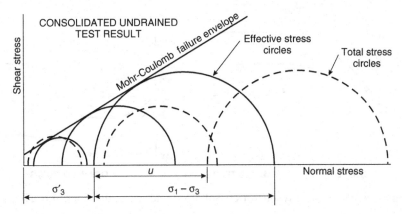

Figure 9.9   Results of a consolidated undrained triaxial test.

on the nature of the material. Drained tests on granular materials (sand and gravel) are easy to perform, but on clays they are not so easy because adequate time must be allowed for drainage to occur. As indicated earlier, consolidated undrained tests are most commonly used to measure c' and ϕ' values for clays.

Typical deviator stress and pore pressure or volume change curves from consolidated undrained and drained tests are shown in Figure 9.10. The following points should be noted:

1. The graphs on the left are from consolidated undrained tests; those on the right are from drained tests.
2. Tests have been done at three cell pressures—high, medium, and low, as labeled in the drawings.
3. In the consolidated undrained tests, the pore pressure is lowest when the cell pressure is lowest and greatest when the cell pressure is highest. This is because at the high cell pressure the particles tend to be forced closer together during shearing—this is resisted by the pore water pressure.

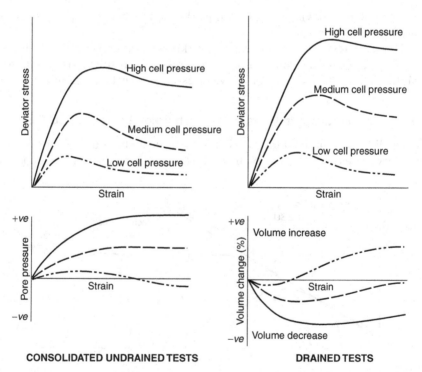

Figure 9.10   Typical results of consolidated undrained and drained triaxial tests.

4. In drained tests, the volume change may be negative or positive, that is, the volume may decrease or increase during loading. In Figure 9.10 all of the volume change graphs show an initial decrease in volume followed by a steady and significant increase. The initial decrease is greatest in the test with the highest cell pressure and vice versa, as would be expected. The increase in volume as the shearing takes place is known as dilatant behavior and such volume increase is called **dilatancy**. Dilatant behavior is a common characteristic of soils, especially sand.

9.2.6   Area Correction in Triaxial Tests

During the loading stage of a triaxial test, the length decreases and the diameter normally increases. There may also be some volume change if the test is drained. This means that the area is steadily changing during a test. A correction is normally made for this change, based on the assumption that the sample retains its cylindrical shape during the test. This is a rather crude approximation as samples deform in various ways, as shown in Figure 9.11. Hence the area correction is a reasonable average, but nothing more.

The area at any stage during the test is determined in the following way:

Let the initial volume, area, and length of the sample be V_0, A_0, and L_0, respectively,

and the volume, area, and length during any stage of the test be V, A, and L, respectively;

also let the volume change and length change during the test be ΔV and ΔL, respectively.

Then we can write $V = V_0 - \Delta V = A_0 L_0 - \Delta V$

and $L = L_0 - \Delta L$.

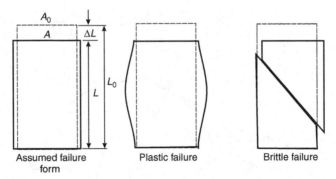

Figure 9.11   Assumed and possible deformation modes during triaxial tests.

The area during the test is then given by

$$A = \frac{V}{L} = \frac{A_0 L_0 - \Delta V}{L_0 - \Delta L} = \frac{A_0 L_0 (1 - \Delta V/(A_0 L_0))}{L_0 - \Delta L}$$
$$= \frac{A_0 (1 - \Delta V/V_0)}{1 - \Delta L/L_0} = \frac{A_0 (1 - e_v)}{1 - e_L} \tag{9.8}$$

where e_v and e_L are the volumetric and axial (vertical) strain, respectively.

Normal convention is that e_v is positive during volume decrease and e_L is positive during compression (length decrease).

9.2.7 Failure Criteria in Terms of Principal Stresses

There are situations where it is useful to express the Mohr–Coulomb failure criteria in terms of principal stresses. The relationship can be obtained by considering the diagram in Figure 9.12

The simplest way to derive the relationship is to examine the radius BD,

$$BD = BC + CD = OB \sin \phi' + CD$$

and then write the values of these in terms of the principal stresses σ_1', σ_3' and the strength parameters c' and ϕ':

$$\frac{\sigma_1' - \sigma_3'}{2} = \left(\frac{\sigma_1' + \sigma_3'}{2}\right) \sin \phi' + c' \cos \phi'$$

Rearranging this gives

$$\sigma_1' (1 - \sin \phi') = \sigma_3' (1 + \sin \phi') + 2c' \cos \phi'$$

Hence

$$\sigma_1' = \sigma_3' \left(\frac{1 + \sin \phi'}{1 - \sin \phi'}\right) + 2c' \frac{\cos \phi'}{1 - \sin \phi'} \tag{9.9}$$

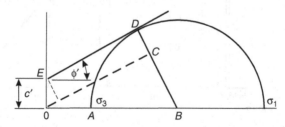

Figure 9.12 Derivation of the Mohr–Coulomb failure criterion in terms of principal stresses.

and the deviator stress is given as

$$\sigma_1 - \sigma_3 = \frac{2\sigma_3' \sin \phi'}{1 - \sin \phi'} + 2c' \frac{\cos \phi'}{1 - \sin \phi'} \qquad (9.10)$$

9.2.8 Determination of Angle of Failure Plane

The failure plane inclination with respect to the plane of the major principal stress can be readily determined by examining a Mohr's circle at failure, as shown in Figure 9.13. Considering the stress state on any plane at an angle α, we find that the only plane on which failure will occur is that defined by the line AP. On this plane the shear stress value is given by τ, which is equal to PD, the available shear strength defined by the Mohr–Coulomb failure line. On planes at any other angle the available shear strength will exceed the shear stress, and failure will not occur.

From the geometry of the figure it is easy to establish the relationship between α and the friction angle ϕ', namely,

$$2\alpha = 90 + \phi' \text{ so that}$$

$$\alpha = 45^{o''} + \frac{\phi'}{2} \qquad (9.11)$$

9.2.9 Worked Example

A series of consolidated undrained triaxial tests has been carried out on a fully saturated clay using samples 40 mm in diameter and 80 mm in height. At failure the information in Table 9.1 was recorded, from which we wish to determine the values of c' and ϕ' of the clay

Sample area $= \pi r^2 = \pi \times (20)^2 = 1256.6$ mm², Sample length $= 80$ mm

We can now systematically calculate the values of effective minor principal stress (σ_3') and the deviator stress ($\sigma_1 - \sigma_3$) needed to determine the values of c' and ϕ'.

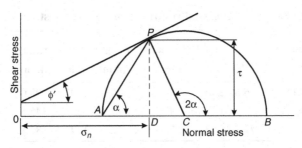

Figure 9.13 Derivation of the failure plane inclination from the Mohr's circle.

Table 9.1 Results of a consolidated undrained triaxial test.

Test Number	Effective Consolidation Pressure (kPa)	Vertical Deformation (mm)	Vertical Force (N)	Pore Pressure (kPa)
1	30	3.1	128.1	6.6
2	80	4.9	194.7	32.3
3	140	7.8	254.0	72.8

The following are the calculations in detail for test 1:

Sample area at failure:

$$A = \frac{\text{volume}}{\text{length}} = \frac{1256.6 \times 80}{80 - 3.1} = \frac{1256.6 \times 80}{76.9} = 1307.3 \, \text{mm}^2$$

Deviator stress:

$$\sigma_1 - \sigma_3 = \frac{\text{load}}{\text{area}} = \frac{128.1 \times 10^{-3}}{1307.3 \times 10^{-6}} \, \text{kPa} = 98.0 \, \text{kPa}$$

Minor principle stress:

$$\sigma_3' = \sigma_3 - u = 30 - 6.6 = 23.4 \, \text{kPa}$$

We can repeat this calculation for the other two samples; the complete results are set out in Table 9.2.

Knowing the values of σ_3' and the deviator stress $\sigma_1' - \sigma_3' = \sigma_1 - \sigma_3$ we can easily plot the Mohr's circles, as is done in Figure 9.14. Drawing a line tangent to these circles gives the following: $c' = 17$ kPa, $\phi' - 28.7°$.

It is often more convenient to plot points rather than circles using the method described later in Section 9.6 and illustrated in Figure 9.20. To do this we need to know the distance to the center of the circle $(\sigma_1' + \sigma_3')/2$

Table 9.2 Complete processed results of the triaxial test.

Sample	Load (kN)	σ_3 (kPa)	u (kPa)	σ_3'	Deformation (mm)	Length (mm)	Area (mm²)	$\sigma_1 - \sigma_3$ (kPa)	$\frac{\sigma_1 - \sigma_3}{2}$ (kPa)	$\frac{\sigma_1' - \sigma_3'}{2}$ (kPa)
1	128.1	30	6.6	23.4	3.1	76.9	1307.3	98.0	49.0	72.4
2	194.7	80	32.3	47.7	4.9	75.1	1338.6	145.5	72.8	120.5
3	254.0	140	72.8	67.2	7.8	72.2	1392.4	182.2	91.1	158.3

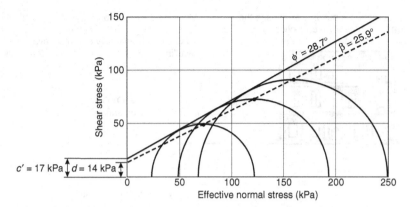

Figure 9.14 Mohr's circles for worked example.

and the radius of the circle $(\sigma_1 - \sigma_3)/2$. These are easily calculated (for the first sample) as follows and are included in Table 9.2:

$$\frac{\sigma_1 - \sigma_3}{2} = \frac{98.0}{2} = 49.0 \, \text{kPa}$$

$$\frac{\sigma_1' + \sigma_3'}{2} = \frac{\sigma_1 - \sigma_3}{2} + \sigma_3' = 49.0 + 23.4 = 72.40 \, \text{kPa}$$

The line drawn through these points has the following values: $d =$ 14 kPa, $\beta = 25.9°$.

From Equations 9.12 and 9.13 given in Section 9.6 we obtain $\phi' = 28.7°$, $c' = 16$ kPa. The c' value does not quite agree with the previous value due to minor errors in the way the lines have been fitted to the data.

9.3 PRACTICAL USE OF UNDRAINED STRENGTH AND EFFECTIVE STRENGTH PARAMETERS

We will now briefly consider the practical situations in which total stress analysis using the undrained strength (s_u) would be appropriate and those in which effective stress analysis using the parameters c' and ϕ' would be appropriate. Figure 9.15 shows two possible types of construction on a layer of uniform saturated clay. The first case is the construction of a building on a surface foundation. The building naturally adds load and increases the confining stress on the soil. The second case involves the excavation of a large body of soil to make space for a motorway. This has the opposite effect—it reduces the load and the confining stress on the soil.

The first case causes consolidation of the soil, as discussed in Chapter 8. Water is "squeezed" out of the soil and the strength increases. The second case has the opposite effect. Reduction in stress initially induces a

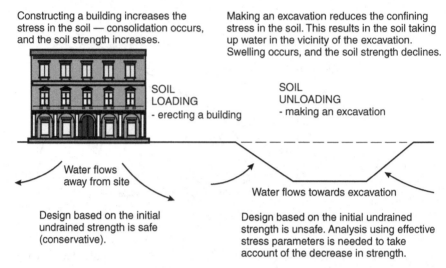

Constructing a building increases the stress in the soil — consolidation occurs, and the soil strength increases.

Making an excavation reduces the confining stress in the soil. This results in the soil taking up water in the vicinity of the excavation. Swelling occurs, and the soil strength declines.

SOIL
LOADING
- erecting a building

SOIL
UNLOADING
- making an excavation

Water flows away from site

Water flows towards excavation

Design based on the initial undrained strength is safe (conservative).

Design based on the initial undrained strength is unsafe. Analysis using effective stress parameters is needed to take account of the decrease in strength.

Figure 9.15   Different loading situations requiring total stress and effective stress analysis.

reduction in the pore pressure in the soil and causes water to seep toward the excavation. As a result, the soil strength decreases. In the first case it is therefore safe to base the foundation design on the initial strength of the soil, that is, the undrained shear strength (S_u). This will increase over time, which will mean an increase in the margin of safety. In the second case the decrease in strength will lower the margin of safety against slip failure into the excavation. To check the new stability situation resulting from the reduced strength, it is necessary to use effective stress analysis. (These situations are analyzed in detail in Section 14.4).

9.4   SHEAR STRENGTH AND DEFORMATION BEHAVIOR OF SAND

The shear strength and deformation behavior of sand is more straightforward than that of clay or silt and will therefore be described first. Shear strength measurements on sand are usually carried out as drained tests since water can flow through sand very quickly, and pore pressures are not likely to be any different in the center of the sample than the values at the ends where drainage is permitted.

Figure 9.16 shows the results of triaxial tests on samples of a sand prepared in a loose and a dense state tested at the same confining pressure. With the dense sample, the deviator stress increases to reach a clear peak value and then decreases to eventually level off at a constant value. The volume shows an initial slight decrease followed by a steady increase and then levels off to a constant value. The loose sample, however, does not

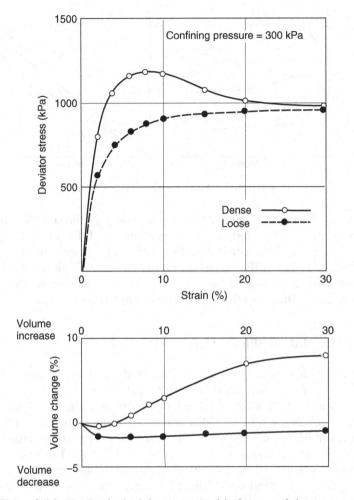

Figure 9.16 Drained triaxial test on sand in loose and dense state.

show a clear peak value of deviator stress. The strength simple rises and levels off to a constant value, which is very close to the value obtained from the dense sample. The volume of the loose sample shows a slight decrease before leveling off to a constant value. In each test, both the deviator stress and volume become constant at large strains. If measurements are made of the density (or void ratio) of the samples once these "constant" or ultimate states are reached, it is found that the values are essentially the same. This ultimate state is known as the **critical state**. In this state, deformation will continue at constant deviator stress and constant volume.

Figure 9.17 shows results of a set of drained shear box tests on identical sand samples plotted as stress versus displacement curves and failure lines in terms of both the peak strength and critical state strength. Sands are by

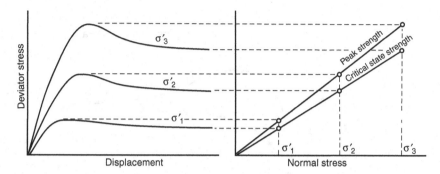

Figure 9.17 Critical state strength of sand.

nature cohesionless, so the failure lines normally go through the origin. Two values of the friction angle ϕ' are thus determined, the peak value (simply designated ϕ') and the **critical state value**, which is commonly designated ϕ'_{crit} or ϕ'_{cv}. The cv suffix refers to constant volume. Typical values of ϕ' for sand are between about 35° and 45°, the lower values normally associated with the loose state and the high values with the dense state.

9.5 RESIDUAL STRENGTH OF CLAYS

Skempton (1964) first proposed the idea of the residual strength of clays in the context of the long-term stability of cut slopes in **stiff fissured clays**. Slips in these clays were observed to occur long after cuttings had been made, sometimes several decades later. Stiff fissured clays, which are relatively common in Europe and North America, are heavily overconsolidated clays containing a random pattern of joints of fissures. The presence of these fissures affects the strength of the clay en masse, and Skempton suggested that a process of progressive failure took place in these slopes, resulting in a drop in strength from the peak strength toward a value referred to as the "residual strength."

Figure 9.18 illustrates the concept of residual strength. Soil has been tested in a device (originally a shear box), which allows continued displacement on a shear plane. Three tests have been done at differing normal stresses and stress–displacement curves of the shape shown obtained. In each test the strength rises to a peak and then decreases. If the test is continued far enough, the strength reaches an ultimate "constant" value; this is termed the residual strength. This behavior appears to resemble that of sands but is fundamentally different in several important aspects. With sands, the critical state is considered to arise because of the creation of a uniform state of stress, deformation, and density. With clays this does not normally occur. Failure planes develop within the soil mass and the platelike shape of the clay particles results in their reorientation to lie flat on the failure plane, forming a "slicken-sided" surface of low strength.

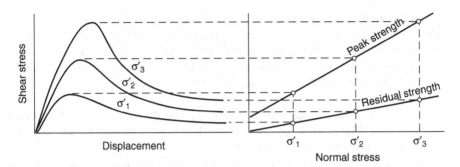

Figure 9.18   Peak and residual strength of soil.

The following important differences between clay and sand behavior should also be noted:

1. The decrease in strength that occurs in sands during shear (except in very loose samples which show only an increase in strength) arises from a change in the denseness of the sand. The particles rearrange themselves to form a new less dense state.

2. The decrease in strength that occurs in clay during shear arises from several causes. First, many natural clays, especially residual clays, contain structural features such as bonds between particles and shear movement results in destruction of these bonds with consequent loss of strength. Second, the platelike shape of the particles causes particle alignment parallel to the shear plane, as already mentioned. Third, some change in denseness may occur, as it does in sands.

3. Particle movement when the critical state is reached in sands is considered to be rolling, while that occurring when the residual strength is reached in clays is considered to be sliding.

4. The term residual strength is intended primarily for clays; if applied to sands it can be assumed to have the same value as the critical state strength.

5. Before reaching the residual state during a strength test on clay, the soil is likely to pass through the critical state, but it is not possible to identify when this occurs. It is thus difficult to create the critical state of a clay in the laboratory, and it is extremely unlikely to develop in field situations.

Two failure plots can be drawn as shown in Figure 9.18 similar to those for sand in Figure 9.17. In this case, however, the lower plot defines the residual strength rather than the critical state strength. This residual strength corresponds to a zero value of c' and a low value of ϕ' (designated ϕ'_r). The value of ϕ'_r for clays of high plasticity is very low—in the range of about $8° - 15°$.

Subsequent work showed that, while Skempton was right in emphasizing the important influence that fissures have on the strength of some clays, his hypothesis that the strength of fissured clays drops over time from the peak value toward the residual value is no longer considered valid. Subsequent research brought out the following important factors:

1. The residual strength measured using shear box tests is substantially higher than the true residual strength (see note below on measurement using the ring shear apparatus).
2. Careful measurement of pore pressures in cuttings in stiff fissured clays showed that in some cases pore pressures were still rising long after the cuttings were completed. This is now believed to be the main cause of the long delay before failures occur.
3. The peak shear strength values used by Skempton were from tests on small "intact" samples. If tests are conducted on very large samples, the values of c' and ϕ' are much lower and are closer to the values obtained by back analysis of actual slips.

The concept of residual strength is, however, still important as it governs the strength of the soil on planes where movement has previously occurred. Skempton (1985, p. 4) states: "Consequently, residual strength is generally not relevant to first-time slides and other stability problems in previously un-sheared clays and clay fills, but the strength of a clay will be at or close to the residual on slip surfaces in old landslides or soliflucted slopes, in bedding shears in folded strata, in sheared joints or faults and after an embankment failure."

9.5.1 Measurement of Residual Strength

Although the direct shear box was used in early studies, it is no longer considered a reliable means of measuring residual strength. The residual strength can only be measured reliably using a device known as the ring shear apparatus. Bishop and co-workers (1971) designed the first ring shear apparatus, which is illustrated conceptually in Figure 9.19.

The device is essentially a shear box curved around to meet up in a complete circle. The soil sample is annular in shape. The displacement possible is therefore unlimited. Tests can be carried out at varying normal stresses as in the conventional shear box; the torque is measured to restrain the upper half of the ring while the lower half is rotated at a constant rate by a motor drive. From the torque the shear stress on the failure plane (annulus) is calculated. The apparatus of Bishop et al. (1971) is a very sophisticated device and ideal for precision measurements. However, setting up the apparatus is an intricate and time-consuming operation. Bromhead (1979) devised a much simpler form of the apparatus which appears to give almost identical results to the Bishop apparatus.

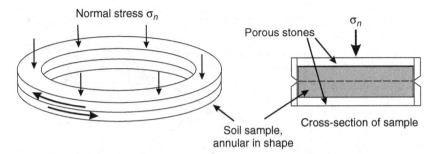

Figure 9.19   Conceptual drawing of the ring shear test.

It is worth observing that the residual strength appears to be an intrinsic property of the soil itself and is independent of the way the soil is prepared for a ring shear test. The same result is obtained regardless of whether the soil is initially undisturbed or remolded in a soft or stiff state.

9.6   STRESS PATH CONCEPT

The term "stress path" refers to a graphical representation of the sequence in which stresses are applied to soil. Stress paths are a useful aid in under-standing the behavior of soils, especially when they are loaded in undrained situations. Figure 4.8 is an example of a stress path; it shows the stress sequence when a soil element is loaded and unloaded vertically with no lateral strain. In that figure the stress path is plotted in terms of the vertical (σ'_v) and the horizontal (σ'_h) effective stresses. While a plot in these terms is always feasible, it is generally more informative to use an alternative plot related directly to the Mohr's circle. The sequence of stresses could be represented by Mohr's circles themselves, but this would present a crowded and confusing picture. To overcome this, conventional practice is to plot the "apex" of the Mohr's circle, that is, the highest point on the circle, as illustrated in Figure 9.20. The distance to the center of the circle is plotted on the horizontal (x) axis and the radius of the circle on the vertical (y) axis. The values of x and y in this plot are thus given by:

$$x = \frac{\sigma_1 + \sigma_3}{2} \text{ and } y = \frac{\sigma_1 - \sigma_3}{2}$$

Stress paths can be plotted using either total or effective stresses.

Figure 9.20a shows the Mohr's circles and associated stress paths in terms of total stress. We can note in passing that in a drained test the effective stress path is identical to the total stress path. Point E is the starting stress state (the cell pressure is given by the distance OE), and as the test continues the circle grows in size until failure occurs when the circle

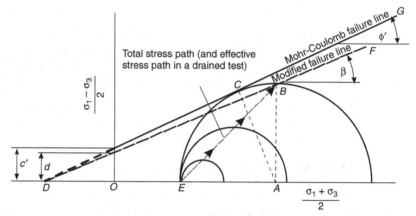

(a) Stress path concept in a triaxial test, and modified Mohr-Coulomb failure line

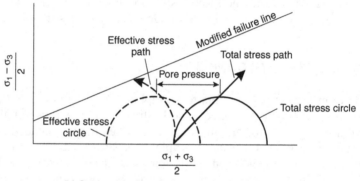

(b) Stress paths in undrained triaxial tests

Figure 9.20 Stress paths in triaxial tests and modified Mohr–Coulomb failure line.

touches the Mohr–Coulomb failure line *DG*. All the points lie on the line *EB*, which thus represents the stress path for loading the soil from *E* to *B*. The line *DF* represents the failure line of the soil for stress paths (or peak strength values) plotted in this way. To determine the parameters c' and ϕ' from a series of tests, it is convenient to plot the failure points (point *B* for the stress path shown) rather than the circles, as it is easier to fit a line to a series of points than to fit a tangent to a series of circles. The relationship between the angle β and the intercept d and the Mohr–Coulomb parameters is a simple geometric one given by

$$c' = d\,\frac{\tan \phi'}{\tan \beta} = d \sec \phi' \qquad (9.12)$$

$$\sin \phi' = \tan \beta \qquad (9.13)$$

Figure 9.20b shows the total and effective stress paths in a consolidated undrained triaxial test. In this case the effective stress path differs from the total stress path by the magnitude of the pore pressure generated during the test. If the pore pressure generated is positive, then both principal stresses decrease, and the Mohr's circle and the stress path move to the left, as shown in the figure. If the pore pressure generated is negative, then the stress path will move to the right of the total stress path.

The change in pore pressure during undrained loading is commonly expressed in terms of a parameter *A*, which is described in the following section.

9.7   PORE PRESSURE PARAMETERS *A* AND *B*

As we have seen in Chapters 7 and 8 and in the discussion above, pore pressure changes in a soil can result from seepage movement of water in the soil as well as from changes in applied total stress when no movement of water occurs. The pore pressure parameters *A* and *B* are used as measures of the change in pore pressure that results from changes in total stress applied to the soil under undrained conditions. These changes in total stress may be simply a change in all-round stress, but they may also be a change in shear stress. The expression used to relate pore pressure change to these total stress changes (Skempton, (1954)) is

$$\Delta u = B\{\Delta \sigma_3 + A(\Delta \sigma_1 - \Delta \sigma_3)\} \tag{9.14}$$

where Δu = change in pore pressure
$\Delta \sigma_3$ = change in minor principal stress
$\Delta \sigma_1$ = change in major principal stress
A, B = pore pressure parameters (or coefficients)

The parameter *B* therefore relates to an all-round increase in stress $\Delta \sigma_3$, while *A* relates to an increase in shear stress given by $\Delta \sigma_1 - \Delta \sigma_3$. If the stress increase is an all-round stress increase, then $\Delta \sigma_1 - \Delta \sigma_3$ is zero and the relationship becomes $\Delta u = B \Delta \sigma_3$.

As we have seen earlier, if the soil is fully saturated, then the pore pressure will increase by the same value as the applied stress, and $B = 1$. When the soil is partially saturated, *B* decreases rapidly as the volume of air in the soil increases. Once the degree of saturation falls to 80 percent, the value of *B* can be expected to be less than 0.2 for many clays.

The parameter *A* relates to changes in shear stress. If the value of σ_3 is kept constant and the soil is fully saturated, then Equation 9.9 becomes

$$\Delta u = A(\Delta \sigma_1 - \Delta \sigma_3)\} = A \Delta \sigma_1$$

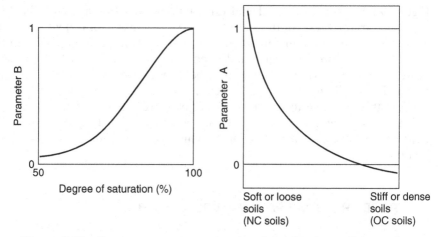

Figure 9.21   Pore pressure parameters A and B related to soil properties.

The value of A is governed by the stiffness or the denseness of the soil. Very soft soils tend to decrease in volume when sheared, so that if conditions are undrained, the pore pressure will rise during shearing. On the other hand, very dense soils tend to dilate (increase in volume) when shear stress is applied, so that if conditions are undrained, the pore pressure will tend to decrease during shearing. The value of A can thus range from less than zero to about 1, though in some very soft soils and soils of high sensitivity A may be greater than 1. For sedimentary soils, A generally ranges from zero for heavily overconsolidated soils to about unity for normally consolidated soils. Figure 9.21 indicates the way the paremeters A and B vary with soil properties.

9.8   SHEAR STRENGTH AND DEFORMATION BEHAVIOR OF CLAY

In the following sections the strength and deformation behavior of clays during consolidated undrained triaxial tests is illustrated and discussed. Examples are given for three types of clays, namely fully remolded clays, undisturbed sedimentary clays, and residual clays. The results are considered to be representative of these materials; however, the range of properties found in natural soils is very large and there are many soils whose behavior may not conform to that illustrated here.

9.8.1   Behavior of Fully Remolded Clay

The behavior of fully remolded clays in consolidated undrained tests is illustrated in idealized form in Figure 9.22. The soil is first mixed to the consistency of a slurry, then a series of normally consolidated samples are

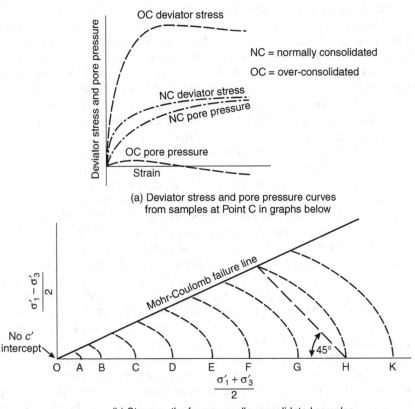

(a) Deviator stress and pore pressure curves
from samples at Point C in graphs below

(b) Stress paths from normally consolidated samples

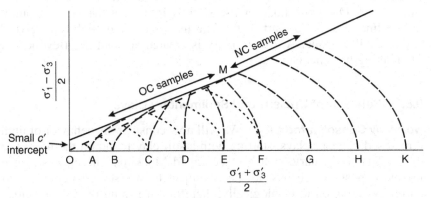

(c) Stress paths from NC samples, and OC samples
after consolidation to Point G and stress reduction

Figure 9.22   Behavior of remolded clay in consolidated undrained triaxial tests.

prepared and tested at consolidation stresses ranging from point A to point K in Figure 9.21b. A typical result for the sample consolidated to stress C is shown in Figure 9.22a in the form of deviator stress and pore pressure curves versus strain. Curves of identical shape are obtained from all the normally consolidated samples. The pore pressure rises by almost the same value as the deviator stress, which means the pore pressure parameter A at failure is close to unity, in accordance with Figure 9.21. The Mohr–Coulomb failure line, shown in Figure 9.22b, goes through the origin, indicating that for normally consolidated clays the cohesion intercept c' is zero.

The influence of overconsolidation is shown by consolidating a number of samples up to the stress level at F, then reducing the consolidation pressure to the values used in the first tests. These samples are now all overconsol-idated, with varying degrees of overconsolidation. The sample at stress E has an overconsolidation ratio (OCR) $= OF/OE$ and is therefore lightly overconsolidated, while the sample at stress A has OCR $= OF/OA$ and is heavily overconsolidated. The deviator stress and pore pressure curves for the overconsolidated sample at point C are also given in Figure 9.22a and can be compared directly with the normally consolidated sample at the same initial stress. The peak deviator stress is now much higher, and there is almost negligible pore pressure. This is to be expected, since overconsol-idation results in a densely packed material which does not tend to decrease in volume when shear stress is applied. A slight positive pore pressure is generated followed by a decrease to become slightly negative.

The stress paths from these overconsolidated samples, seen in Figure 9.22c, are quite different from the normally consolidated samples, the difference becoming greater as the OCR increases. For heavily over-consolidated samples the parameter A is close to zero, in accordance with Figure 9.21. Overconsolidation also slightly influences the Mohr–Coulomb failure line. The initial part of the line moves upward slightly to give the soil a small c' value. A discontinuity is created at point M. Beyond point M, nothing has changed.

9.8.2 Behavior of Undisturbed Sedimentary Clays

Normally Consolidated Clay We will now consider the behavior of undis-turbed sedimentary clays, starting with a soft clay found on the north coast of the Thames estuary, near a place called Mucking. The clay is a recent marine deposit which has not been subject to stresses greater than those currently acting on it. Geologically, therefore, it is a normally consolidated clay. The behavior of the clay in a series of consolidated undrained triax-ial tests at different consolidation pressures is illustrated in Figures 9.23 and 9.24. Full details of these tests can be found in Wesley (1975).

Figure 9.23 shows curves of deviator stress and pore pressure versus strain for a sample consolidated at the in situ stress state (21 kPa) as well as at a much lower (3.5 kPa) and a much higher stress state (150 kPa). The

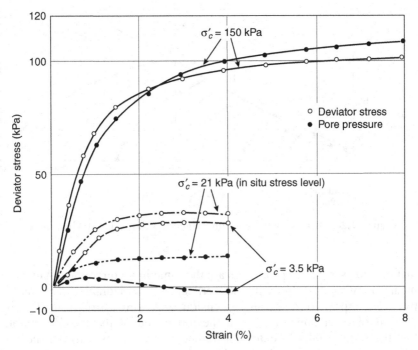

Figure 9.23 Stress–strain curves from undrained triaxial tests on Mucking clay.

behavior is fairly similar to that of the remolded samples in Figure 9.22. At a high consolidation pressure the soil behaves as a normally consolidated soil, with the pore pressure rising by the same value as the deviator stress. Similarly at a very low stress the soil behaves as an overconsolidated clay, with very little pore pressure increase.

There is, however, a significant difference in the behavior at the in situ stress level. The stress paths in Figure 9.24 show normally consolidated behavior only at consolidation pressures above about 60 kPa. The average in situ effective stress at the depth from which these samples were taken was 21 kPa, which is well below 60 kPa. This suggests an OCR of about 3. This illustrates the hardening processes that sedimentary soils undergo after deposition. As stated earlier (Section 8.3.3), there are many clays that are geologically "normally consolidated" but there are very few, if any, that behave as normally consolidated clays in terms of their shear strength or consolidation characteristics.

Further illustrations of the behavior of the Mucking clay during testing at its in situ stress state are given in Figures 9.25–9.27. These tests were carried out in the hydraulic stress controlled triaxial apparatus described in detail by Bishop and Wesley (1975). Figure 9.25 shows the results of triaxial tests on samples consolidated isotropically to their average in situ effective stress and also to their anisotropic K_0 stress state. The term **isotropic** means

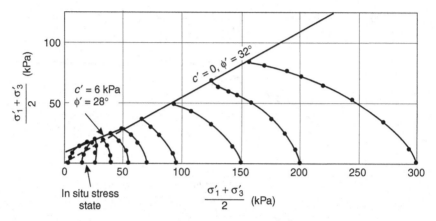

Figure 9.24 Stress paths from undrained triaxial tests on Mucking clay.

that the stresses or soil properties are the same in all directions. **Anisotropic** means they are different in the vertical and horizontal directions. Both triaxial compression tests and extension tests were carried out. Extension tests involve an upward pull on the top platen of the triaxial sample so that the sample fails in extension rather than compression. All these tests are therefore measures of the undrained shear strength (s_u) of the soil. The value of s_u is half the deviator stress, which in Figure 9.25 is plotted as $\sigma_V - \sigma_H$.

The graphs of deviator stress and pore pressure versus strain in Figure 9.25a show that the peak value of deviator stress is not greatly influenced by whether the initial stress state is isotropic or anisotropic, but the value in the extension tests is significantly lower than in the compression tests. The reason for this is evident from the stress paths in Figure 9.25b. These show that the pore pressure generated in extension tests is different to that in compression tests, resulting in lower effective stresses in the latter. Failure of all samples occurs close to the same Mohr–Coulomb line established in Figure 9.24 from triaxial compression tests.

The influence of the test method on the undrained shear strength is further illustrated in Figure 9.26, which shows stress paths and failure values from a further range of test types. These include compression and extension triaxial tests on horizontal samples and plane strain compression tests on vertical and horizontal samples, all plotted in terms of σ_1' and σ_3' rather than σ_V' and σ_H' as was done in Figure 9.25. This is done to relate them all in the same Mohr–Coulomb failure line. **Plane strain** means no horizontal deformation occurs in the direction at right angles to the direction of shear. A special apparatus is needed for these tests. Four samples were tested by each method, and the stress paths shown in the figure are averages from these. The difference in the shape of the stress paths is a reflection of the

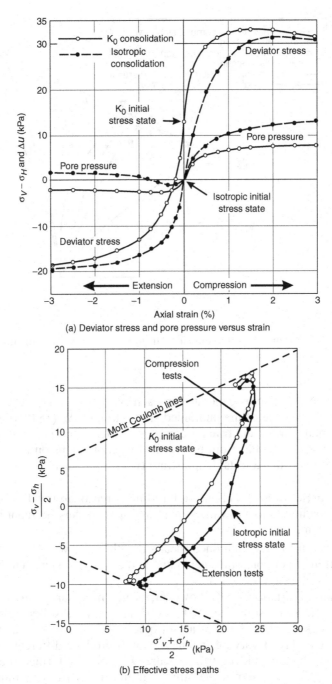

Figure 9.25   Undrained triaxial compression and extension tests at the in situ stress state.

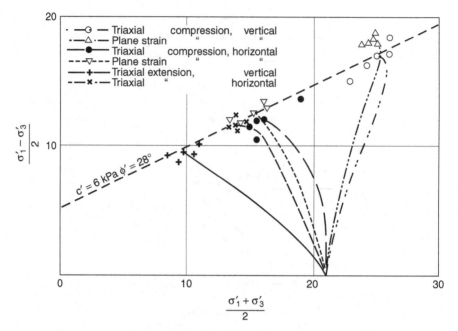

Figure 9.26 Mohr–Coulomb envelope from undrained tests at in situ stress level (Mucking clay).

different pore pressures generated in each type of test. The tests show that the maximum value of the undrained shear strength S_u (18 kPa) is from the vertical plane strain and triaxial compression tests and is about double the minimum value (9 kPa) from the vertical triaxial extension tests.

In summary, these tests demonstrate two important factors:

(a) A single Mohr–Coulomb failure line is reasonably valid for all types of loading, and the values of c' and ϕ' can thus be regarded as soil constants.

(b) The value of the undrained shear strength varies widely, because of the different pore pressures generated in the type of test used to measure it.

The large variation in s_u, however, is not necessarily due only to the type of test. Some soils, especially soft, normally consolidated clays such as the Mucking clay, have an anisotropic structure, so that identical tests on the same clay at varying inclinations will produce different results.

This is already evident from Figure 9.26 and is further illustrated in Figure 9.27, which shows the stress–strain curves from undrained triaxial tests on samples trimmed with vertical, 45°, and horizontal orientation. The curves clearly show a steady reduction in strength from the vertical to the horizontal sample. The reason for this is that the soil is less stiff in the

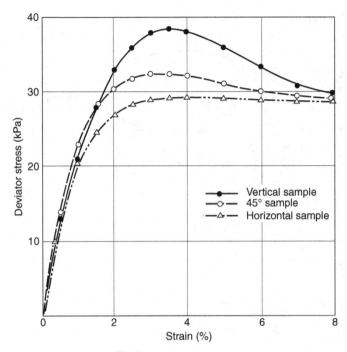

Figure 9.27 Undrained triaxial compression tests at different orientations (Mucking clay).

horizontal direction than the vertical direction and higher pore pressures are generated when the sample is loaded horizontally. The maximum and minimum values of s_u from the vertical and horizontal samples are respectively about 18 and 14 kPa. The maximum value of s_u in these tests is therefore only about 30 percent greater than the minimum value, whereas from the full range of tests in Figure 9.26 it was about 100 percent. Thus, this 100 percent range is due partly to the intrinsic structural anisotropy of the soil and partly due to the type of test.

Overconsolidated Clays The behavior of a heavily overconsolidated clay during undrained triaxial testing is illustrated in Figures 9.28 and 9.29. These tests, carried out on London clay, are described fully by Bishop et al. (1965). Figure 9.28 shows the curves of deviator stress and pore pressure versus strain. The most significant feature of these tests is the sharp loss of strength that occurs when they are strained beyond their peak (failure) stress. Only the sample consolidated at the highest pressure does not show a sharp drop in strength, although it still shows some loss of strength with increasing strain.

Figure 9.29 shows the stress paths during the tests and the approximate Mohr–Coulomb failure line. It is clear that this line is not linear, showing

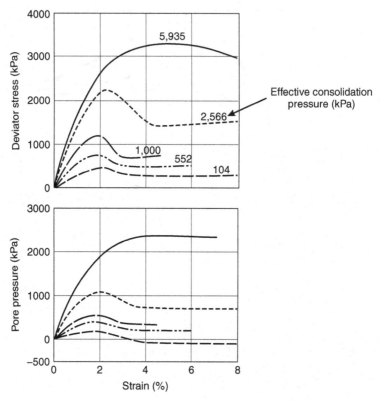

Figure 9.28 Consolidated undrained triaxial tests on London clay. (After Bishop et al., 1965.)

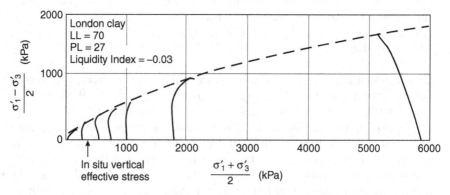

Figure 9.29 Stress paths from triaxial tests on London clay. (After Bishop et al., 1965.)

a steady decrease in the value of the friction angle with increasing stress level. It is noteworthy also that the stress path at the highest consolidation pressure shows that the soil is still not behaving as a normally consolidated soil, despite the fact that this stress is much higher than the stress the soil is likely to have experienced at any time in the past. The behavior of this heavily overconsolidated clay therefore does not conform closely to that of the remolded soil shown in Figure 9.22. This again illustrates that stress history alone is not an adequate explanation of the behavior of natural soils. The structure created by aging effects is equally important.

9.8.3 Behavior of Residual Soils

The behavior of two particular residual soils in consolidated undrained tests is illustrated in Figures 9.30 to 9.31. The undisturbed samples of both materials were taken as block samples within about 3 m of the ground surface. Details of the two soils are as follows:

  Soil a: This is a tropical red clay found widely in Java, Indonesia. These clays are derived from the weathering of andesitic volcanic materials and contain a high proportion of the clay mineral halloysite. They are generally very fine grained and of moderate to high plasticity. Their natural water contents are normally close to their plastic limits and they are of negligible sensitivity.

  Soil b: This is a clayey silt derived from the weathering of a sandstone formation found in and around the city of Auckland in New Zealand. Properties within the formation vary widely from low to very high plasticity soils. The samples tested here were of low plasticity, with a liquidity index of 0.5 and a sensitivity of about 10. Their undrained shear strength was approximately 100 kPa.

Figures 9.30 and 9.31 show the results of consolidated undrained triaxial tests on tropical red clay. Both undisturbed and compacted samples were tested at a range of consolidation pressures.

The deviator stress and pore pressure curves in Figure 9.30 suggest that the soil tends to behave as an overconsolidated clay at low stress levels, since the pore pressure generated during testing is relatively small. The pore pressure in the samples consolidated to 50 kPa is approaching zero at 8 percent strain when the tests were terminated. At higher consolidation pressures the pore pressures are higher in relation to the deviator stress. The stress paths in Figure 9.31 confirm that the soil behaves as an overconsolidated clay at low consolidation stress levels. At higher stress levels the stress paths show behavior tending toward normally consolidated behavior, but only over the initial sections.

As the samples approach failure, a tendency to dilate is evident, and the pore pressure steadily declines. This behavior is normally associated

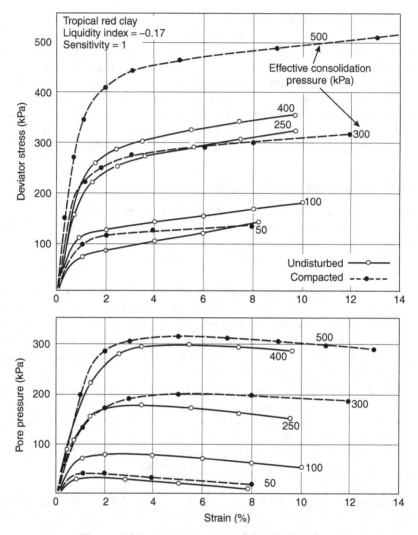

Figure 9.30 Triaxial tests on tropical red clay.

with dense materials and is therefore not surprising. With a liquidity index less than unity, this clay clearly exists in a dense state. Structure is not a significant influence on this clay as its behavior is the same after remolding and recompacting as in its undisturbed state.

Similar results from the second soil are shown in Figures 9.32 and 9.33. The second soil, with a liquidity index of 0.5, is a much less dense material than the first soil and the test results reflect this. The deviator stress curves from the undisturbed samples reach a peak at a strain of about 2 percent and then show a small but steady decline. The pore pressure curves show that

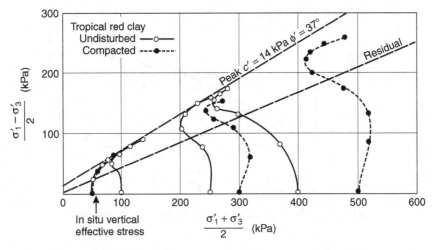

Figure 9.31 Triaxial test stress paths from tropical red clay.

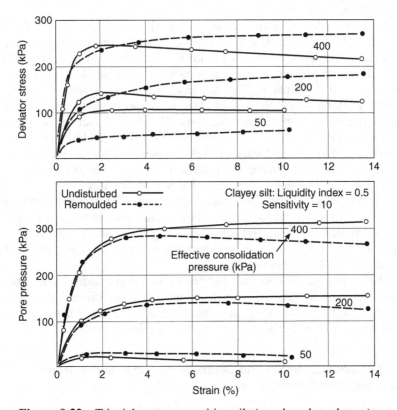

Figure 9.32 Triaxial tests on sensitive silt (weathered sandstone).

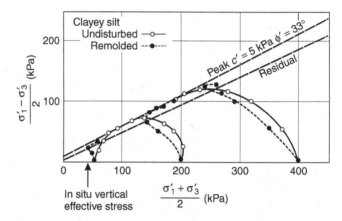

Figure 9.33 Triaxial test stress paths from sensitive silt.

the soil is tending to decrease in volume (rather than dilate as occurred with the red clay), and the pore pressure consequently shows a slight increase with strain.

After remolding and reconstituting the samples, the behavior is significantly different. The soil is more "ductile" and there is no longer a peak deviator stress, except at the lowest consolidation pressure of 50 kPa. The other samples show a slow steady increase in deviator stress until the tests were terminated. This increase is due to a small rise in effective stress resulting from the slight decline in pore pressures. Figure 9.33 shows the stress paths from these tests. These show that in its undisturbed state the soil behaves as an overconsolidated clay at low stresses and as a normally consolidated clay at high stresses. It also shows that the pore pressures generated during the tests at the higher consolidation pressures are substantially greater than was the case with the red clay. The figure also shows a significant difference between the undisturbed and remolded samples. The shape of the undisturbed stress paths indicate a tendency of the soil to collapse over the full stress range, whereas the remolded stress paths show the soil tending to dilate when it reaches the point of failure on the Mohr–Coulomb line.

The difference in behavior between these two residual soil types thus arises because the red clay is a dense unstructured soil and suffers no loss of strength during shearing, or remolding, whereas the silt is a less dense, structured, soil, which loses strength as shearing proceeds.

9.8.4 Failure Criterion and Determination of c' and ϕ' from Consolidated Undrained Tests

The above triaxial tests show that a considerable range of stress paths is possible, depending on the soil type and the consolidation pressure; this range is

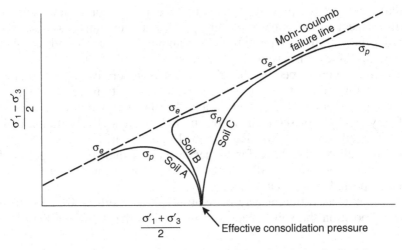

Figure 9.34   Possible stress paths in consolidated undrained triaxial tests.

summarized in an idealized fashion in Figure 9.34 and is applicable to both sedimentary and residual soils. For simplicity, a single Mohr–Coulomb failure line is adopted, although in practice this would not be the case because we are dealing with three different soils. Soil A is a soft, normally consolidated clay, or a highly sensitive soil. Soil B is a moderately stiff soil, and soil C is a dense silt, or a heavily overconsolidated clay.

These stress paths show that there is some ambiguity with regard to the stress state at which failure actually occurs and thus the stress state to be used in determining the effective strength parameters c' and ϕ'. The peak failure stress during the test is the peak value of the deviator stress; this is denoted by point σ_p on each curve. However, the point at which the stress paths first touch the Mohr–Coulomb line is the point at which the shear stress first equals the available shear resistance, and therefore this more correctly indicates the failure envelope and the values of c' and ϕ'. These points are denoted σ_e on each curve. The use of points σ_p rather than points σ_e results in lower values of c' and ϕ', although with most soils the difference is not great. The point at which the stress path first reaches the Mohr–Coulomb failure line corresponds to the maximum value of the principal stress ratio σ_1'/σ_3', and this can be used as a criterion to determine the values of c' and ϕ', rather than the peak deviator stress.

9.9   TYPICAL VALUES OF EFFECTIVE STRENGTH PARAMETERS FOR CLAYS AND SILTS AND CORRELATIONS WITH OTHER PROPERTIES

It is not feasible to give a comprehensive account of the full range of c' and ϕ' values found in fine-grained soils, but it is useful to have some "indicators" of the probable values to be expected for various soils. First, with respect to the cohesion intercept c', it is generally the case that the harder the soil, the higher will be the value of c'. A crude indication of the range of values for several soil types is as in Table 9.3:

Second, with respect to the ϕ' value, it is generally highest in silts and clays of low plasticity and decreases as the plasticity increases. Figure 9.35 shows the relationship between ϕ' and the plasticity index for a wide range of clays, mainly of sedimentary origin. Some clays clearly do not conform to the trend illustrated in Figure 9.35. In particular, Mexico City clay and the Attapulgite clay have far higher ϕ' values than those expected from their plasticity indices. Allophane clays also have much higher ϕ' values than expected from Figure 9.35.

A somewhat different way of relating ϕ' to Atterberg limits is to focus on the position the soil occupies in relation to the A-line on the plasticity

Table 9.3 Typical values of cohesion intercept c

Soil Type	Cohesion Intercept c' (kPa)
Soft, normally consolidated clays	Normally close to zero, but could approach 10 kPa
Firm to stiff clays, including residual soils with s_u in the range 70–150 kPa	From about 10 to 25 kPa
Stiff to hard clays, especially heavily overconsolidated clays	From about 25 to 100 kPa
Compacted clays	Normally between about 12 and 25 kPa

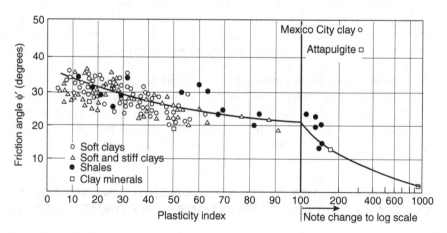

Figure 9.35 Values of the peak friction angle ϕ' for clay of various compositions plotted against their plasticity index. (After Terzaghi et al., 1996.)

chart. As described in Section 3.3, the distance of a soil above or below the A-line is normally a good indicator of its likely engineering characteristics. Figure 9.36 shows the friction angle plotted against distance above or below the A-line from a rather limited range of soils. This distance is represented by ΔPI, defined by $\Delta PI = PI - 0.73(LL - 20)$. The data are from the author's own files and involve both sedimentary and residual soils, all with relatively high liquid limits. It is seen that this graph shows a reasonably well defined correlation.

Various attempts have been made over the years to relate the residual angle ϕ'_r to index properties: the clay fraction, liquid limit, or plasticity index. These attempts showed that no general relationship exists between the residual angle ϕ'_r and any of these parameters. An improved relationship of more general validity is obtained by using the above plot in terms of ΔPI; this is shown in Figure 9.37.

Correlations such as those in Figures 9.36 and 9.37 are not intended as a means of avoiding the measurement of the friction angle. Measurement of ϕ' by triaxial testing is time consuming but not particularly difficult, so if reliable values are required, such tests should be carried out. In the case of the residual friction angle ϕ'_r, the advent of the Bromhead (1979) ring shear apparatus has made the measurement of residual strength almost

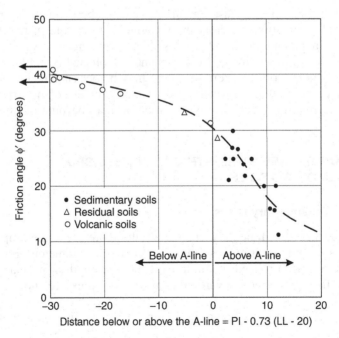

Figure 9.36 Values of the friction angle ϕ' plotted against distance above or below the A-line on the plasticity chart.

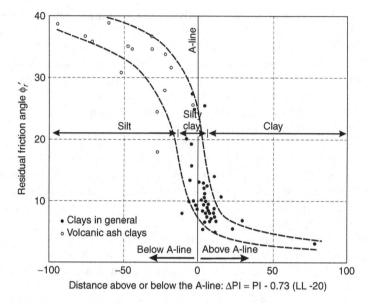

Figure 9.37 Residual friction angle ϕ_r' plotted against distance above or below the A-line on the plasticity chart. (From Wesley, 2003.)

as straightforward as measurement of the Atterberg limits. The principal benefit of correlations such as those above is their role in building up as complete a picture of a soil as possible. Checking the way the soil fits in with known trends in soil behavior and established correlations between properties is part of this exercise. There may be occasional situations where the Atterberg limit is known but direct measurement of the friction angle is not feasible, in which case the correlation may become of direct use.

9.10 UNDRAINED STRENGTH OF UNDISTURBED AND REMOLDED SOILS

9.10.1 Sedimentary Clays

For undisturbed sedimentary clays and silts there are some empirical correlations relating undrained strength to other soil parameters. For example, Skempton (1957) related the undrained shear strength of normally consolidated soils to the effective vertical consolidation pressure and the Atterberg limits.

His relationship is shown in Figure 9.38 and is summarized in the expression:

$$\frac{s_u}{\sigma'} = 0.11 + 0.0037 \, (\text{PI}) \tag{9.16}$$

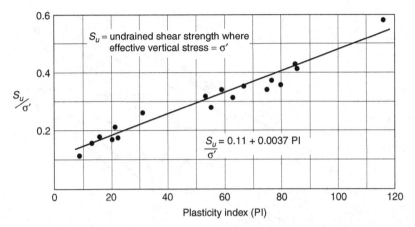

Figure 9.38   Undrained shear strength for normally consolidated clays (Skempton, 1957).

where s_u = undrained shear strength
$\quad\quad \sigma'$ = effective overburden pressure
$\quad\quad$ PI = plasticity index

The undrained shear strength of normally consolidated soils can also be related theoretically to their consolidation pressure provided the soils are "unstructured" and their properties are directly related only to their stress history. From Equation 9.9 we have the following relationship:

$$\sigma_1' = \sigma_3' \left(\frac{1 + \sin \phi'}{1 - \sin \phi'} \right) + 2c' \frac{\cos \phi'}{1 - \sin \phi'}$$

We can rearrange this to give us the deviator stress in terms of the major principal stress σ_1':

$$\sigma_1 - \sigma_3 = \sigma_1' - \sigma_3' = \frac{2\sigma_1' \sin \phi'}{1 + \sin \phi'} - \frac{2c' \cos \phi'}{1 + \sin \phi'}$$

The undrained strength s_u is half the deviator stress (Figure 9.8), and for a normally consolidated clay, the value of c' is zero, so that

$$s_u = \frac{\sigma_1' \sin \phi'}{1 + \sin \phi'}$$

For normally consolidated clays it is reasonable and convenient to assume that the pore pressure parameter $A = 1$. This means that the major principal stress σ_1' does not change during the test, and σ_c' is the same as the major principal stress σ_1' at failure. This is illustrated in Figure 9.39.

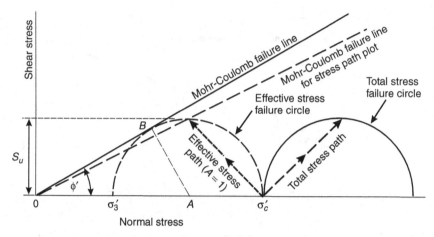

Figure 9.39 Undrained shear strength (S_u) related to consolidation pressure for normally consolidated clay.

We can therefore write:

$$s_u = \left(\frac{\sin \phi'}{1 + \sin \phi'} \right) \sigma'_c$$

so that we have

$$\frac{s_u}{\sigma'_c} = \frac{\sin \phi'}{1 + \sin \phi'} \qquad (9.17)$$

This gives $s_u/\sigma'_c = 0.33$ for $\phi' = 30°$ and $s_u/\sigma'_c = 0.25$ for $\phi' = 20°$.

These values agree with the values from Figure 9.37 for medium-plasticity clays. However, the trend suggested by this theoretical analysis (Equation 9.17) does not agree with the trend in Figure 9.37 (Equation 9.16) since an increasing plasticity index is generally accompanied by a decrease in ϕ' value (at least for sedimentary soils), which would mean a decrease in the ratio S_u/σ'_c with increasing plasticity index. Field measurements indicate the opposite trend. This is possibly due to a greater influence of interparticle bonds in the higher plasticity clays.

9.10.2 Remolded Soils

For fully remolded soils it is possible to relate undrained shear strength to the Atterberg limits and the liquidity index of the soil. The ratio of undrained shear strength at the plastic limit and liquid limit is considered to be close to 100 (Schofield and Wroth, (1968)). Actual values proposed have generally been about 200 and 2 kPa. Recently, Sharma and Bora (2003) have recommended 170 and 1.7 kPa. The above values and the curve relating

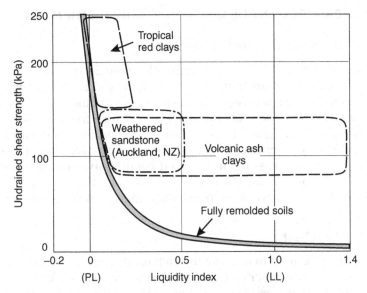

Figure 9.40   Undrained shear strength versus liquidity index for fully remolded soils and values for several undisturbed residual soil groups.

undrained shear strength with liquidity index are shown in Figure 9.40. This curve represents the lower limit of shear strength at which a soil can exist at a particular value of liquidity index. Most soils will exist in nature with higher undrained shear strength than that given by this curve. Only undisturbed soils that show no loss of strength on remolding (i.e., nonsensitive soils) will lie on this line.

9.10.3   Residual Soils

Also shown in Figure 9.40 are approximate limits for the undrained shear strength and liquidity index of three particular residual soils with which the author is familiar. As expected, there is no relationship between these values and those for remolded soils. The undrained shear strength of undisturbed residual soils normally lies well above the strength of the fully remolded soil—a fact that arises because of the contribution which the structure of the material makes to its strength. It appears that the undrained shear strength of most fine-grained residual soils is seldom less than about 70 kPa, with the possible exception of "black cotton" clays, and is normally above 100 kPa. Black cotton clays are a variety of residual soil found in poorly drained areas and contain a large proportion of the swelling clay mineral montmorillonite. They are not particularly soft, though probably softer than most residual soils. The problem with black cotton soils is that they are prone to very large volume changes with water content changes, and also their effective strength parameters are generally very low.

9.11 MEASUREMENT OF UNDRAINED SHEAR STRENGTH

9.11.1 Unconfined Compression test

We saw in Section 9.2.4 and Figure 9.8 that undrained triaxial tests can be used to measure the undrained shear strength of clay. It is apparent from Figure 9.8 that the test results are the same regardless of the confining (cell) pressure. It is therefore not necessary to use a triaxial cell to measure the undrained shear strength and a simple compression test on an unconfined cylindrical sample can be used. The compressive strength measured in this way is known as the **unconfined compressive strength**, and it will be apparent from Figure 9.8 that the undrained shear strength is half the unconfined compressive strength provided the soil is fully saturated.

9.11.2 Vane Test

Another method used to measure the undrained shear strength of clays is the vane test, which is described in detail in Section 10.6.4. The test is widely used in the field to measure the undrained shear strength of clays, especially soft, normally consolidated clays, partly because it is an in situ test and is relatively simple to perform and partly because obtaining good-quality undisturbed samples of soft clays can be very difficult. Vane tests in the field are usually carried out with the help of a drilling rig to push the vane into the soil and withdraw it.

REFERENCES

Bishop, A. W., and L. Bjerrum. 1960. The relevance of the triaxial test to the solution of stability problems. In *Proceedings of the Research Conference on Shear Strength of Cohesive Soils*, pp, 437–501. Reston, VA: American Society of Civil Engineers.

Bishop, A. W., G. E. Green, V. K. Garga, A. Andresen, and J. D. Brown. 1971. A new ring shear apparatus and its application to the measurement of residual strength. *Geotechnique*, Vol. 21, No. 4 pp 273–328.

Bishop, A. W., D. L. Webb, and P. I. Lewin. 1965. Undisturbed samples of London clay from the Ashford Common shaft: Strength — Effective stress relationships. *Geotechnique*, Vol. 15, No. 1 pp. 1–31.

Bishop, A. W., and L. D. Wesley. 1975. A hydraulic triaxial apparatus for controlled stress path testing. *Geotechnique*, Vol. 25 No. 4, pp. 657–70.

Bromhead, N. E. 1979. A simple ring shear apparatus. *Ground Engineering* (July), pp 40–44.

Schofield, A. and P. Wroth. 1968. *Critical Soil Mechanics*. London: McGraw - Hill.

Sharma, B., and P. K. Bora. 2003. Plastic limit, liquid limit, and undrained shear strength of soil — Reappraisal. *Proc. ASCE J. Geotech. Geoenviron.l Eng.,*. Vol. 129, No. 8, pp. 774–777.

Skempton, A. W. 1954. The pore pressure coefficients A and B. *Geotechnique*, Vol. 4, No. 4, pp. 143–147.

_____. 1957. Discussion on the planning and design of the new Hong Kong airport. *Proceedings Institution of Civil Engineers*, Vol. 7, pp. 305–307.

_____. 1964. The long term stability of slopes. *Geotechnique*, Vol. 14 No. 2, pp. 77–101.

_____. 1985. Residual strength of clays in landslides, folded strata and the laboratory. *Geotechnique*, Vol. 35, No. 1, p. 4.

Terzaghi, K., R. B.Peck, and G.Mesri. 1996. *Soil Mechanics in Engineering Practice*, 3rd ed. New York: John Wiley and Sons.

Wesley, L. D. 1975. Influence of stress path and anisotropy on the behaviour of a soft alluvial clay. PhD Thesis, Imperial College, University of London.

_____. 2003. Residual strength of clays and correlations using Atterberg Limits. *Geotechnique*, 53, No 7, 669–672.

EXERCISES

1. The following readings were taken during a shear box test on a sample of sand, the box being 60 mm square:

Vertical Load (kg)	Peak Shear Force (kN)
20	0.147
40	0.250
60	0.368
80	0.520

Determine the angle of shearing resistance of the sand. **(33°)**.

2. The following results were obtained from a drained triaxial test on a dense sand:

Confining Pressure (kPa)	Peak Deviator Stress (kPa)
100	370
200	720
300	1400

(a) Determine ϕ' by constructing a Mohr failure envelope. **(39°)**

(b) Determine the theoretical inclination of the failure plane during these tests. **(64.5°)**

3. The angle of shearing resistance for a certain sand is found from drained shear box tests to be 37°. If a sample of the same sand at the same density is subjected to a drained triaxial test, determine the expected value of the deviator stress $(\sigma_1 - \sigma_3)$ at failure when the cell pressure is 60 kPa. **(182 kPa)**

4. In a consolidated undrained triaxial test on a clay, the deviator stress and pore pressures were measured at failure. From the following results, determine the cohesion intercept and angle of shearing resistance (in terms of effective stress). (**10 kPa, 32°**)

Cell Pressure (kPa)	Maximum Deviator Stress (kPa)	Pore Pressure at Failure (kPa)
60	103	33
120	140	74
180	190	115

5. A silty clay has shear strength parameters $c' = 18$ kPa, $\phi' = 36°$. A consolidated undrained triaxial test is carried out using an effective consolidation pressure of 50 kPa. If the pore pressure parameter $A = 0.25$ at failure, determine the deviator stress and the values of σ'_3 and σ'_1 at failure. (**124.5 kPa, 18.9 kPa, 143.4 kPa**)

6. A series of consolidated undrained triaxial tests was carried out on a firm clay using samples 50 mm in diameter and 100 mm in length. The following results at failure were obtained:

Test Number	Effective Consolidation Pressure (kPa)	Vertical Deformation (mm)	Vertical Force (N)	Pore Pressure (kPa)
1	50	3.7	305.4	15.0
2	100	6.9	460.8	30.6
3	200	11.8	751.7	70.9
4	300	17.6	952.3	139.9

Determine the values of c' and ϕ' of the clay. (**26 kPa, 30°**)

CHAPTER 10

SITE INVESTIGATIONS, FIELD TESTING, AND PARAMTER CORRELATIONS

10.1 OVERVIEW

Site investigations are carried out to obtain information needed for planning and designing geotechnical engineering projects. This includes an overall picture of the geology and the soil layers (stratigraphy) at the site as well as values of the various soil parameters needed for design purposes. Most of the information on the geology and stratigraphy is obtained from the site investigation, while soil parameters for design are mainly obtained from laboratory tests, although some are obtained from in situ soil tests during the site investigation. The predominant method used in most site investigations is drilling boreholes, from which cores or samples of the soil are recovered for visual examination, and laboratory testing.

10.2 DRILLING

The objective in drilling boreholes is to recover samples of the soil that can be examined visually so that a careful record can be made of the soil types and depths at which they are encountered. This record of the soil "profile," made over the full depth to which the borehole is taken, is called the bore log. It requires the presence on-site of someone trained in identifying soil types and describing them in a systematic manner, in accordance with established soil classification methods. These methods were described in

Chapter 3. For such a record to give a true picture of the soil profile, it is essential that the samples recovered from the borehole retain as much of their natural characteristics as possible, that is, they should be "undisturbed" samples.

With machine drilling, it is possible to obtain "cores" of essentially undisturbed material in almost all cohesive or cemented materials, ranging from soft clay to hard rock. However, with granular materials, especially clean sands and gravels, it is extremely difficult to obtain intact samples, and normally only completely disturbed samples of these materials can be recovered. Most site investigation drilling is done using machine rigs, although for small projects and inaccessible sites there is still a place for manual "hand auguring" of boreholes.

10.2.1 Hand Auguring

There are a variety of tools available for hand auguring boreholes. An example of the most common type is shown in Figure 10.1. Many of these augurs are known as posthole augurs, as this is the use for which they are normally intended. The drilling rods for hand auguring are normally 1 m in length and are added in stages as drilling proceeds. Hand auguring is only suitable for soft- to medium-strength clays and can easily be carried out to a depth of about 5 m.

Beyond this depth, the length of the rods makes withdrawing the augur from the borehole difficult unless a derrick-and-winch system is available. With such gear it is possible to go considerably deeper. In some parts of the world hand auguring is done down to depths of 10– to 15 m.

10.2.2 Machine Drilling

Figure 10.2 shows the principles of operation of a machine-driven rotary drilling rig. This involves attaching various tools for retrieving soil samples to the end of drilling rods and pushing or drilling them into the ground. The rig has the capacity to undertake the following essential operations:

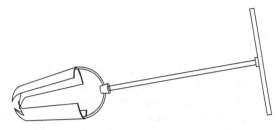

Figure 10.1 Example of a hand augur (Iwan type).

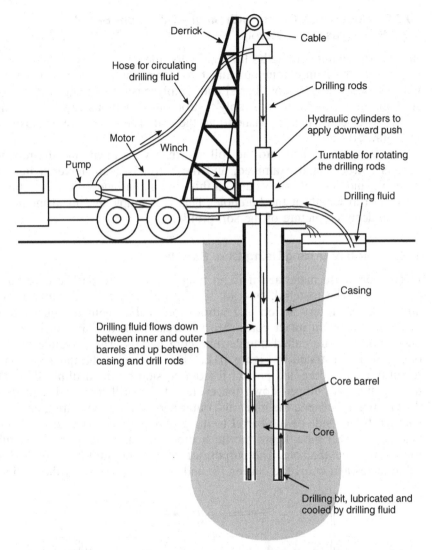

Figure 10.2   Rotary drilling rig and core barrel operation (not to scale).

(a) Provide a vertical downward push to the drill rods. This is normally done using a hydraulic system.

(b) Rotate the drill rods. This is done using a rotating device (a turntable) to which the rods can be clamped.

(c) Pump water or drilling fluid down the hollow space at the center of the drilling rods. The purpose of this is to lubricate and cool the cutting tool at the end of the rods.

The rotary drilling rig can be used for a range of different types of drilling procedures, as described in the following sections.

10.2.3   Continuous Coring with Single-Tube Core Barrel (Also Known as Open Barrel)

In soft- to medium-strength clays and silts, the easiest way to put down a borehole and obtain continuous intact core over the full length of the hole is to use a single-tube core barrel and simply push it straight into the soil. A single-tube core barrel is simply a strong tube with a sharp cutting edge at the lower end which is pushed into the soil. The procedure is illustrated in Figure 10.3.

The barrel is pushed into the soil and then withdrawn, and the retained core is extruded and examined, and the barrel is again lowered into the borehole and another core taken. In this way a continuous core is obtained over the full length of the borehole. This is a simple and ideal method to gain an accurate picture of the soil profile.

10.2.4   Rotary Drilling Using Core Barrels

In hard or dense materials, it is no longer possible to push a core barrel directly into the material. Instead, special core barrels with cutting teeth are used, and rotary action and lubrication enable them to penetrate the material and retain intact cores. Such barrels normally have an inner and outer tube and are called double-tube core barrels. The principle of their operation is illustrated in Figure 10.2. The function of the inner tube is to retain the intact core and protect it from erosion by the drilling fluid. The drilling fluid is pumped down the center of the drill rods and then flows down the gap between the inner and outer barrels, past the cutting edge and back up the outside of the barrel to the ground surface. The retained core, held in the inner tube, keeps the inner tube stationary while the outer tube rotates and cuts the core. The purpose of the drilling fluid is twofold. First, it lubricates and cools the cutting bit and, second, it transport the cuttings,

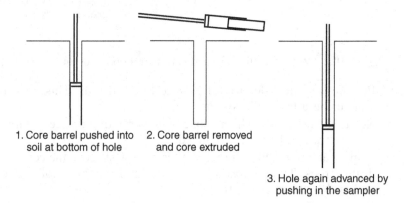

1. Core barrel pushed into soil at bottom of hole
2. Core barrel removed and core extruded
3. Hole again advanced by pushing in the sampler

Figure 10.3   Continuous coring with a single-tube core barrel (also known as an open barrel).

as suspension in the drilling fluid, up to the ground surface. Drilling fluid is normally water, but additives may be used to cope with particular drilling situations.

A wide range of core barrels is available; the choice of the type of barrel to be used in any particular formation requires considerable expertise and experience. With hard soils and soft rocks, the core can often be retained in the core barrel by friction or adhesion alone between the core and the inside of the inner tube. However, with hard materials and solid rock, this is not normally possible and a device known as a "core catcher" is placed at the bottom end of the inner tube to prevent the core falling out when the core barrel is withdrawn from the hole.

10.2.5 Wash Drilling

Wash drilling is a simple and primitive method of putting down a borehole. Water is pumped down the drill rods to exit at the bottom with the aid of a chisel or "fishtail" cutting bit. Rotary action is normally used, though in loose material this may not be necessary. The drilling fluid brings the cuttings back up to the ground surface. Wash drilling is not a good method to obtain a reliable picture of the soil profile, as only "washings" are recovered, that is, the fine material suspended in the drilling fluid. Its use in sands and gravels is inevitable, as these materials cannot easily be recovered by other methods.

10.2.6 Percussion Boring

Percussion boring is used in materials that are not amenable to core drilling or wash drilling. Such materials are typically very broken and shattered rock and gravels. Core barrels will not cut intact cores in these materials, and they are too coarse to be brought directly back up to the surface in the drilling fluid. Percussion drilling involves the use of cable tools to drive devices into the soil or rock. These devices are mainly various types of chisels, which break up the material and enable the borehole to advance. The material may be brought up to the surface by the drilling fluid or a device known as a bailer (or sand pump) may also be used to recover the material from the bottom of the hole.

10.3 UNDISTURBED SAMPLING USING SAMPLE TUBES

Undisturbed samples are most commonly taken from boreholes, in sample tubes, during the drilling operation. The base of the borehole is cleared of any debris resulting from the drilling operation, and a sample tube is then attached to the drilling rods (using a holding device) and pushed into the base of the hole. To ensure minimum disturbance of the sample, the sample tube should be of minimum wall thickness; hence the term **thin-walled sample tubes** is commonly used for such samplers. A diagrammatic view of the cutting end of such a sample tube is shown in Figure 10.4a.

The degree of disturbance is governed by the area ratio A_r of the tube defined as

$$A_r = \frac{D_e^2 - D_i^2}{D_e^2} \tag{10.1}$$

It is generally considered that A_r should not be greater than 10%. Some sample tubes have threaded ends which take screw-on cutting "shoes." These often have area ratios in excess of 25% and are therefore far from ideal tubes. It is also desirable that the inner diameter D_i be slightly smaller at the cutting end than over the remainder of the tube, as indicated in Figure 10.4. Apart from the quality of the tube, the sampling operation depends on the way the hole has been drilled and cleaned out prior to taking the sample. If the drilling operation is poorly done, the soil at the bottom of the hole may be disturbed even before the sampling operation starts.

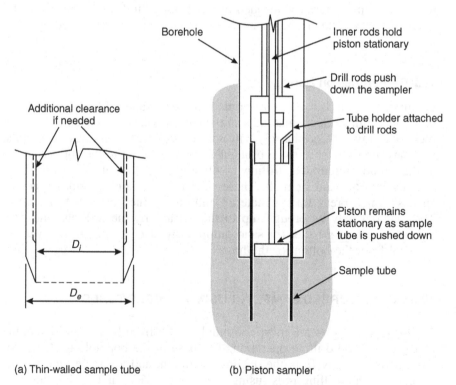

(a) Thin-walled sample tube   (b) Piston sampler

Figure 10.4 Thin-walled sample tube and the piston sampler.

In some situations, a more sophisticated device, known as a piston sampler, is used for obtaining undisturbed samples, especially from soft clay. Its mode of operation is illustrated in Figure 10.4b. The drill rods used have a set of inner rods to which a piston is attached. This piston is placed at the open end of the sample tube and the inner and outer drill rods "locked" together while the device is lowered into the hole. The piston ensures that mud or other debris at the bottom of the hole is displaced before the sampling operation begins. With the device firmly at the bottom of the hole, the inner rods are held in a fixed position, and the outer rods are used to push down the sample tube.

10.4   BLOCK SAMPLING

Samples obtained from boreholes are never without some degree of disturbance, and it is often very uncertain just what the degree of disturbance is in any particular sampling technique. The disturbance may have occurred in advancing the hole during the drilling operation, prior to the actual taking of the sample, or it may occur during the sampling operation itself. To overcome this uncertainty, block samples are a very attractive alternative and are the ideal procedure for minimizing disturbance. Block samples are simply carved out of the material by hand. The procedure is illustrated in Figure 10.5.

Block samples have enormous advantages over samples taken in sample tubes, although they also have some quite severe limitations. Their advantages are:

(a) They can be obtained with a minimum of disturbance.
(b) They can be much larger than tube samples.
(c) Their location can be selected carefully after visual inspection of the site to ensure that they are obtained from the material of interest.

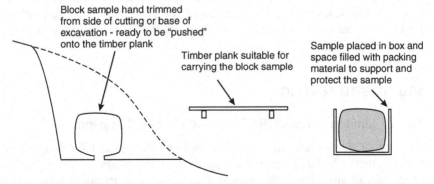

Block sample hand trimmed from side of cutting or base of excavation - ready to be "pushed" onto the timber plank

Timber plank suitable for carrying the block sample

Sample placed in box and space filled with packing material to support and protect the sample

Figure 10.5   Concept for trimming and carrying block samples.

They have the disadvantage of limitations of access, except close to the surface. Only when a mechanical digger is available or an excavation has been made for a highway or a construction site is it possible to get access to deep soil layers. However, it is often the case that the soil close to the surface is of greatest interest, so the difficulty of access may not be a major constraint.

To prevent the soil from drying out, it is very important to have ready plastic sheeting or similar watertight wrapping cloth. Carefully wrapping the sample in such cloth will help prevent it from both falling apart and drying out. It is good practice to place the sample in a properly prepared box, so that packing material can be placed around it to protect it and hold it together. Once the sample is back in the laboratory, it should be further protected from drying out by wrapping it in damp rags on top of the plastic wrapping and then applying a further layer of plastic sheeting.

10.5   INVESTIGATION PITS (OR TEST PITS)

Investigation pits are either hand dug or machine excavated holes put down for the purpose of examining soil conditions beneath the surface and for providing an opportunity to obtain block samples. If they are hand dug (only likely to be the case where labor is cheap), they can be made with a neat circular cross section, like water supply wells, in which case they will generally be very unlikely to collapse. Machine dug pits, however, cannot normally be formed with a neat circular cross section and the sides cannot be relied on to remain stable. Such pits therefore require proper lateral bracing before people can enter them.

Investigation pits provide a very much clearer picture of soil conditions than can be obtained from boreholes because direct inspection of the soil in its undisturbed state is possible. They also provide a better opportunity to select and obtain block samples. As already observed, it is often the case that the layers closest to the surface are of greatest interest to the geotechnical engineer, and test pits are by far the best way to investigate these layers. While test pits and block sampling are widely used for research purposes, they are not used as much as they should be in routine geotechnical engineering.

10.6   IN SITU TESTING

10.6.1   Limitations of Drilling and Undisturbed Sampling

Because of the limitations of undisturbed sampling and laboratory testing, various methods have been developed over the years to determine soil properties in situ. The best known of these are penetrometer tests and the shear vane test. There are two main types of penetrometer tests, both of which have been in use for many years. These are the standard penetration test (SPT) and the Dutch cone penetrometer test (CPT).

10.6.2 Standard Penetration Test (Dynamic Test)

This test grew out of the use of a split-"spoon" sampler for obtaining "undisturbed" samples. The sampler is shown in Figure 10.6; it is seen to be a small-diameter, thick-walled sampler and therefore not a suitable device for obtaining high-quality undisturbed samples. The sampler is attached to drill rods in the normal way, lowered to the bottom of a borehole, and driven into the soil using a falling weight hammer. In the course of its use, it was recognized that in addition to obtaining a sample, information could be obtained on the hardness of the soil by counting the number of blows to drive in the sampler. The dimensions of the device and the weight and

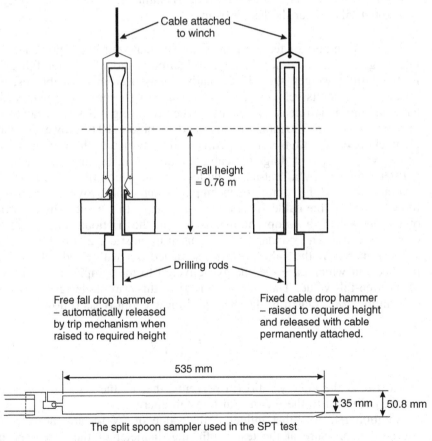

Figure 10.6 Split-spoon sampler and drop hammer used in SPT.

drop height (fall) of the hammer were therefore standardized to give the test universal acceptance.

The test requires a borehole and is carried out as the borehole is advanced. It is most commonly used in sands, although it can be carried out in any soil, including clay. In sands the test requires a casing to prevent collapse of the hole. The essentials of the test are illustrated in Figure 10.6

The test is only "standardized" to a certain degree; unfortunately, important parts of the test are not standardized. Those features that are standardized are the following:

- The dimensions of the sampler.
- The hammer weight is 65 kg (140 lb).
- The fall height of the hammer is 760 mm.
- The penetration distance is $3 \times 150\,\text{mm} = 450\,\text{mm}$.
- The blow count is taken over each 150 mm of penetration.
- The test result is expressed as the N number, where N is the blow count taken over the last 300 mm.

The most important feature that is not standardized is the procedure for releasing the drop hammer. Ideally, all hammers should be "free fall"; that is, the falling weight should be totally released from the cable used to raise it prior to its release. This is the case with some hammers which have an automatic release mechanism when the weight has been raised the required height of 760 mm, as illustrated in Figure 10.6. However, many hammers operate with the cable permanently attached to the weight, so that as the weight drops it drags the cable along with it. In this case, the cable is often operated by a "cat-head" pulley around which the cable is wound several turns, and the friction between pulley and cable provides the traction force to pull on the cable. This procedure means that a substantial portion of the energy of the drop hammer is lost in the frictional drag from the operating cable. The standard N value used in most correlations with other soil parameters is the value from tests carried out with fixed cables (U.S. practice), in which case the energy delivered to the sampler is about 60% of the free-fall value. The value obtained in this way is designated $(N_1)_{60}$ (or simply N_{60}), related to the measured blow count N as follows:

$$(N_1)_{60} = \left(\frac{\text{ER}}{60}\right) N \tag{10.2}$$

where ER is the energy ratio (in percent), that is, the ratio of the actual energy delivered to the theoretical free-fall energy.

Additional factors that influence the value of $(N_1)_{60}$ are the effective overburden pressure at the test depth, the diameter of the borehole, and the presence of any fine material in the sand being tested. To account for influence of the overburden pressure, a standard effective overburden pressure of 100 kPa (1 ton/ft^2) has been adopted. The corrected value of N_1 is related to the value N, uncorrected for overburden pressure, as follows:

$$N_1 = C_N N \qquad (10.3)$$

where C_N is a correction factor.

Figure 10.7 shows values of C_N for a range of overburden pressures. The curve is an approximate average from a number of relationships proposed in the soil mechanics literature.

For tests carried out in very fine sand, or sand containing silt, below the water table, it is considered that the measured N value will be influenced by the increased resistance due to negative excess pore pressure caused during driving and unable to dissipate instantaneously. The correction is recommended for N values greater than 15, given by

$$N' = 15 + \tfrac{1}{2}(N - 15) \qquad (10.4)$$

where N' and N are the corrected and uncorrected values, respectively.

Unless otherwise stated, the N values quoted in other parts of this book, especially those used in correlations with other soil properties (Section 10.7), are actually N_{60} values, that is, the values obtained by the test procedure normally used in the United States. The frequency with which SPTs are carried out is a matter of choice and depends on the objective of the tests and the nature of the material under investigation. Common intervals are 1–2 m.

Advantages and disadvantages of the SPT are as follows:

Advantages

- The test is widely used.
- It can be used in all sorts of materials, including quite hard or dense materials.

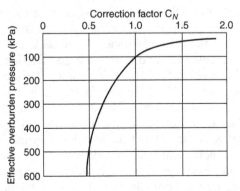

Figure 10.7   Standard penetration test correction factors for overburden pressure.

- Well-established correlations exist with other soil properties, especially relative density, liability to liquefaction, and undrained shear strength.

Disadvantages

- The test requires a borehole, which means the soil may be disturbed by the drilling operation before the test is done.
- It is a rather "crude" dynamic test.
- Some important aspects of the test have not been standardized, especially the method of driving the sampler (i.e., dropping the weight). Both fixed-cable and free-fall ("trip") hammers are used.

Skempton (1986) gives a very comprehensive analysis of the SPT and the many factors that influence its execution and interpretation.

10.6.3 Dutch Static Cone Penetration Test CPT

A conical point is pushed into the ground using a rig at the ground surface. The rig is equipped with a mechanical or hydraulic jack and measuring devices as well as a means of resisting the reaction from the downward "pushing" force; this can be either gravity weight or screw anchors. The cone, which has a cross-sectional area of $10\,\text{cm}^2$, can be either mechanical or electrical. When the device was first developed in Holland, the cone was entirely mechanical, but in recent years cones with electronic transducers have become widely available and used. Some cones measure only point resistance while others measure both point resistance and skin friction. The latter, which may be either mechanical or electrical, make use of a independent sleeve located immediately behind the conical point. Modern devices may also measure pore pressure.

Figure 10.8 shows both the original mechanical device and the modern electronic version. The mechanical device makes use of hollow rods 1 m in length, inside which are small-diameter solid rods. When the outer rods are pushed down, the device telescopes into its fully telescoped position, ready for a test to be carried out. The test is done by pushing on the inner rods. Initially, only the tip moves, which it does over a distance of about 4 cm, during which the cone resistance is measured. The tip then picks up the friction sleeve behind it and both cone and friction sleeve move down together, allowing a second reading to be made. The frictional resistance (or adhesion) on the sleeve is the difference between the total resistance and the cone resistance already measured. These stages of the test are illustrated in Figure 10.8a. The values of resistance are measured using a hydraulic cell and Bourden gauge on the thrust device at the surface. This thrust device was originally a mechanical jack, but devices today are usually hydraulic, often mounted on heavy vehicles to provide an adequate reaction load.

The electronic devices contain transducers that independently measure both the point resistance and the friction (or adhesion) on the friction sleeve

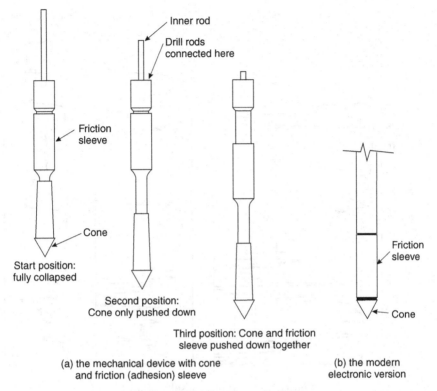

Figure 10.8   Dutch static CPT.

and make use of a cable up the center of the rods to transmit the values to recording gear at the surface.

In soft clays, tests can be carried out to great depths, not uncommonly up to 30 or 40 m. The depth achievable in any situation is governed by the hardness or denseness of the material being tested. In dense sands and gravels the force required to advance the penetrometer is likely to be very great and severely limit the depth to which the test can be taken. Readings with the mechanical device have traditionally been taken at 20-cm intervals; with the electronic device readings can be virtually continuous. The standard rate of penetration is 2 cm/s. The results are presented as graphs of cone resistance and skin friction versus depth, as illustrated in Figure 10.9

It is also common practice to plot the ratio of friction to cone resistance, as shown in Figure 10.9. This latter plot can be used as an indication of soil type. Clays have relatively high friction ratio, while sands have a relatively low friction ratio. The cone resistance values are normally reliable and repeatable. However, the sleeve friction is a less reliable measurement, and determination of soil type from the friction ratio is therefore a fairly approximate procedure.

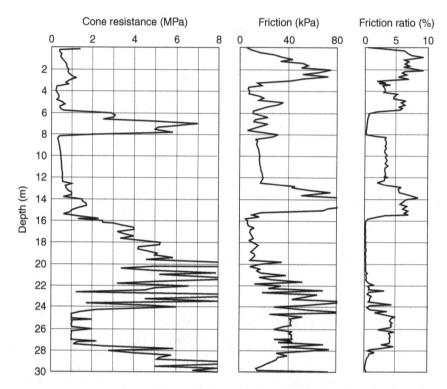

Figure 10.9   Typical results of a CPT.

Advantages and disadvantages of the Dutch CPT are as follows:

Advantages

- The test does not require a borehole, so that undisturbed soil is tested.
- An almost continuous record of resistance can be obtained over the full depth of penetration.
- The test is widely used in many countries.
- The test is simple and quick to perform.
- The equipment is very portable.

Disadvantages

- The penetrometer cannot penetrate hard or dense material.
- It is an empirical test and requires correlations with other soil properties.

A comprehensive account of the CPT and correlations with other soil parameters is given by Lunne, Robertson, and Powell (1997).

10.6.4 Shear Vane Test

The shear vane test can be used in both the field and the laboratory to measure the undrained shear strength of clays. The vane consists of four blades attached to a central rod as illustrated in Figure 10.10. It is pushed carefully into the soil and then a torque applied in a steady, controlled fashion, until a peak value of torque is reached. At this point the vane rotates and shears the soil as a cylindrical mass, as the figure indicates. The shear strength on this cylindrical surface is directly related to the applied torque and the vane dimensions. By calculating the moment coming from the shear resistance on the ends and curved surface of the cylinder, it is easily shown that the undrained shear strength is given by

$$S_u = \frac{T}{2\pi R^2 H \left[1 + \dfrac{2R}{3H} \right]} \tag{10.5}$$

where T is the peak value of the applied torque.

   The above relationship assumes that the peak shearing resistance occurs at the same time on all surfaces. This is not strictly correct, as the displacement on the end surfaces is proportional to the radius, and thus peak resistance may occur at slightly different times. However, the contribution coming from the two ends is actually very small, so that the influence of varying displacement on the end surfaces will be very minor. A wide range of vane sizes is available, normally with a height-to-diameter ratio of 1.5–2. Small laboratory vanes may have a diameter of only 1 cm, while some field vanes may have a diameter up to 20 cm.

   When the vane is used in the field, it can be pushed directly into the soil from the ground surface for tests close to the surface. For deeper tests,

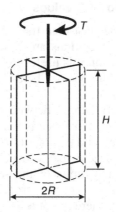

Figure 10.10   Diagrammatic view of the shear vane device.

boreholes are necessary, and the test is carried out by pushing the vane into the soil at the base of the borehole. The test is only suitable for soft to medium-strength clays; with stiff to hard clays it is not possible to push the vane into the soil or carry out the test without damage to the vane. Various corrections and correlations for field vane tests have been proposed over the years; for a useful overview of these the reader is referred to Chandler (1987).

10.7 CORRELATIONS BETWEEN IN SITU TEST RESULTS AND SOIL PROPERTIES

Penetrometer tests are useful for two reasons. First, they give a visual picture of the relative hardness or strength of the materials making up the soil profile. The cone penetrometer test in particular, because of its closely spaced (or continuous) readings, gives an excellent visual picture of the soil layering and variations within layers. Second, the tests provide numerical values that can be correlated with other parameters of direct interest to geotechnical engineers, such as strength, compressibility, and relative density. Many such correlations are to be found in the soil mechanics literature, some of which are presented in the following sections. These correlations are very useful, especially in the case of sands, where undisturbed sampling is extremely difficult, if not impossible, in most situations. While these correlations are very valuable, it should be appreciated that they may also be very approximate. Both SPTs and CPTs measure resistance at failure states in the soil, rather than stiffness or compressibility of the soil, and while stiffness can be expected to have a general relationship to ultimate strength, this is not always the case, and correlations with stiffness will always be approximate.

10.7.1 SPT N Values and CPT Values

Because the SPT is an older test than the CPT and is very widely used in the United States, many correlations are based on SPT values rather than on CPT values. For this reason it is useful to have a correlation between the two tests, so that SPT correlations can be used when the field data involves only CPTs.

Figure 10.11 shows correlations between CPT cone resistance values and SPT N values. The cone resistance is normalized with respect to atmospheric pressure ($\sim$100 kPa), and the ratio of this parameter to the N value is plotted against the mean particle size. Correlation curves from two sources are shown, together with the limits of the scatter involved in developing the correlations. The Robertson et al. curve is probably the more reliable of the two curves, although there is not a lot of difference between them. It is very clear that the correlation is strongly influenced by the particle size of the material tested.

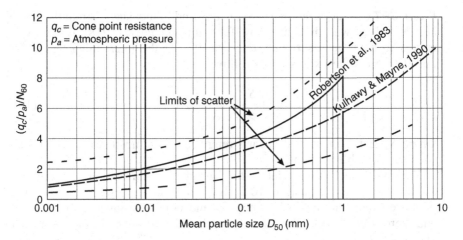

Figure 10.11 Relationship between CPT cone resistance and SPT N value for varying mean particle size. (After Robertson et al., 1983, and Kulhawy and Mayne, 1990.)

10.7.2 Undrained Shear Strength of Clay

From CPT The cone resistance can be likened to the ultimate bearing capacity of the clay for undrained loading. This is described in detail in Chapter 12, and the bearing capacity equation has the form

$$q = S_u N_c + \gamma D \qquad (10.6)$$

where q is the ultimate bearing capacity of the soil (i.e., the maximum vertical stress it can withstand), S_u is the undrained shear strength, N_c is a bearing capacity factor, γ is the unit weight of the soil, and D is the depth. For a simple surface foundation, the value of N_c can be determined theoretically. However, the cone in the penetrometer test behaves somewhat differently from a simple foundation because of side effects and shape influence, and the value of N_c can only be determined by empirical measurements. Rearranging the above formula, the undrained shear strength is related to the cone resistance as follows:

$$S_u = \frac{q_c - \sigma_v}{N_k} \qquad (10.7)$$

where q_c is the cone resistance, σ_v is the total vertical stress, and N_k is an empirical factor. The factor N_k has been measured for a wide range of soils, giving values ranging generally between about 12 and 20. This range in values probably reflects the different methods used for measuring the undrained strength as well as the degree of sample disturbance associated with the methods. A reasonable average value for N_k is 15.

From SPT   A correlation between the undrained shear strength of clay and N values was first given by Terzaghi and Peck (1948) and does not appear to have been seriously questioned over the intervening years. This correlation can be expressed as follows:

$$\left(\frac{S_u}{p_a}\right) = 0.06\,N \tag{10.8}$$

where S_u is the undrained shear strength, p_a is atmospheric pressure, and N is the SPT blow count.

We can check this relationship using the correlation illustrated in Figure 10.11 and Equation 10.2. Ignoring the influence of overburden pressure, the relationship between S_u and cone resistance is $S_u = q_c/15$, and from Figure 10.11 the relationship between cone resistance and N for clay is

$$\left[\frac{q_c}{p_a}\right] = 1.0\,N\,(\text{approx.})$$

We can therefore write

$$\left[\frac{15\,S_u}{p_a}\right] = N \qquad \text{and hence} \qquad \left[\frac{S_u}{p_a}\right] = 0.067\,N$$

which is in reasonable agreement with the Terzaghi and Peck expression in Equation 10.8.

10.7.3   Relative Density of Sand

From CPT   The cone resistance from CPTs in sand is clearly dependent on the relative density and the vertical effective stress on the sand. Many investigations of this relationship have been carried out using "calibration chambers" in the laboratory. These tests produce relationships of the sort shown in Figure 10.12, which is an approximate mean of graphs presented by Lunne and Christofferson (1983), Baldi et al. (1989), and Jamiolkowski et al. (1988).

Figure 10.12 is based on tests on clean, hard-grained, quartz sand and is therefore only valid for such sand. For mixed-grained or soft-grained sands the relationships may be quite different (see Robertson and Campanella, 1983; Wesley, 2006). There is a further factor that influences the values of cone resistance in sand, and this is the stress history and the age of the sand. Some sands may have experienced past overburden stresses greater than their preset values and are thus overconsolidated in a manner similar to overconsolidated clays. Overconsolidation and aging of sands alter their properties, especially the relationship between SPT or CPT values and

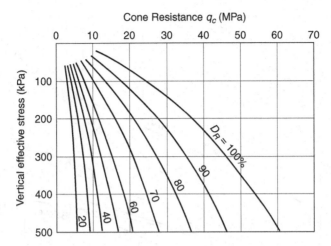

Figure 10.12 Relationship between relative density and cone resistance for varying vertical effective stresses.

other soil properties. Baldi et al. (1989) present graphs similar to those in Figure 10.12 that take account of these effects.

10.7.4 Stiffness Modulus of Sand

Figure 10.13 shows a number of correlations between N values, or CPT cone resistance values, and the Young's modulus of sand. Not all of these correlations were originally presented in the form shown in Figure 10.13, and some approximations have been made in converting them to the form in this figure. In particular, some of the correlations are based on N values, and some on cone resistance values. To plot them on the same chart, it has been assumed that $(q_c/p_a)/N = 4$, which is a midrange value for sand (see Figure 10.11). This means $q_c(\text{MPa}) = 0.4N_{60}$. In addition, some of the correlations are in terms of the constrained modulus rather than Young's modulus; to convert these to Young's modulus, an assumed value of Poisson's ratio of 0.2 has been used. Relationships are given for both normally consolidated and overconsolidated sands.

It is seen that there is good agreement between the graphs for normally consolidated sands but much less agreement for overconsolidated sands. It is clear, however, that the effect of overconsolidation or aging is much greater on the stiffness of the sand than on the penetration resistance.

In conclusion, the point made earlier is emphasized, namely that penetrometer tests measure ultimate resistance values, which means the soil is in a failure state when the measurements are made. Thus any correlations with other parameters such as relative density or compressibility are

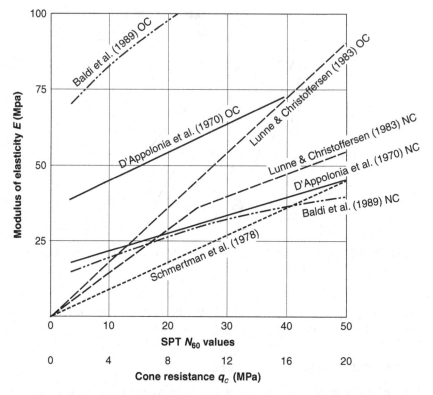

Figure 10.13 Young's modulus E from CPTs and SPTs.

approximations only. In addition many of the correlations are based on hard-grained quartz sand and will not be valid for soft-grained sands.

REFERENCES

Baldi, G., R. Bellotti, V. N. Ghionna, M. Jamiolkowski, and C. F. C. Lo Presti. 1989. Modulus of sands from CPTs and DMTs. *In Proceedings of the Twelfth International Conference on Soil Mechanics and Foundation Engineering*, Rio de Janeiro (August 13–18, 1989), Vol. 1, pp. 165–70.

Chandler, R. J. 1987. The in situ measurement of the undrained shear strength of clays using the field vane. In *Proceedings of the International Symposium on Laboratory and Field Vane Strength Testing*, ASTM, Tampa, FL (January 1987), pp 13–44.

D'Appolonia, D. J., E. D'Appolonia, and R. G. Bristette. 1970. Discussion on paper: Settlement of spread footings on sand. *Journal of Soil Mechanics and Foundation Division ASCE*, Vol. SM2, pp. 754–764.

Jamiolkowski, M., V. N. Ghionna, R. Landellotta, and E. Pasqualini. 1988. New correlations of penetration tests for design practice. *In Proceedings of the International Symposium on Penetration Testing (ISOPT-1)*, Vol. 1, Orlando, FL (March 1988), pp. 263–296.

Kulhawy, F. H., and P. H. Mayne, P. H. 1990. *Manual on Estimating Soil Properties for Foundation Design*. Palo Alto, CA: Electric Power Research Institute.

Lunne, T., and H. P. Christofferson. 1983. Interpretation of cone penetrometer data for offshore sands. In *Proceedings of the Fifteenth OTC Conference*, Vol. 1, Houston, TX (May 2–5, 1983), pp. 181–192.

Lunne, T., P. K. Robertson., and J. J. M. Powell. 1997. *Cone Penetration Testing in Geotechnical Practice*. New York: Routledge.

Robertson, P. K., and R. G. Campanella. 1983. Interpretation of cone penetrometer tests: Part I: Sand. *Can. Geotech.l J.*, Vol. 20, No. 4, pp. 718–733.

Robertson, P. K., R. G. Campanella, and A. Wightman. (1983). SPT-CPT correlations. *Journal of Geotechnical Engineering*, ASCE, 109 (11), 1449–59.

Schmertman, J. H., J. D. Hartman, and P. R. Brown. 1978. Improved strain influence factor diagrams. *Journal of the Geotechnical Engineering Division*, Vol. GT8, pp. 1131–1135.

Skempton, A. W. 1986. Standard penetration test procedures and the effects in sands of overburden pressure, relative density, particle size, ageing and oversconsolidation. *Geotechnique*, Vol. 36, No. 3, pp. 425–447.

Terzaghi, K., and R. B. Peck. 1948. *Soil Mechanics in Engineering Practice*. New York: John Wiley and Sons.

Wesley, L. D. 2006. Geotechnical characteristics of a pumice sand. In *Proceedings of the Second International Workshop on Characterisation and Engineering Properties of Natural Soils*, Vol 4, pp. 2449–2472. London: Taylor and Francis Group.

CHAPTER 11

STABILITY CONCEPTS AND FAILURE MECHANISMS

11.1 BASIC CONCEPTS

In Chapter 9, we introduced the three major design situations addressed by geotechnical engineers (Figure 9.1) that are governed by the shear strength of the soil, namely foundations of structures, retaining walls, and earth slopes. In each situation, part of the soil mass tends to fail by sliding on a surface (or surfaces) within the soil mass. These surfaces are known as slip surfaces or failure surfaces. In some situations the failure surface may be straight, or approximately straight, but in general it is more likely to be curved, though not necessarily of a predetermined shape. The manner in which the soil mass tends to fail is termed a **failure mechanism**. Analysis of these failure mechanisms is essentially an exercise in static equilibrium and involves two basic steps:

(a) Identification of the failure mechanism by which the soil is most likely to fail. In some cases, the appropriate (or governing) failure mechanism may be self-evident or its form may be constrained by theoretical considerations. It may also be predetermined by particular geological factors and soil conditions. In other cases it may be uncertain and can only be determined by a trial-and-error process.

(b) Analysis of the soil mass involved in the failure mechanism. This is a matter of static equilibrium, involving the weight of the soil mass, the shear strength of the soil on potential failure surfaces, and any external forces involved. Seepage and pore pressure are included in the analysis as factors governing the shear strength of the soil.

The principles of the analysis are essentially the same in each case, although the form of the analysis and the forces involved are somewhat different. The soil mass is considered as a free body in equilibrium under the action of a set of forces. These forces can be divided into those tending to cause failure (**destabilizing forces**) and those tending to maintain stability (**resisting forces**). These methods of analysis are commonly referred to as **limit equilibrium methods,** as they basically consider the equilibrium of a soil mass on the point of failure or collapse. The purpose of this chapter is to introduce the concepts and methods that are common to the three situations, in particular the ideas of failure mechanism, stability analysis, and safety factors, and to bring out their differences. Each case is described in detail in Chapters 12–14. To illustrate the basic approach involved, we will consider the situation illustrated in Figure 11.1, which shows a tank or building on the surface of a uniform clay layer.

The force tending to cause failure is the pressure (σ) coming from the weight of the structure. We will assume that the pressure is uniform, which would normally be the case for a tank containing a liquid but may not be the case with other structures. If we have no prior knowledge of the most likely failure mode, we can adopt a possible failure mode such as the simple circular arc failure shown in the figure, with its center at point d. Failure involves movement of the block of soil $abcda$ along the surface abc with rotation about the point d.

We can now analyze the stability of the soil block by taking moments about the point d.

The disturbing moment is given as

$$\sigma B \left(\frac{B}{2} \right) = \frac{\sigma B^2}{2}$$

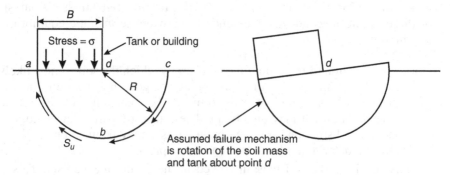

Figure 11.1 Possible failure of a surface load by slip movement (shear) along a circular arc.

and the resisting moment as

$$S_u \pi RR = \pi B^2 S_u$$

since $B = R$.

If we assume the soil is on the point of failure, these two must be equal and the maximum value of the stress σ is given by

$$\sigma = 2\pi S_u = 6.28 S_u$$

We do not know that the assumed failure mechanism is the correct one, and we can further investigate the circular arc mechanism by examining failure about points other than d. It is readily shown by trial and error or mathematically (for a mathematical solution, see Budhu, 2000, pp. 323–325) that the lowest value of the stress σ is given by

$$\sigma = 5.52 S_u$$

We will see shortly in considering the bearing capacity of soils that this is still not the lowest value that can be obtained by limit equilibrium methods. The true value, together with the failure mechanism that leads to it, is described in Section 11.3. In this simple example we determined the maximum pressure (the bearing capacity) the soil could support in terms of its strength S_u. Alternately, for known values of the pressure σ and the strength S_u, we could have determined the safety factor. To illustrate the use of safety factors, we will next consider the stability of slopes and then return to bearing capacity and earth pressure on retaining walls.

11.2 STABILITY OF SLOPES

The force causing instability (the destabilizing force) in this case is simply the weight of the soil itself, and the resisting forces come from the shear strength of the soil. External forces are not normally involved. The way in which failure will occur is uncertain and cannot normally be predetermined. For slopes of homogeneous soil, failure generally occurs on surfaces that approximate to circular arcs, as indicated in Figure 11.2.

It is possible, by analyzing the rotational equilibrium of the soil mass, to estimate the shearing stress that acts on any assumed potential failure surface. It is also possible to determine the maximum available shear strength of the soil on the same surface from a knowledge of the shear strength parameters of the soil. The actual method for doing this analysis is described in detail in Chapter 14. Failure will occur if the shear stress equals or exceeds the available strength. The shearing stress that actually acts on the failure surface in a stable slope is obviously less than the maximum

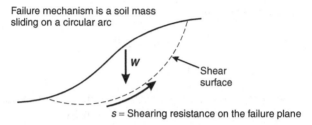

Failure mechanism is a soil mass
sliding on a circular arc

w

Shear
surface

s = Shearing resistance on the failure plane

Figure 11.2 Failure mechanism for slope failure.

strength available; if this were not the case, the slope would have failed. The strength actually operating on the failure plane must equal the shearing stress and is termed the **mobilized shear strength**.

The objective of the analysis in this case is to determine whether the slope is stable and what its margin of safety is. By carrying out the equilibrium analysis, a comparison can be made between the mobilized strength and the available strength. The stability of the slope is then described in terms of the **factor of safety**:

$$\text{Factor of Safety} = \frac{s}{s_m} \tag{11.1}$$

where s is the available strength and s_m is the strength needed to maintain stability (**mobilized shear strength**).

The use of the factor of safety (or safety factor) has come in for considerable criticism since its first use. D. W. Taylor, writing in 1948, states: "Much criticism has been levelled in the past at improper use of factors of safety and the incomplete definitions that have sometimes been given for such factors. However, any quantitative stability analysis must make use of some measure of the degree of safety" (Taylor 1948, p. 414). The criticism continues today, but Taylor's observation is still valid. It is important, however, to both define the safety factor in as rigorous a manner as possible and recognize its limitations. The preferred definition, in accordance with Equation 11.1, is as follows:

> The **factor of safety** is that factor by which the shear strength of the soil must be reduced in order to bring the soil mass into a state of limiting equilibrium, that is, to bring the soil to the point of failure (Bishop, 1964).

Critics of the use of the factor of safety argue that a statistical or probability approach is preferable and that a measure such as probability of failure would be a better concept than factor of safety. However, while this approach has some merit, the factor of safety appears more practical and is still in general use.

The definition of factor of safety given above still has shortcomings that we should be aware of. In particular, it does not define the means by which the shear strength of the soil is to be measured or expressed. As we have seen in Chapter 9, the undrained strength of soil is not the same as the effective stress strength. Hence the safety factor of any slope will be different when it is expressed in terms of the effective stress strength from what it would be in terms of the undrained strength.

11.3 BEARING CAPACITY

Bearing capacity is the term used to describe the ability of the soil to support a load bearing on the surface of the soil or in some cases a load embedded at depth below the surface. Such a load would normally be the foundation of a building or bridge but could also be a storage tank or a soil embankment. Foundation design normally consists of two components. The first is to ensure the stability of the foundation, which is governed by the strength and therefore the bearing capacity of the soil. The second is to ensure that the settlement of the foundation is within acceptable limits. This is governed by the compressibility of the soil, as discussed in Chapter 8.

The force situation is now quite different from that of the soil slope described above. There is no disturbing force coming from the soil weight, as the soil surface is level and no unbalanced forces are involved. The disturbing force in this case is the external load, and the objective of the analysis is to determine the maximum load that the foundation can support. We have already considered a simple circular arc failure mechanism and noted that it did not give the correct theoretical answer. The true theoretical failure mechanism is determined using theory of plasticity concepts and is illustrated conceptually in Figure 11.3. It is described in greater detail in Chapter 12. It involves a fixed wedge of soil immediately below the foundation and zones of radial shear and "passive" earth pressure on each

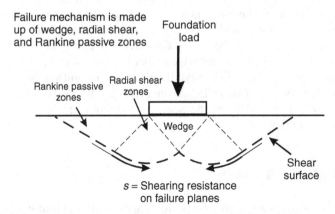

Figure 11.3 Failure mechanism for bearing capacity failure.

side, as shown. The soil strength is assumed to be fully mobilized and leads to the solution:

$$\text{Maximum bearing pressure } \sigma_m = (2 + \pi)S_u = 5.14S_u$$

While soils do not strictly conform to plasticity theory, the use of the theory is adequate for practical purposes. For estimating bearing capacity, we therefore do not have to make assumptions about the failure mechanism, as was the case with soil slopes; we adopt the theoretical solutions that are available to us from the work of others in this field.

The maximum load the foundation will support is known as the **ultimate bearing capacity**, or the failure load of the soil. The term "ultimate bearing capacity" is well established in soil mechanics and geotechnical engineering but is not entirely satisfactory as it is not consistent with the use of the term "ultimate" in other contexts. The term ultimate is often used to denote the strength remaining when a material has been loaded and deformed beyond its peak strength. In estimating the ultimate bearing capacity, it is assumed that the soil is on the point of failure; that is, the shear strength is its peak, or failure value, rather than its ultimate value.

The factor of safety in foundation design is commonly applied to the ultimate value to give us the **allowable bearing capacity**. Although the term "allowable bearing capacity" (or the design bearing capacity) is commonly used in this context, it would perhaps be more appropriately called the "allowable bearing pressure" or the "design bearing pressure" since the term "capacity" is normally taken to imply a maximum value. Thus

$$\text{Allowable bearing capacity} = \frac{\text{Ultimate bearing capacity}}{\text{Factor of safety}} \qquad (11.2)$$

This definition of safety factor is not the same as the definition used in Equation 11.1 for slope stability. We will see in Chapter 12 that the use of the safety factor in this way is not as straightforward as this relationship suggests. With deep foundations, the situation is complicated, because the weight of the soil itself becomes (in effect) a component of the bearing capacity and has implications with respect to how the safety factor should be defined and applied. For this situation, the definition in Equation 11.1 appears sounder than that in Equation 11.2, as will be explained in Chapter 12. This situation highlights the need for the safety factor to be adequately defined and understood.

11.4 RETAINING WALLS

A retaining wall is a wall built to hold up a mass of soil that would otherwise collapse. This case is different again from the two previous cases. The mechanism of failure is not certain but normally takes the form of the failure of a wedge-shaped block of soil tending to slide toward the wall, as shown in Figure 11.4. The failure surface often approximates to a straight line. The destabilizing force is the weight of the mass of soil behind the wall. The resisting forces are now made up of two components. One comes from the shear strength of the soil itself on the potential failure plane, and the other is the supporting force P provided by the wall. The principal objective of the analysis is to determine the magnitude of the force P necessary to maintain equilibrium, that is, the force needed to hold up the soil. This provides the basis for the design of the wall.

In some situations, especially those involving a level surface behind the wall, the inclination of the failure plane can be determined from theoretical concepts. However, in most situations, especially those involving sloping ground behind the wall and nonuniform pore pressures in the soil, it is necessary to determine the most probable failure plane by a trial-and-error process. It is usually sufficiently accurate to analyze only planar surfaces and to vary their inclination.

The factor of safety is applied in a somewhat different way again from the two previous cases. As with the estimation of the ultimate bearing capacity, the soil strength on the failure plane is assumed to be fully mobilized in the equilibrium analysis of the soil wedge. The factor of safety now comes into the design of the wall itself. The wall can fail in several ways and each requires analysis and an appropriate safety factor. The three possible failure mechanisms are as follows:

(a) The wall may slide horizontally (sliding failure).
(b) The wall may overturn (overturning failure).
(c) The bearing capacity of the soil supporting the wall may be exceeded, especially at the toe of the wall where the pressure on the soil will be greatest (bearing capacity failure).

Stability analysis is then carried out with respect to each of these possible failure modes and safety factors are incorporated into the analysis. For

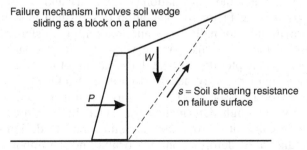

Figure 11.4 Failure mechanism for earth pressure on a retaining wall.

example, if it is considered that a safety factor of 2 is appropriate with respect to sliding failure, then the wall is designed to withstand a force of $2P$, where P is the force necessary to maintain equilibrium. The safety factors adopted for design against sliding failure may not necessarily be the same as for overturning or bearing capacity failure.

11.5 FURTHER OBSERVATIONS

11.5.1 Safety Factors, Load Factors, and Strength Reduction Factors

The question of the safety factor has already been discussed in some detail, but there are some additional matters that should be appreciated. The use of the safety factor does not imply that the soil is not overstressed (i.e., in a state of failure) anywhere in the soil mass. With retaining walls, the method of design assumes that the soil is actually in a state of failure all along the shear plane on which the soil wedge is tending to slide. This would be expected to be the case with flexible retaining walls but not with rigid walls. In the soil slope, if the safety factor is small, it is probable that some zones within the slope will be overstressed and the soil will yield. This does not mean that the soil mass will fail, although this might happen if the safety factor is low and the soil is highly sensitive. With foundation loads, it is common practice to use quite high safety factors, usually greater than 2 and possibly as high as 3. This helps to minimize settlement of the foundation and means that it is unlikely that any zones within the soil are overstressed.

Modern design practice, especially for structural design, makes use of what is commonly called "limit state" design. In place of safety factors, the method uses strength reduction factors that are applied to material strengths and load factors that are applied to the loads. The application of this method to geotechnical situations is not straightforward and is not covered here.

11.5.2 Questions of Deformation Versus Stability

As mentioned earlier, the design of foundations for structures involves questions of both stability and deformation (settlement), and deformation is more often the controlling factor in foundation design than is stability. In contract, deformation is normally less important with both slopes and retaining walls but in some situations may be very critical. The stability concepts discussed in this chapter do not take deformation into account. With natural hill slopes or with properly designed earth embankments, deformation is seldom an important issue and in many cases is not considered in assessing their stability, although deformation may well be monitored during construction to check that performance is conforming to design expectations. With retaining walls, deformation may or may not be important. Deformation of a free-standing gravity retaining wall may be of no consequence, while the deformation of a wall supporting an excavation alongside existing buildings may be a critical consideration.

REFERENCES

Bishop, A. W. 1964. Soil Properties, Imperial College MSc(Eng) course lecture notes.

Budhu, M. 2002. *Soil Mechanics and Foundations*. New York: John Wiley and Sons.

Taylor, D. W. 1948. *Fundamentals of Soil Mechanics*. New York: John Wiley and Sons.

CHAPTER 12

BEARING CAPACITY AND FOUNDATION DESIGN

12.1 BEARING CAPACITY

In designing foundations for buildings or other structures, the following two basic issues need to be addressed:

1. The bearing capacity of the soil, that is, its ability to support the load placed on the foundation. As we have seen in Chapter 11, this is an issue governed by the shear strength of the soil.
2. The settlement of the foundation that may occur. This depends on factors other than the shear strength of the soil, as discussed in Chapter 8.

In this chapter we are concerned only with the bearing capacity of the soil. If a foundation is constructed on a soil and the load on it steadily increases, the soil below it will deform, allowing the foundation to settle, as illustrated in Figure 12.1. If the soil is relatively hard, or dense as in the case of a granular material, the load deformation curve will be similar to curve G_1 in the figure. If, however, it is relatively softor loose, the curve will be more like curve G_2. If the curve is like G_1, there is clearly an upper limit to the load the foundation will carry. This is known as the failure load or the **ultimate bearing capacity** (q) of the soil, as discussed in Chapter 11. If the curve is similar to G_2, then the failure load is not so clear. Normal practice is to take the point of maximum curvature (point A) as the ultimate bearing capacity.

The form of the failure mechanism beneath a surface foundation can be determined analytically if the assumption is made that soil behavior

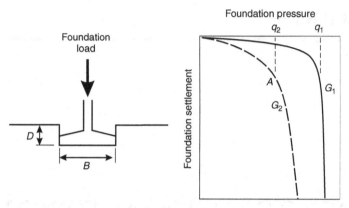

Figure 12.1 Load, settlement, and ultimate bearing capacity of a foundation.

approximates that of a plastic material with shear strength made up of a cohesive component having a constant value and a frictional component proportional to the normal stress. Its strength can therefore be expressed as, $s = c + \sigma \tan \phi$, where c and ϕ are respectively the cohesion and friction angle of the soil.

The approximate shape of the failure mechanism is illustrated in Figure 12.2. By analyzing this mechanism, Terzaghi (1943) obtained the general expression for the bearing capacity of a soil, namely

$$q = cN_c + \gamma DN_q + \frac{1}{2}\gamma BN_\gamma \qquad (12.1)$$

where B and D are the width and depth of the foundation,

γ is the unit weight of the soil, and

N_c, N_q, and N_γ are bearing capacity factors dependent on the value of the friction angle ϕ.

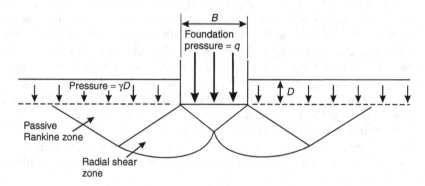

Figure 12.2 Failure mechanism assumed in the Terzaghi expression for bearing capacity.

The bearing capacity is thus made up of three components, each dependent on distinct aspects of the failure mechanism. The first term, cN_c, is the contribution from the cohesive strength of the soil. The second term, γDN_γ, is the contribution coming from the depth of soil above the level of the foundation base, which acts as a surcharge load above the Rankine passive pressure zone. Finally, the term $\frac{1}{2}\gamma BN_\gamma$ is the component that comes from the self-weight of the soil below the level of the base of the foundation. The values of these three bearing capacity factors are shown in Figure 12.3.

Terzaghi's analysis involved certain simplifying assumptions, as does all such analysis in the field of soil mechanics, and various workers in this field have refined the analysis somewhat in the intervening years. The principal changes were due to Meyerhof (1963). However, the changes to Terzaghi's bearing capacity factors are not great, as Figure 12.3 indicates. These bearing capacity factors are for an infinitely long (strip) footing of width B. For rectangular or circular footings the form of the failure mechanism is different and alters the value of the ultimate bearing capacity. Exact solutions are not available, but the adjustments proposed by Terzaghi and Peck (1967) are still widely used and believed to be conservative. They involve shape factors applied to each term of Equation 12.1, resulting in the following expressions:

$$\text{Square footing: } q = 1.2cN_c + \gamma DN_q + 0.4\gamma BN_\gamma \qquad (12.2)$$

$$\text{Circular footing: } q = 1.2cN_c + \gamma DN_q + 0.3\gamma BN_\gamma \qquad (12.3)$$

For rectangular footings of breadth B and length L, the shape factors should be obtained by linear interpolation between the values for the infinitely long footing and a square footing. For example, for the cohesive

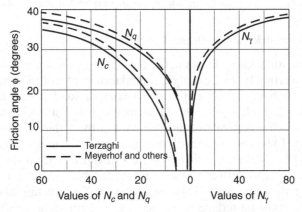

Figure 12.3 Bearing capacity factors for a strip footing.

component, the shape factor changes from 1.0 for an infinite footing ($B/L =$ 0) to 1.2 for a square footing ($B/L = 1$), so that for intermediate values of B/L the shape factor is given by $1 + 0.2B/L$.

The above formulas are based on the analysis of a theoretical material having a shear strength that can be expressed in terms of a cohesive component c and a frictional component governed by the friction angle ϕ. The formulas can therefore be applied either in terms of effective stress or in terms of total stress.

12.1.1 Bearing Capacity in Terms of Effective Stress

When expressed in terms of effective stress, the formulas retain the form given in Equations 12.1–12.3, with $c = c'$ and $\phi = \phi'$. However, the value of the soil unit weight (γ) should now be selected taking account of the level of the water table. If the water table is deep compared to the width of the footing, the bulk unit weight is appropriate, but if the water table is at the ground surface, then the submerged unit weight of the soil ($\gamma' = \gamma - \gamma_w$) should be used. For intermediate depths, interpolation between the above limits is required.

12.1.2 Bearing Capacity in Terms of Total Stress (Undrained Behavior)

When applied to an undrained situation, the value of c is the undrained shear strength of the soil, and the value of ϕ is zero, in which case $N_q = 1$ and $N_\gamma = 0$. Equation 12.1 then becomes

$$q = s_u N_c + \gamma D \qquad (12.4)$$

This is for a strip footing and requires adjustment for other footings using the shape factor described above for N_c. Skempton (1951) has presented the values of N_c in graphical form, as shown in Figure 12.4.

12.1.3 Eccentric and Inclined Loads

Some foundation loads do not act at the center of the foundation, and others do not act vertically. For example, foundations subject to earthquake forces are likely to have a horizontal component of load acting some distance above ground level, so that the resultant force is both eccentric and acting at an angle to the vertical. To take account of the influence of eccentric load application, Meyerhof's construction is commonly used. This is illustrated in Figure 12.5, where the load acts at a distance e from the center of the foundation. A new "effective" area is created such that the load acts at the centroid of this new area. This means that for the rectangular area shown the width B must be reduced by twice the eccentricity e, and the new width is given by $B' = B - 2e$.

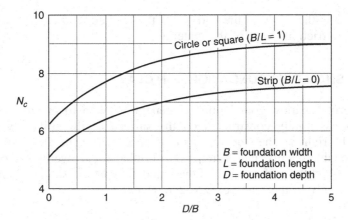

Figure 12.4 Values of N_c for bearing capacity in terms of undrained strength (Skempton, 1951).

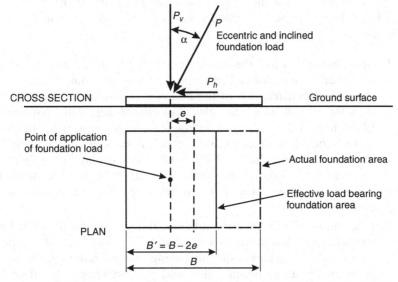

Figure 12.5 Eccentric and inclined loading. (After Meyerhof, 1963.)

Meyerhof also proposed the following adjustments to the bearing capacity factors to take account of inclined loading:

$$i_c = i_q = \frac{N_c(\text{inclined loading})}{N_c(\text{vertical loading})} = \frac{N_q(\text{inclined loading})}{N_q(\text{vertical loading})} = \left(1 - \frac{\alpha^\circ}{90^\circ}\right)^2 \tag{12.5}$$

$$i_\gamma = \frac{N_\gamma(\text{inclined loading})}{N_\gamma(\text{vertical loading})} = \left(1 - \frac{\alpha^\circ}{\phi^{\prime o}}\right)^2 \tag{12.6}$$

where α is the angle to the vertical of the applied load and ϕ' is the friction angle of the material.

12.2   SHALLOW FOUNDATIONS ON CLAY

The term "shallow foundation" is generally taken to mean a foundation that is essentially constructed on the surface of the soil. It will almost never be right on the surface, as the surface layer generally tends to be of questionable quality and is subject to shrinkage and swelling effects which would affect the performance of the foundation. Most "shallow" foundations are probably in the range of 0.5–2 m deep.

12.2.1   Use of Undrained Shear Strength

Foundations on clay are normally designed on the basis of undrained shear strength, at least when the clay is fully saturated and of medium to high plasticity. There are several good reasons for this, the principal being the following:

1. The assumption that the application of the foundation load will often be closer to an undrained situation than a drained one. In other words, the rate of construction of the building is assumed to be more rapid than the rate at which the clay will consolidate. This assumption is likely to be true with most sedimentary clays, especially in the case of overconsolidated clays. With residual soils, it is less likely to be true on a general basis. With some residual clays consolidation is slow and may take years, while with others it may be much more rapid and involve only a few months or even weeks; the range is enormous.

2. Regardless of rate of consolidation, the use of the undrained shear strength for design purposes is almost always a safe approach. The bearing capacity estimated using the undrained shear strength is normally a minimum value and will increase over time when the foundation load is applied and consolidation of the clay takes place.

3. In many parts of the world, foundations are likely to be subject to seismic loads. The application of such loads is naturally very rapid, and there will be no time for consolidation to occur. Design for seismic loads must therefore be on the basis of undrained strength.

12.2.2   Application of Factor of Safety

The use of the factor of safety has been introduced and discussed in Chapter 11, and its possible use for estimating allowable bearing capacity from ultimate bearing capacity is introduced, that is,

$$\text{Allowable bearing capacity} = \frac{\text{Ultimate bearing capacity}}{\text{Factor of safety}}$$

This use is satisfactory for shallow foundations on materials of high strength, especially those with high friction angles. However, with low-strength materials, especially soft clays, this use can result in anomalous results. The ultimate bearing capacity of clay in terms of undrained strength is

$$q = s_u N_c + \gamma D \quad \text{(Equation 12.4)}$$

For soft clay and a foundation at some depth beneath the surface, the depth term γD may well exceed the strength term $s_u N_c$, and applying the safety factor to the value of q may result in an allowable bearing pressure less than γD, in other words less than the natural stress at depth D before the construction of the foundation. This is clearly not sensible. It is therefore desirable to apply the safety factor (F) only to the net bearing capacity, that is, the increase in stress above that which originally applied at depth D. In this way the allowable bearing capacity is given by

$$q_{\text{all}} = \frac{s_u N_c}{F} + \gamma D \tag{12.7}$$

The safety factor is now the ratio of the mobilized shear stress to the strength available, and this use of the safety factor is in keeping with the original definition given by Equation 11.1.

The value of the safety factor used in foundation design is generally large, in the range of 2.5–3.0 for normal dead and live loads. An important reason for this is to help minimize settlements. A generous safety factor tends to ensure that settlement arises predominantly from consolidation of the soil, and not from the deformation required to mobilize the strength of the soil.

12.2.3 Bearing Capacity Versus Settlement Tolerance in Design of Foundations

The design of a foundation requires an adequate margin of safety against bearing capacity failure (a strength issue) and an assurance that its potential settlement will not exceed the tolerance limits of the structure (a compressibility issue). With most buildings, especially those used for human occupation, such as office blocks, homes, schools, and hospitals, the tolerance to settlement is quite small, and the foundation design is more often governed by settlement considerations than by bearing capacity issues.

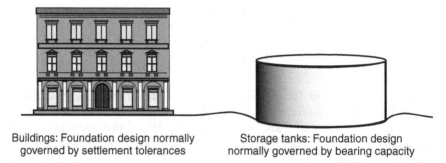

Buildings: Foundation design normally governed by settlement tolerances

Storage tanks: Foundation design normally governed by bearing capacity

Figure 12.6   Bearing capacity versus settlement in the design of foundations.

Some factories also have very low settlement tolerances and the same situation applies. It is often the case therefore, when designing foundations, that sophisticated refinement of the bearing capacity of the soil is not an issue. Foundations are likely to be dimensioned to give low applied pressures in order to limit settlements, or piles may be used in place of surface foundations because of settlement concerns.

There are some clear exceptions to the above generalization as illustrated in Figure 12.6. Large storage tanks are a typical example. Many large storage tanks are made of steel and are able to tolerate large settlements, especially between their perimeter wall and the center of the steel floor. Desirable locations for such tanks are often level sites close to port facilities, which frequently consist of soft, normally consolidated clays with low shear strength and high compressibility. In these situations it is inevitable that bearing capacity will be the dominant consideration in design. Such tanks are often designed with relatively low safety factors against bearing failure and with the capacity to tolerate large settlements.

12.2.4 Worked Examples

Example 1: A steel water storage tank with a height of 10.5 m and a diameter or 20 m is to be built on the surface of soft to firm clay having an average undrained shear strength of 30 kPa. The depth of water to be stored in the tank is 10 m. The area is subject to earthquakes and the tank is to be designed to withstand a horizontal acceleration of $0.25g$. Determine the safety factor with respect to bearing capacity failure for static and earthquake loading, assuming the tank is full to 10 m depth. The tank is shown in Figure 12.7.

Static Case   The ultimate bearing capacity of the soil is determined from Equation 12.4 and Figure 12.4:

$$q_{\text{ult}} = s_u N_c + \gamma D = s_u N_c = 30 \times 6.2 = 186\,\text{kPa}.$$

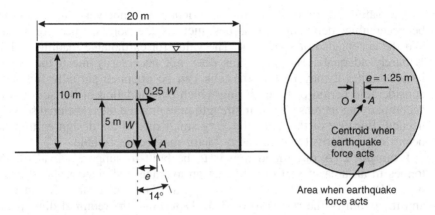

Figure 12.7   Water storage tank subject to static and earthquake loading (worked example 1).

The applied pressure is given as $10 \times 9.8 = 98\,\text{kPa}$. We will take this as $100\,\text{kPa}$, making a small allowance for the weight of the tank itself.

Hence the safety factor is $186/100 = 1.86$.

Earthquake Case   The earthquake acceleration means that in addition to the gravity force there is an earthquake force acting horizontally with a magnitude of $0.25W$, where W is the weight of the tank plus the water in it. These forces act at the center of gravity, which is 5 m above ground level, as indicated in the figure. The inclination and eccentricity of the applied force are therefore as follows:

$$\text{Force inclination: } \alpha = \tan^{-1}(0.25/1) = 14°$$

$$\text{Force eccentricity: e} = 0.25 \times 5.0 = 1.25\,\text{m}$$

We can use Meyerhof's method to determine the safety factor. From the eccentricity and the tank dimensions we calculate the new area, which has point A as its centroid. From geometric analysis, this new area is $270.2\,\text{m}^2$. The original area was $\pi R^2 = \pi \times 10^2 = 314.2\,\text{m}^2$. Hence the new bearing pressure is now $314.2 \times 100/270.2 = 116.3\,\text{kPa}$.

We must now determine the ultimate bearing capacity by applying the appropriate factors to take account of the load inclination. This is given by Equation 12.5, namely,

$$\frac{N_c\,(\text{inclined loading})}{N_c\,(\text{vertical loading})} = \left(1 - \frac{\alpha°}{90°}\right)^2$$

For $\alpha = 14°$, the factored N_c value is $6.2 \times 0.713 = 4.42$.

The ultimate bearing capacity is $4.42 \times 30 = 133\,\text{kPa}$.

The new safety factor equals $133/116.3 = 1.14$.

The safety factor of 1.86 for static loading is not very high but may be acceptable, depending on issues such as the consequences of failure and the acceptance of some risk. The value for earthquake loading of 1.14 is barely adequate. However, this does not necessarily mean the design is not viable. It may be that the tank can be operated partially full for a considerable period of time, during which consolidation of the soil and an accompanying increase in shear strength may provide the additional security needed. Because of the generally very small risk of the design earthquake occurring, a relatively small safety factor may be acceptable.

Example 2: A raft foundation is to be built to support a load of 550 tonnes in an area of soft clay having an average undrained shear strength of 18 kPa and a unit weight of 15.8 kN/m^3. The water table is 1 m deep and the depth of the raft is to be 3 m. Determine the required dimensions of the raft, assuming it to be square, to give it a safety factor of 3.

The ultimate bearing pressure is given by $q = s_u N_c + \gamma D$ (Equation 12.4), but as explained above, we need to be careful in applying the safety factor for deep foundations, especially on soft clay. The safety factor should only be applied to the shear strength component of the bearing capacity. The allowable bearing pressure is therefore

$$q_{all} = \frac{s_u N_c}{F} + \gamma D = 18 \times 6.2/3 + 15.8 \times 3 = 84.6 \,\text{kPa}$$

The total load equals $550 \times 1000 \times 9.81 \,\text{N} = 550 \times 9.81 \,\text{kN} = 5395 \,\text{kN}$.

The required area is $5395/84.6 = 63.8 \,\text{m}^2$, and the required dimension is $\sqrt{63.8} = 8.0 \,\text{m}$. In applying the formula for bearing capacity, no allowance was made in selecting the parameter N_c for the depth of the foundation. The D/B ratio is now $3/8 = 0.375$. From Figure 12.4 we can see that the value of N_c is about 6.5 rather than the 6.2 we adopted and we could refine our calculation if we felt this was necessary.

Because this is a total stress (undrained) estimate, the water table does not influence the calculation. We could repeat this calculation taking account of the buoyancy effect of the water table, but the effect is to reduce both the foundation pressure and the allowable bearing pressure by the same amount, and the end result is identical.

12.3 SHALLOW FOUNDATIONS ON SAND

12.3.1 Use of Bearing Capacity Theory

Sands have large friction angles usually between about 37° and 45°. The use of conventional bearing capacity theory for these materials leads to very large values of bearing pressures, except in the case of very narrow foundations directly on the ground surface. Examination of the Terzaghi charts (Figure 12.3) shows that the values of the factors N_c, N_q, and N_γ

increase very rapidly once ϕ' approaches $40°$, and the values of bearing capacity determined using them tend to be unrealistically high. In practice these large values could only be achieved with very large displacement (settlement) of the foundation. Control of settlement thus frequently becomes the principal design consideration for foundations on sand, just as we saw was the case with foundations on clay. Because direct measurements of the compressibility of sand are extremely difficult, either in the field or the laboratory, methods for estimating settlement in sands are essentially empirical.

12.3.2 Empirical Methods for Foundations on Sand

Two empirical methods for estimating settlement of foundations on sand were covered in Section 8.10; these can be used as a basis for determining acceptable bearing pressures. For circular foundations on uniform sands it is also possible to estimate the settlement by using elastic theory solutions, as described in Section 8.2. For nonuniform sands and foundations of variable shape, elastic theory can still be used to determine the stresses in the ground beneath the foundation and the settlement estimated by adding up the compression of appropriate sublayers. The modulus value needed for such estimates can be determined from the correlations given in Section 10.7.4.

A further simple design method is provided by the graphs shown in Figure 12.8, put forward by Terzaghi and Peck (1967). This gives the allowable bearing pressure for foundations of varying width that will ensure settlement does not exceed 25 mm. The sand density is expressed by the N value from the standard penetration test. The graphs in Figure 12.8 are

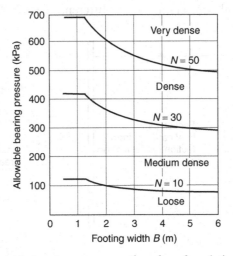

Figure 12.8 Allowable bearing pressure of surface foundations on sand to limit settlement to 25 mm. (After Terzaghi and Peck, 1967.)

not intended to accurately predict settlement, only to ensure that it does not exceed 25 mm; they will therefore tend to give conservative values of allowable pressure. The graphs are based on the assumption that the water table is deeper than $2B$, where B is the width of the foundation. If the water table is at the surface, then the allowable bearing pressure can be taken as half the values in the figure.

12.4 PILE FOUNDATIONS

12.4.1 Basic Concepts and Pile Types

Piles are a very common form of foundation and are used to transfer surface loads to deeper layers in situations where the soil has insufficient strength to support surface foundations. Piles may also be used to resist lateral (horizontal) loads or uplift forces. Piles may be either **end-bearing piles** or **friction piles**, as illustrated in Figure 12.9, or a combination of these. End-bearing piles are taken down to a hard layer and derive their load-carrying capacity from the strength of the hard layer. Friction piles do not extend down to a hard layer. They derive their load-carrying capacity primarily from "skin friction," that is shear resistance on the shaft surface of the piles. The terms "skin friction" and "shaft adhesion" are both used to designate this component of bearing capacity. Most piles are a combination of end-bearing and shaft adhesion, although in some situations one of the components may be so small as to be ignored for design purposes. The majority of piles in practice derive most of their load-bearing capacity from end bearing, though there are some situations, namely piles in soft, normally consolidated clays, where shaft friction is the major component.

The ways in which these two components of load develop as the load is applied to the pile are significantly different. The shaft resistance is fully

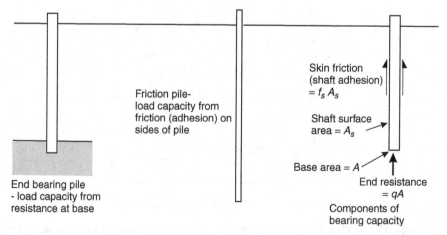

Figure 12.9 End-bearing and friction piles.

mobilized at very small deflections; it generally peaks and then decreases slightly. The end resistance takes much greater deflection to be fully mobilized and does not decline after reaching its peak value. This behavior is illustrated in Figure 12.10 for a pile which derives its bearing capacity in approximately equal proportions from end-bearing and skin friction.

There are many ways of constructing piles. The principal ones are illustrated in Figure 12.11 and are described briefly in the following sections.

Driven Precast Piles These are the most common type of pile and can be made of steel, reinforced concrete, or timber. They are driven into the ground using a pile-driving machine, or "rig," which in its simplest form consists of a heavy weight and a winch system which raises the weight and drops it onto the top of the pile, thus driving it into the ground. Modern rigs use more sophisticated driving methods which make use of steam or diesel engines to provide the driving energy.

Cast In Situ Piles These piles are formed by first driving a "shell" or "casing" with a blocked-off base or "plug" to prevent soil entering the casing. When the casing has reached the required depth, a reinforcing cage is lowered into it and it is filled with concrete. The casing is then withdrawn, leaving behind the "plug" at its base. Strong vibrations are commonly applied to the casing to assist the withdrawal process.

Bored Piles Bored piles are formed by first drilling a hole and then lowering a reinforcing cage into it and filling it up with concrete. Depending on the nature of the soil, a casing may be used over part or all of the hole to prevent collapse of the sides.

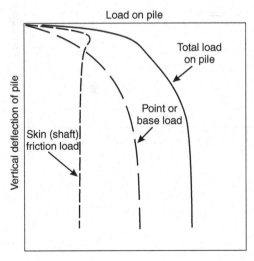

Figure 12.10 Load distribution on a pile versus deflection of the pile.

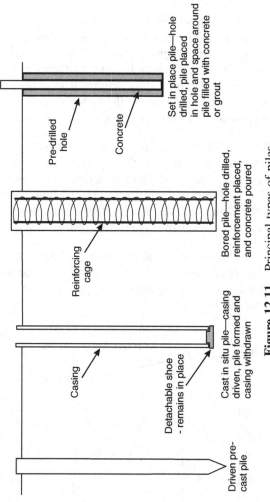

Figure 12.11 Principal types of piles.

Driven pre-cast pile

Cast in situ pile—casing driven, pile formed and casing withdrawn

Casing

Detachable shoe - remains in place

Bored pile—hole drilled, reinforcement placed, and concrete poured

Reinforcing cage

Set in place pile—hole drilled, pile placed in hole and space around pile filled with concrete or grout

Pre-drilled hole

Concrete

Set-in-Place Piles   These are piles that are precast or preformed and are lowered into a prebored hole. The gap between the pile and the side of the hole is then filled with concrete. This method is often used for relatively low capacity piles, such as timber piles for pole houses, or piles for forming retaining walls.

12.4.2   Pile-Bearing Capacity — Basic Formula and Methods of Estimation

As indicated above in Figure 12.8, the vertical load capacity of a single pile is made up of two components and can be expressed as follows:

$$Q = qA + f_s A_s \tag{12.8}$$

where Q = ultimate (peak) load capacity (bearing capacity)
A = cross-sectional area of pile
A_s = embedded surface area of pile
q = ultimate (peak) end-bearing pressure
f_s = ultimate skin friction (shaft adhesion) per unit area on sides of pile

Estimation of pile-bearing capacity is essentially an exercise in estimating q and f_s. There are a variety of ways of determining these, either as separate components or in total; the principal ones are discussed next.

Use of Bearing Capacity Theory   In principle, the basic bearing capacity formula

$$q = cN_c + \gamma DN_q + \frac{1}{2}\gamma BN_\gamma$$

can be used to estimate the end resistance but is of no assistance in determining the skin friction. The use of this formula is acceptable for piles in clay when the design can be based on the undrained shear strength. However, there are severe limitations in applying the formula to piles in sand. First, the formula gives unrealistically high values for the term γDN_q. For example, for $\phi' = 40°$, $\gamma = 20 \, \text{kN/m}^3$, $D = 10 \, \text{m}$, $N_q = 100$ (approx), which gives $\gamma DN_q = 20 \, \text{MPa}$, which is an extremely high value. Second, field tests show that the end resistance does not conform to the above equation. The increase in end-bearing capacity with depth is less than the formula suggests and reaches a limiting (peak) value when $L \geq 10B$ (approximately), where L is the length and B the width or diameter of the pile.

Empirical Methods Based on Field Tests The use of empirical correlations between field tests and other soil properties is the most common method for estimating pile-bearing capacity. These will be discussed in more detail in Sections 12.4.3 and 12.4.4, dealing specifically with piles in either clays or sands.

Pile-Loading Tests Full-scale pile-loading tests can be undertaken for large projects in uncertain soil conditions when considerable cost savings may be possible and the expense involved in loading tests becomes acceptable. These tests are normally carried out using high-capacity hydraulic jacks to apply the load in a series of controlled stages. A reaction source to jack against is necessary and is one of the factors contributing to the high cost of pile-loading tests. A number of adjacent piles can be utilized for this purpose or a large stack of weights ("kentledge") provided on a specially built platform. If appropriate instrumentation is used, these tests can distinguish between the contributions to bearing capacity coming from end-bearing and skin friction. Pile-loading tests clearly provide the most reliable determination of pile capacity.

Methods Based on Behavior During Driving The simplest of these methods involves making use of the distance the pile moves downward per hammer blow. The height and weight of the hammer are readily known and the theoretical energy input per blow can easily be estimated. This energy is absorbed by the pile and is related to the pile resistance and the distance the pile moves. This distance is commonly called the set or the set/blow. A range of formulas have been developed making use of this "energy balance"; they are commonly known as pile-driving formulas. The best known and most widely used is probably the Hiley formula. The formulas are all fairly crude and should therefore be used with considerable caution. However, while they may not give very accurate estimates of pile capacity, the measurement of the set is very useful in controlling or checking the consistency of behavior between piles on a particular site.

In recent years, more sophisticated methods involving a pile-driving analyzer (PDA) have been developed. This makes use of wave propagation theory to analyze the response of the pile to the wave generated by the hammer blow at the surface. Strain and acceleration at the top of the pile are measured during driving using strain transducers and accelerometers attached to the top of the pile. The results are processed by a computer system which yields an estimate of the load-carrying capacity of the pile.

12.4.3 Bearing Capacity of Piles in Clay

Bearing capacity of piles in clay is normally based on undrained shear strength, in which case Equation 12.8 can be written as

$$Q = N_c s_u A + \alpha s_u A_s \qquad (12.9)$$

where N_c = end-bearing capacity factor
$\qquad s_u$ = undrained shear strength of clay
$\qquad \alpha$ = reduction factor (ratio of skin friction to undrained shear strength)

Point Resistance The first term in the above expression is the point resistance. The value of N_c is normally taken as 9, in accordance with the value indicated in Figure 12.4 for a deep circular foundation. The undrained shear strength can be determined in a variety of ways, including laboratory tests or in situ field measurements, as described in Section 10.5. It should be noted that the cone resistance from the cone penetration test (CPT) in clay cannot be used directly as a measure of the point resistance of a pile. As described in Chapter 10, the conversion factor from cone point resistance to undrained shear strength is about 15, much greater than the theoretical value of 9.

Skin Friction The main unknown in Equation 12.9 is the value of α, the coefficient governing the skin friction acting on the side surfaces of the pile. A number of attempts have been made to provide empirical tables or graphs from which to obtain the value of α. These attempts are based on the analysis of data from actual pile load tests at sites where s_u has been reliably measured. This analysis shows that the value of α is close to unity in soft clays (i.e., the skin friction equals the undrained shear strength) but decreases steadily as the soil becomes harder. These correlations therefore relate α to the undrained shear strength. Several well-known correlations are shown in Figure 12.12 (from McClelland, 1974).

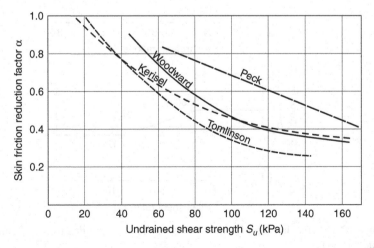

Figure 12.12 Skin friction reduction factors for piles in clay. (After McClelland, 1974.)

More recently, Semple and Rigden (1984) have reviewed available data and concluded that α is dependent on the pile dimensions as well as the value of s_u. They express α as the product of two coefficients, F and α_p.

Thus

$$\alpha = \alpha_p F$$

where α_p = coefficient dependent on s_u/σ'_v (σ'_v = effective vertical stress)
F = coefficient dependent on L/D ratio (L = pile length, D = pile diameter)

The values of these two coefficients are shown in Figure 12.13.

These graphs indicate that for a clay of particular undrained shear strength the skin friction factor decreases as the ratio of pile length to pile diameter increases as well as decreases as the soil becomes harder. For a very long pile, the skin friction is only 70 percent of that for a short pile in the same soil.

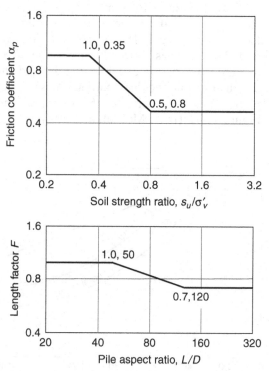

Figure 12.13   Further skin friction reduction factors for piles in clay (Semple and Rigden 1984).

12.4.4   Bearing Capacity of Piles in Sand

As already mentioned, conventional bearing capacity theory does not yield reliable results when applied to piles in sand. Empirical methods, or correlations, based on the results of standard penetration tests (SPTs) or CPTs are widely used in practice.

Point Resistance   The Dutch CPT was originally devised in Holland to function as a "model pile" with the hope that the point resistance from the test would give a direct measure of the point resistance to be expected with an actual pile. Field testing has shown this to be more or less correct (for sand).
Thus

$$Q_b = Aq_c$$

where Q_b = end-bearing capacity of pile
q_c = cone resistance from CPT tes.
A = area of base of pile

If SPTs are used, then a correlation between N and q_c can be used to convert N values to equivalent q_c values. In accordance with Figure 10.11, it is common practice to take $q_c = 400N$ (where q_c is in kilopascals).

Skin Friction   For piles in sand, the end resistance is usually much greater than the contribution from skin friction, and in some cases the latter may be ignored. Meyerhof (1976) suggested the correlations given in Table 12.1 (for both end resistance and skin friction) based on the results of SPTs. The SPT values have been converted to CPT values using the correlation given in the previous section ($q_c = 400N$). Meyerhof suggests these correlations

Table 12.1   Driven Piles in Sands: Values of End Resistance and Skin Friction from SPT or CPT Values

	Ultimate (Peak) End Resistance (kPa)		Ultimate (Peak) Skin Friction (kPa)	
	Sand and Gravel	Silty Sand and Silt	Average Size Pile	Small-Displacement Piles
SPT N value	$40N\dfrac{L}{D} \le 400N$	$30N\dfrac{L}{D} \le 300N$	$2N$	N
Dutch CPT point resistance q_c	$\dfrac{q_c}{10}\dfrac{L}{D} \le q_c$		Dense Sand $\dfrac{q_c}{200}$	Loose Sand $\dfrac{q_c}{400}$

for conventional driven precast piles. For bored piles, both end resistance and skin friction are lower than for driven piles. Meyerhof suggests that the end resistance and skin friction are about one-third of the values for driven piles.

It is sometimes suggested that a more "fundamental" theoretical approach should be used to estimate the skin friction on piles. In theory, the skin friction is given by

$$f = K\sigma'_v \tan \phi'_p \tag{12.10}$$

where σ'_v = effective vertical stress
K = ratio of effective horizontal to effective vertical stress
ϕ'_p = friction angle between pile and soil

This equation is theoretically sound, but there is great uncertainty regarding the value of K and some uncertainty about ϕ'_p. For driven piles, the driving process displaces and "densifies" the soil, tending to increase the horizontal stress. The K value could be as high as 2 or 3. On the other hand, for bored, cast-in-place piles K may be only 0.5, but the actual value is uncertain. Thus the method is difficult to apply in practice despite its theoretical appeal.

12.4.5 Pile Group Behavior

The load-bearing capacity of a pile group is not necessarily the sum of the capacity of the individual piles. This is because the piles no longer act as individual piles unless the spacing between piles is very large, which is not normally the case. The capacity of the pile group may be greater or less than the sum of the individual piles, depending on the soil conditions and the spacing of the piles. The situation is different for piles in clay and sand, as described in the following sections.

Piles in Sand (or Gravel) and End-Bearing Piles Generally In this case, the load-bearing capacity of a pile group can be expected to be at least equal to the sum of the bearing capacity of each pile. In practice, piles driven at close spacing into a sand layer will result in some compaction or "densifying" of the sand and thus increase the capacity of each individual pile and of the group. In practice, this effect is not easily quantifiable and is ignored.

Piles in Clay For piles in clay, it is necessary to consider the bearing capacity of the pile group as a whole as well as the capacity of the individual piles. The block of soil formed by the pile group is treated as a single entity and analyzed as a foundation, as indicated in Figure 12.14. Here, M and N are the width and length of the pile group.

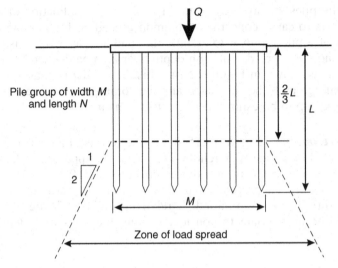

Figure 12.14   Estimation of bearing capacity and settlement of a pile group in clay.

The bearing capacity of a single pile is given by

$$Q_i = 9s_u A + \alpha s_u A_s$$

The bearing capacity of the pile group, treating it as a solid block, is given by

$$Q_g = 9s_u A_g + A_p s_u$$

where A_g = base area of group
A_p = perimeter area of block

Thus

$$Q_g = 9s_u MN + 2(M + N)L s_u \qquad (12.11)$$

Alternatively, the group capacity can be taken as the sum of individual piles to give $Q_g = nQ_i$ where n is the number of piles and Q_i the bearing capacity of a single pile.

The lesser value of Q_g given by the two calculations is taken as the true value.

Settlement of Piles   For most routine design situations, the settlement of piles is sufficiently small to be of little or no practical significance. With large-diameter piles carrying very heavy loads settlement may be of concern, although this is the exception rather than the rule. Pile groups, especially piles in clay, may result in sufficient consolidation of the deeper clay layers to cause concern. A common procedure for estimating the settlement of a pile group is to assume the load acts at two thirds of the pile length and then spreads out with depth over an area defined by 1:2 boundary lines, as shown in Figure 12.14. This allows the increase in stress due to the pile group to be calculated and the compression of the layer beneath the pile group to be estimated in the usual manner.

Worked Example Figure 12.15 shows the results of a CPT and soil types as indicated by a nearby borehole. Determine an appropriate depth and the ultimate bearing capacity of a driven concrete pile, 0.4m by 0.4m square.

Selection of an appropriate depth is not straightforward in this case as the soil conditions are quite varied. If relatively light loads are to be carried, it might be appropriate to consider driving the piles to a depth of only

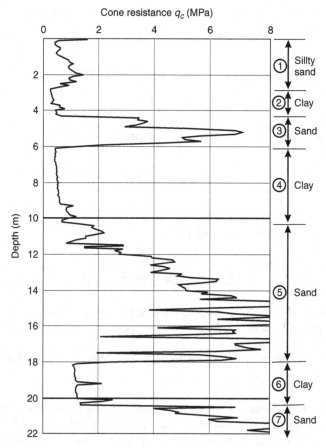

Figure 12.15 Soil conditions for worked example.

Table 12.2 Estimation of pile bearing capacity

Layer	Soil Type	Thickness (m)	q_c (MPa)	$s_u = q_c/15$ (kPa)	Factor α	Skin Friction (kPa)	Shaft (m²) Area(m²)	$F_S A_S$ (kN)
1	Sand	2.9	0.6	—	—	1.5	4.64	7.0
2	Clay	1.4	0.3	20	0.95	19	2.24	42.5
3	Sand	1.8	4.0	—	—	10	2.88	28.8
4	Clay	4.2	0.5	33.3	0.83	28	6.72	188.2
5	Sand	4.7	5.0	—	—	12.5	7.52	94.0

Total: 360.5 kN

5 m so they would be founded in the sand layer found between 4 and 6 m. However, if reasonably high bearing capacity is required, it would be much more appropriate to found the piles in the second sand layer encountered between 12 and 18 m. We will adopt a depth of 15 m, which is sufficiently deep into the layer but not in danger of significant influence from the clay layer found immediately below it.

Calculation of point resistance:

We can take the average cone resistance as 5 MPa and

Cross-sectional area $= 0.4 \times 0.4 = 0.16\,\text{m}^2$

End bearing capacity $= 0.16 \times 5000\,\text{kN} = 800\,\text{kN}$

Calculation of skin friction:

Table 12.2 sets out the calculation, making use of the relationships given in Figure 12.11 (for the clay layers) and Table 12.1 (for the sand layers).

For the sand layers, we have assumed $f_s = q_c/400$, in keeping with Table 12.1 for a loose sand. Figure 10.12 indicates this to be the case. The ultimate bearing capacity of the pile is the sum of the end resistance and skin friction: $800 + 360.5 = 1160.5\,\text{kN}$.

12.4.6 Lateral Load Capacity of Piles

Most piles are required to resist some lateral load, in addition to their vertical load. In some situations, the lateral load may be negligible in comparison to the vertical load; in other situations the reverse may be true. Piles used to construct retaining walls, for example, are only subject to the horizontal (or lateral) load coming from the earth pressure. Pile design with respect to lateral load must meet the following requirements:

1. The pile must have adequate lateral load capacity to resist the applied load without danger of failure.

2. The deformation (or lateral deflection) of the pile under the applied load must be within acceptable limits.

With many piles, only the first requirement is of significance; this is the case with conventional pile retaining walls. These are free-standing structures and any deformation occurring is unlikely to have undesirable effects. Their design therefore does not normally involve a consideration of lateral deflection. Timber pile retaining walls are a common example of this type of wall, although concrete piles are also used in the same way. The term "pole" is normally used for such piles.

Piles used to support buildings are primarily vertical load-carrying members, and their lateral load capacity is usually a secondary consideration. Wind and earthquake activity, however, can be sources of considerable lateral force that needs to be taken account of in design. It is unlikely that deflection is an important issue in most cases. For these reasons only lateral load capacity will be considered here; deflection issues will not be covered.

Pile Behavior Under Lateral Loads Consider the behavior of a pile subjected to a steadily increasing horizontal load as shown in Figure 12.16.

As the force on the pole increases, the pole is pushed toward the soil and pressure builds up on the side of the pole being pushed against the soil. To maintain static equilibrium, pressure also builds up on the opposite side. One of two possible failure modes eventually occurs. Either the pole will break, as indicated, or the pile will be pushed sideways through the soil. A pole which fails by breaking is termed a **long pile** and a pole which fails by moving through the soil (i.e., causes failure of the soil) is termed a **short pile**. For long piles the ultimate lateral load capacity is governed by

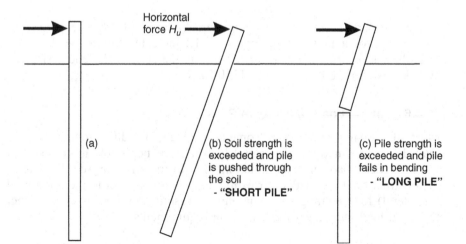

Figure 12.16 Failure modes of a pile under lateral load.

the strength of the pile, while for short piles it is governed by the strength of the soil. Broms (1965) gives a comprehensive account of the behavior of laterally loaded piles and presents methods that can be used in their design. Broms's treatment is largely followed here. It should be noted that a pole can be both a long and short pile since it can be designed so that the pile will be on the point of breaking at the same time that it is on the point of being pushed sideways through the soil.

Ultimate Lateral Load Capacity   To calculate the ultimate lateral load of a pile, it is necessary to know the following:

- Dimensions and strength of the pile, in particular its length and diameter, and its ultimate bending moment
- Soil properties and the way they relate to the pressure which the soil will exert on the pile

Figure 12.17 illustrates the expected soil pressure distribution acting on short and long piles in clay and sand when subjected to lateral load. In the case of clay, the behavior is assumed to be undrained. According to Broms (1964), experimental tests show that the net maximum pressure on the pile rises to between $8s_u$ and $12s_u$, where s_u is the undrained shear strength of the clay. The average value is about $9s_u$. This is comparable to a value of $9s_u$ for the base resistance of a deep circular footing. Some similarity in the values is expected as the pile acts like a deep strip footing being pushed sideways through the soil.

To maintain moment equilibrium, the soil pressure on the lower part of the pile will be in the opposite direction to that higher up, as the figure

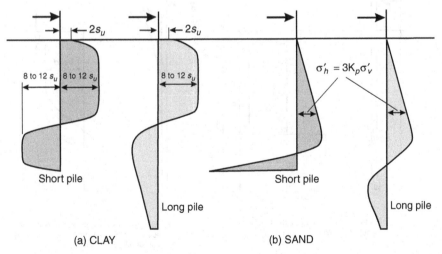

Figure 12.17   Soil pressure on piles in clay and sand under lateral load.

indicates. If the pile behaves as a short pile, the clay will be brought to a failure state both along the upper part of the pile and over the lower section on the reverse side. If, however, the pile behaves as a long pile, the pile will fail by breaking before it has fully mobilized the strength of the soil.

In the case of sand, the pressure distribution is significantly different. Experimental work shows that the lateral pressure increases linearly with depth below the surface, at least for some distance. Experimental work also shows that the average magnitude of this lateral pressure is given by:

$$\sigma'_h = 3K_p \sigma'_v$$

where K_p is the passive pressure coefficient of the sand, the meaning of which is described in Chapter 13.

$$\sigma'_h \text{ and } \sigma'_v$$

are the effective horizontal and vertical pressures, respectively.

Near the base of the pile, the pressure acting on the pile reverses in direction, in the same way as it did with clay. The pressure distributions in Figure 12.17 make analysis somewhat difficult, and to overcome this, Broms suggested some simplifying assumptions for both clay and sand. These are presented and analyzed in the following sections.

Analysis of Piles in Clay   Figure 12.18 illustrates the simplifying assumptions of Broms. In this figure, d is the pile diameter, f is the distance

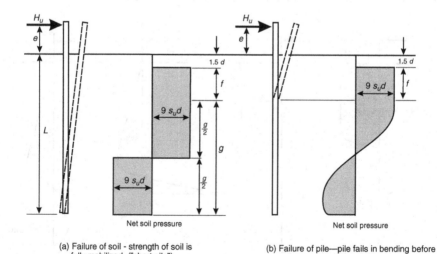

(a) Failure of soil - strength of soil is fully mobilized ("short pile")

(b) Failure of pile—pile fails in bending before soil strength is fully mobilized over lower part of pile ("long pile")

Figure 12.18   Simplifying assumptions made for analysis of piles in clay.

below the start of the pressure zone to the point of maximum moment, and the remaining terms are self-explanatory. To take account of the lower lateral pressure near the surface, the pressure over the top $1.5d$ is completely ignored. This also makes some allowance for softening or even shrinkage of the soil away from the pile that may occur near the ground surface. Analysis of the moment equilibrium of the upper part of either pile gives us a relationship between the horizontal force, the pile diameter and moment, and the strength of the soil. Analysis of the static equilibrium of the short pile gives us a relationship between the horizontal force, the pile dimensions (both diameter and length), and the strength of the soil.

The depth f defines the location of the maximum moment, and since the shear force here is zero, it follows that

$$H_u = 9s_u df \quad \text{and} \quad f = \frac{H_u}{9s_u d} \qquad (12.12)$$

Taking moments about the point of maximum moment gives

$$M = H_u \left(e + 1.5d + f\right) - (9s_u df)\frac{f}{2} \quad \text{so that} \quad M = H_u \left(e + 1.5d + \frac{f}{2}\right) \qquad (12.13)$$

These expressions are common to both long and short piles, as they only involve the upper part of the pile where the pressure and equilibrium situations are identical. Substituting for f gives

$$M = H_u \left(e + 1.5d + \frac{H_u}{18s_u d}\right) \qquad (12.14)$$

This is a simple expression relating the moment in the pile to the horizontal force H_u. It is valid for short and long piles and does not involve L, the pile length. We can rearrange this to obtain H_u in terms of the moment:

$$H_u^2 + 18S_u d \left(e + 1.5d\right) H_u - 18s_u dM = 0$$

Solving for H_u gives

$$H_u = 3s_u d \left\{ \sqrt{9\left(e + 1.5d\right)^2 + \frac{2M}{s_u d}} - 3\left(e + 1.5d\right) \right\}$$

and dividing through by $S_u d^2$ and replacing M with the ultimate (failure) moment M_y give

$$\frac{H_u}{s_u d^2} = 3 \left\{ \sqrt{9\left(\frac{e}{d} + 1.5\right)^2 + \frac{2M_y}{s_u d^3}} - 3\left(\frac{e}{d} + 1.5\right) \right\} \qquad (12.15)$$

This dimensionless equation relates the ultimate load the pile can take (H_u) to the ultimate bending moment capacity of the pile (M_y) and thus gives the ultimate capacity of a long pile. This relationship is presented by Broms in the form of a chart suitable for use in design. This is shown in Figure 12.20a. It should be noted, however, that while these charts are useful, their scale is such that they are not very accurate for some pile combinations of size and soil strength, and it is often desirable to make use of the equations rather than the charts to obtain sufficient accuracy.

For short piles we need a similar relationship involving the length of the pile, but not its moment, since it is assumed that failure will not occur in the pile. We can do this by considering horizontal force equilibrium and moment equilibrium of the pile. To simplify this analysis, it is convenient to slightly alter the force equilibrium diagram, as is done in Figure 12.19.

The block of pressure on the right side of the pile is divided into two blocks; the upper one has the same dimension as the pressure block on the left side of the pile. This thus creates two pressure components—one central component of height β and two equal components of height x acting in opposite directions. The point O is at the center of the central component. The analysis is further simplified by putting $e' = e + 1.5d$ and $L' = L - 1.5d$.

Resolving horizontally gives

$$H_u = 9s_u d\beta \text{ and } \beta = \frac{H_u}{9s_u d} \tag{12.16}$$

Moments about O give

$$H_u \left(e' + x + \frac{\beta}{2} \right) = 2 \times 9s_u d . x \left(\frac{x}{2} + \frac{\beta}{2} \right)$$

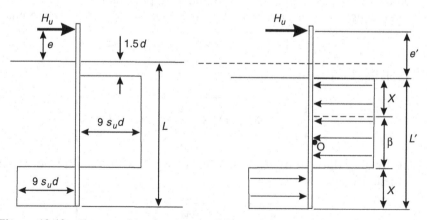

Figure 12.19 Force and earth pressure diagram for static equilibrium analysis of short piles.

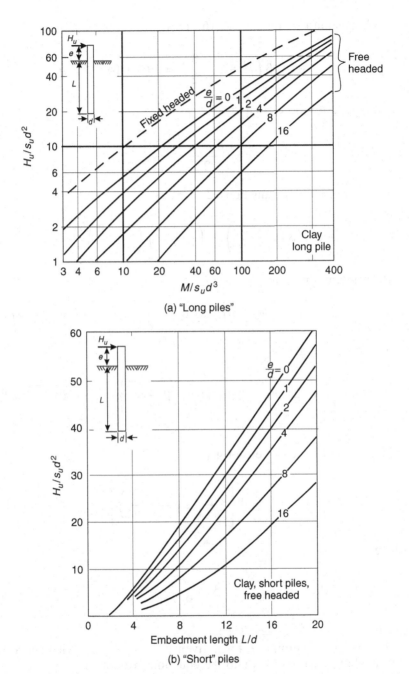

Figure 12.20 Charts for estimation of lateral load capacity of piles in clay. (After Broms, 1964.)

And substituting Equation 12.16 gives

$$\beta\left(e' + x + \frac{\beta}{2}\right) = x\,(x + \beta) \tag{12.17}$$

Now $2x + \beta = L'$ so that

$$x = \frac{L' - \beta}{2} = \frac{L'}{2} - \frac{\beta}{2}$$

Substituting this in Equation 12.17 gives

$$\beta\left(e' + \frac{L'}{2}\right) = \left(\frac{L' - \beta}{2}\right)\left(\frac{L' + \beta}{2}\right)$$

Rearranging this gives

$$L'^2 - 2\beta L' - \left(4e'\beta + \beta^2\right) = 0 \tag{12.18}$$

Solving for L' gives

$$L' = \frac{1}{2}\left\{-(-2\beta) \pm \sqrt{4\beta^2 - 4\left(-4e'\beta - \beta^2\right)}\right\} \beta \pm \frac{1}{2}\sqrt{8\beta^2 + 16e'\beta}$$

Neglecting the negative root gives

$$L' = \beta\left\{1 + \sqrt{2\left(1 + \frac{2e'}{\beta}\right)}\right\} \tag{12.19}$$

where

$$\beta = \frac{H_u}{9s_u d}$$

and $L' = L - 1.5d \quad e' = e + 1.5d$

We can thus determine L for a given value of H_u and vice versa. If we put the above relationships into the equation, we obtain

$$L - 1.5d = \frac{H_u}{9s_u d}\left\{1 + \sqrt{2\left[1 + \frac{2(e + 1.5d)}{H_u/9s_u d}\right]}\right\} \tag{12.20}$$

Or

$$\frac{L}{d} - 1.5 = \frac{H_u}{9s_u d^2} \left\{ 1 + \sqrt{2 \left[1 + \frac{2(e/d + 1.5)}{H_u/9s_u d^2} \right]} \right\} \qquad (12.21)$$

This is the dimensionless form used by Broms for his chart, presented in Figure 12.20b.

Analysis of Piles in Sand Figure 12.21 shows the simplifying assumptions for sand. The two blocks of pressure on each side of the short pile are replaced with a simple triangle of pressure on one side and a concentrated force at the base on the other side. This is a common assumption made in the analysis of sheet piles and is encountered again in Chapter 13. Broms adopts the assumption that the net soil resistance acting on the pile is given by $\sigma'_h = 3\sigma'_v K_p$ where σ'_v is vertical pressure and K_p is the Rankine passive pressure coefficient given by

$$K_p = \frac{1 + \sin \phi'}{1 - \sin \phi'} \text{ (explained in Chapter 13).}$$

The factor of 3 is based on experimental data and in effect states that the soil resistance is three times the normal passive pressure, an effect which arises from the 3-D nature of the pole–soil interaction.

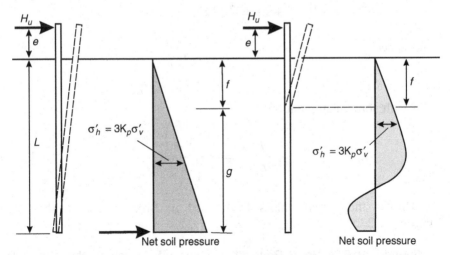

(a) Failure of soil—soil strength is fully mobilized ("short pile")

(b) Failure of pile—pile fails in bending before strength of soil is fully mobilized over lower part of pile ("long pile")

Figure 12.21 Simplifying assumptions for analysis of piles in sand.

As with the pile in clay, the relationship between pile-bending moment and the applied force H_u is the same for both the long and the short pile. Taking f as the depth to the point of maximum bending moment, the following relevant equations are easily derived:

$$H_u = \frac{3}{2}\gamma dK_p f^2 \quad \text{and} \quad M = H_u\left(e + \frac{2}{3}f\right)$$

Eliminating f from these equations leads to the expression

$$(M - eH_u)^2 = \frac{8}{27}\frac{H_u^3}{\gamma dK_p} \tag{12.22}$$

For a long pile the moment M will be the ultimate (failure) bending moment, M_y, of the pile and thus we can write

$$\left(M_y - eH_u\right)^2 = \frac{8}{27}\frac{H_u^3}{\gamma dK_p} \tag{12.23}$$

This is a cubic which can be solved for H_u in terms of M_y, e, γ, d, and K_p. Broms has produced a solution in dimensionless form and presented it in his chart, which is given in Figure 12.22a.

For **short piles**, the value of H_u is easily obtained by taking moments about the pile tip:

$$H_u\,(e + L) = \frac{1}{2}\left(3\gamma K_p Ld\right) \times L \times \frac{L}{3}$$

and hence

$$H_u = \frac{0.5\gamma dL^3 K_p}{e + L}$$

and with some manipulation

$$\frac{H_u}{\gamma K_p d^3} = \frac{0.5\left(\dfrac{L}{d}\right)^2}{\dfrac{e}{L} + 1} \tag{12.24}$$

This is the dimensionless form used in Broms's charts, shown in Figure 12.22b.

Influence of Pile Head Restraint   Some piles, such as those used for retaining walls (and considered above), are not restrained at the top and are termed **free headed**. Other piles, especially those used to support buildings, are connected at the ground surface to large foundation blocks or pile "caps"

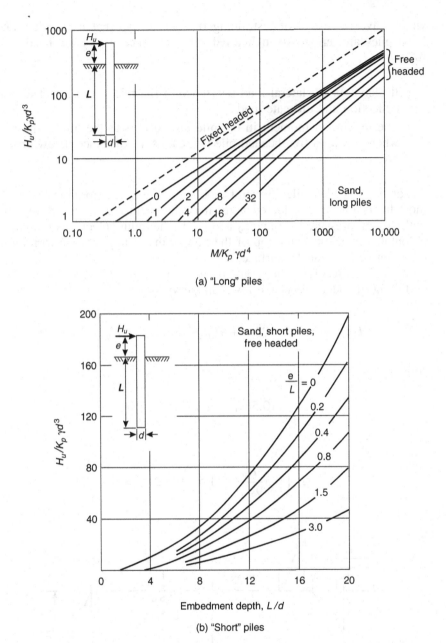

(a) "Long" piles

(b) "Short" piles

Figure 12.22 Charts for estimation of lateral load capacity of piles in sand. (After Broms, 1964.)

which have the effect of restraining the pile top so that it cannot rotate. Such piles are termed **fixed headed**. The influence of a fixed head is as follows:

- It increases the lateral load resistance of the pile with respect to both deflection and ultimate load.
- The maximum moment in the pile now occurs at the top of the pile, whereas with a free-headed pile it occurs some distance below ground level.

For fixed-headed piles, several modes of failure are possible. Only the most likely one is considered here; it is shown in Figure 12.23, together with the soil pressure in both clay and sand. The failure mechanism involves a double "hinge" near the top of the pile, with the upper hinge forming at the point of fixity at the surface.

With clays, resolving horizontally gives $H_u = 9s_u \, df$.

Taking moments about the lower hinge yields

$$H_u (f + 1.5d) = 2M_y + 9s_u df \frac{f}{2} = 2M_y + H_u \frac{f}{2}$$

so that

$$H_u (0.5f + 1.5d) - 2M_y = 0$$

or

$$H_u \left(0.5 \frac{H_u}{9s_u d} + 1.5d \right) - 2M_y = 0$$

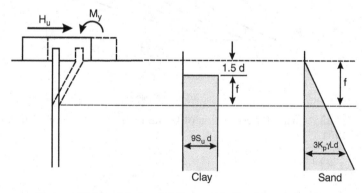

Figure 12.23 Mode of failure with fixed-headed piles.

Rearranging this gives

$$H_u^2 + 27s_u d^2 H_u - 36s_u d M_y = 0$$

and solving for H_u gives

$$H_u = \frac{1}{2}\left[-27s_u d^2 \pm \sqrt{(27s_u d^2)^2 + 4 \times 36s_u d M_y}\right]$$

Neglecting the negative root, dividing through by $S_u d^2$, and rearranging give

$$\frac{H_u}{s_u d^2} = 13.5\left[\sqrt{1 + \frac{16M_y}{81s_u d^3}} - 1\right] \tag{12.25}$$

This is the dimenionless form used by Broms in his charts. With sands, resolving horizontally gives

$$H_u = \frac{1}{2}\left(3\gamma f d K_p\right) \times f = \frac{3}{2}\gamma d K_p f^2$$

and taking moments about the lower hinge yields

$$H_u f = 2M_y + \frac{3}{2}\gamma d K_p f^2 \times \frac{f}{3} = 2M_y + H_u\frac{f}{3}$$

Hence

$$M_y = \frac{1}{3}H_u f \quad or \quad f = 3\frac{M_y}{H_u} \text{ and}$$

$$\left(\frac{3M_y}{H_u}\right)^2 = f^2 = \frac{H_u}{3/2\gamma d K_p}$$

Rearranging this gives

$$M_y^2 = \frac{2}{27}\frac{H_u^3}{\gamma d K_p}$$

so that

$$\left(\frac{M_y}{\gamma d^4 K_p}\right)^2 = \frac{2}{27}\left(\frac{H_u}{\gamma d^3 K_p}\right)^3 \tag{12.26}$$

This is the dimensionless form used by Broms in his charts.

Worked Example   A retaining wall is to be built using timber piles (often termed poles rather than piles). Each pole is required to withstand a horizontal force of 15 kN acting 0.6 m above ground level. The soil consists of clay having an undrained shear strength (s_u) of 60 kPa. Timber poles are available with the following characteristics:

Pole diameter (mm)	200	250	300	350	400
Ultimate (failure) bending moment (kNm)	12.9	35.3	43.7	69.3	103.5

It is preferable for a pole to fail as a short pile rather than as a long pile, as this mode of failure is very unlikely to occur suddenly. It will be preceded by warning signs as the pole starts to lean. Failure as a long pile may well be sudden without any warning signs. We will therefore use a safety factor of 2.5 with respect to the pole strength and 2.0 with respect to the soil strength. Note that although we are designing the pole to fail as a short pile, we must still ensure it has the necessary strength to perform as a long pile. The design method involves a trial process to determine the required diameter of the pole.

Let us try a 250-mm pole. The allowable bending moment in the pole is 35.3/2.5 =14.1 kNm. The allowable or mobilized undrained shear strength of the soil is 60/2 = 30 kPa.

We will first do the design using the Broms charts:

$$\frac{H_u}{s_u d^2} = \frac{15}{30\,(0.25)^2} = 8.0$$

and $e/d = 0.6/0.25 = 2.4$. From Figure 12.18a we obtain

$$\frac{M}{s_u d^3} = 35$$

so that $M = 35 \times 30 \times (0.25)^3 = 16.4$ kN-m. This exceeds the allowable value of 14.1 kNm, so we must try the next size up, that is, the 0.30-m-diameter pole. The allowable moment is now $43.7/2.5 = 17.5$ kNm.

$$\frac{H_u}{s_u d^2} = \frac{15}{30\,(0.30)^2} = 5.6$$

and $e/d = 0.6/0.3 = 2.0$. From Figure 12.18a we obtain

$$\frac{M}{s_u d^3} = 21$$

so that $M = 21 \times 30 \times (0.3)^3 = 17.0$, which is less than the allowable value of $17.5\,\text{kNm}$, so the 0.30-m size is satisfactory. We can now go on to determine the required embedment depth using Figure 12.18b.

$$\frac{H_u}{s_u d^2} = 5.6$$

and $e/d = 2.0$, and from the chart we obtain $L/d = 5.0$ (approximately). This gives $L = 5.0 \times 0.3 = 1.5\,\text{m}$.

We will now repeat the analysis using the equations to obtain a more exact solution:

The bending moment in the 250 mm dia. pole is given by Equation 12.14:

$$M = H_u \left(e + 1.5d + \frac{H_u}{18 s_u d}\right) = 15(0.6 + 0.375 + 0.111) = 16.3\,\text{kNm}$$

This exceeds the allowable value in the pole so we must try the next size, 300 mm. The allowable moment in this pole is $43.7/2.5 = 17.5\,\text{kNm}$.

The bending moment in the pole is now $15(0.6 + 0.45 + 0.093) = 17.1\,\text{kNm}$. This is less than the allowable, so we can adopt this size.

The required pole embedment depth is given by Equation 12.20, which we can write as

$$L - 1.5d = \beta \left\{1 + \sqrt{2\left[1 + \frac{2(e + 1.5d)}{\beta}\right]}\right\} \quad \text{where } \beta = \frac{H_u}{9 s_u d}$$

This gives $L = 1.48\,\text{m}$.

We find that in this case there is good agreement between the results from the charts and those using the formulas. However, this is somewhat fortuitous, as it depends on our judgment in using the charts, especially in interpolating values on the logarithmic scales.

It should be noted that in practice some timber piles are driven into the ground while others are concreted into predrilled holes. In the latter case the above calculation will be conservative, as the concrete increases both the effective diameter of the pile as well as its bending moment capacity. The calculations could be refined to take account of this.

REFERENCES

Broms, B. B. 1965. Design of laterally loaded piles. *ASCE Journal of Soil Mechanics and Foundations Division*, 91 No (SM3),. pp 79–99.

McClelland, B. 1974. Design of deep penetration piles for ocean structures. *ASCE J. Geotech. Eng. Div.*, Vol. GT7, pp. 705–747.

Meyerhof, G. G. 1963. Some recent research on the bearing capacity of foundations. *Can. Geotech.l J.l*, Vol. 1, No. 1, pp. 16–26.

———. 1976. Bearing capacity and settlement of pile foundations. *Proceedings ASCE*, Vol. 82, No. SM1, pp. 1–19.

Semple R. M., and W. J. Rigden. 1984. Shaft capacity of driven piles in clay. In *Analysis and Design of Pile Foundations*. Reston, VA: American Society of Civil Engineers.

Skempton, A. W. 1951. The bearing capacity of clays. *Proceedings Building Research Congress*, Vol. 1, pp. 180–189.

Terzaghi, K. 1943. *Theoretical Soil Mechanics*. New York: John Wiley and Sons.

Terzaghi, K., and R. B. Peck. 1967. *Soil Mechanics in Engineering Practice* (2nd ed.). New York: John Wiley and Sons.

EXERCISES

1. Determine the allowable bearing capacity of a strip foundation 0.6 m wide and 1.0 m deep in clay having the following properties: $s_u = 85$ kPa, unit weight $= 16.2$ kN/m^3. Assume the clay is fully saturated and use a safety factor of 3. (**209 kPa**)

2. A structure is supported by a large surface foundation 20 m square and 1 m deep on clay having a unit weight of 18.5 kN/m^3 and $s_u = 220$ kPa. The structure has a total weight of 10,500 tonnes with its center of gravity at the center of the foundation 7.5 m above ground level. The structure must be able to resist a horizontal earthquake force equivalent to $0.3g$. Determine the safety factors under static loading and during earthquake loading. (**2.85, 1.86**)

3. A circular structure 8 m in diameter and weighing 800 tonnes is to be built on soft clay. The clay has a unit weight of 16.2 kN/m^3 and undrained shear strength (S_u) of 18 kPa. Determine the depth at which the structure would need to be founded to have a safety factor of 3. (**6.82 m**)

4. Figure 12.24 shows the result of a CPT carried out at a site consisting of clay to a depth of about 13 m followed by sand layers down to at least 20 m. Determine the ultimate bearing capacity of a 0.4-m^2 concrete pile driven to a depth of 17 m. (**1480 kN**).

5. Please note that this is a "joint" question requiring knowledge of material from both Chapters 12 and 13.

   A timber pole retaining wall has been built to a height of 2 m to retain level ground. The retained soil is essentially noncohesive with properties $\gamma = 18$ kN/m^2, $c' = 0$, $\phi' = 32°$. The poles are 0.25 m in

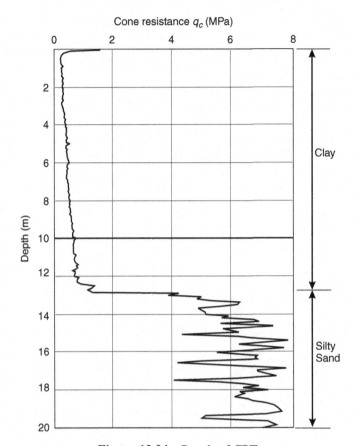

Figure 12.24   Result of CPT.

diameter and have an ultimate bending moment capacity of 40 kNm. They are spaced at 1.2-m intervals and are embedded in clay to a depth of 1.5 m. The clay has an undrained shear strength of 60 kPa.

Determine the safety factors with respect to failure as a long pile (pile fracture) and failure as a short pile (sideways push through the soil). **(1.84, 2.50)**

CHAPTER 13

EARTH PRESSURE AND RETAINING WALLS

13.1 COULOMB WEDGE ANALYSIS

As discussed in Chapter 11, the assumed failure mechanism governing the force that acts on a retaining wall is a wedge-shaped soil mass sliding on an inclined failure surface. In the Coulomb method of analysis, the assumption is made that the failure surface is a plane. This concept is illustrated in Figure 13.1.

The disturbing force is the weight of the soil, and the resisting forces are made up of the soil shear strength on the failure plane plus the force provided by the wall. It is assumed that the soil shear strength is fully mobilized. For simplicity, Figure 13.1 is for a vertical frictionless wall and a level ground surface behind the wall. It is also assumed that the soil has strength parameters c' and ϕ', and there are no pore pressures or seepage effects involved.

The polygon of forces acting on the soil wedge, assuming the failure plane is at an angle α to the horizontal, is also shown in the figure. The shear strength of the soil is made up of a cohesive and a frictional component. The magnitude and direction of the cohesive component are known, but only the direction of the frictional component is known. The magnitude of the force P_a is obtained as follows:

$$\frac{FD}{DC} = \tan\left(\alpha - \phi'\right)$$

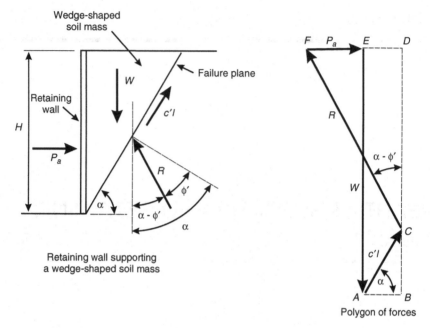

Figure 13.1 Coulomb method for determining active earth pressure on a retaining wall.

$$FD = FE + ED = P_a + c'l \cos \alpha$$

$$DC = DB - CB = W - c'l \sin \alpha$$

where l is the length of the failure surface. Therefore

$$P_a + c'l \cos \alpha = \left(W - c'l \sin \alpha \right) \tan \left(\alpha - \phi' \right) \qquad (13.1)$$

From the geometry of the wedge (assuming the soil unit weight as γ) we can also write

$$W = \frac{\gamma H^2}{2 \tan \alpha} \qquad (13.2)$$

$$l = \frac{H}{\sin \alpha} \qquad (13.3)$$

Hence

$$P_a = \left(\frac{\gamma H^2}{2 \tan \alpha} - c'H \right) \tan \left(\alpha - \phi' \right) - \frac{c'H}{\tan \alpha} \qquad (13.4)$$

In the situation in Figure 13.1, the proximity of the frictionless wall and the ground surface means that the horizontal and vertical stresses approximate to principal stresses, so that the inclination of the failure plane can be taken as $\alpha = 45 + \frac{\phi'}{2}$. Substituting this into Equation 13.4, we obtain

$$P_a = \left[\frac{\gamma H^2 \tan\left(45 - \frac{\phi'}{2}\right)}{2 \tan\left(45 + \frac{\phi'}{2}\right)} \right] - c'H \left[\tan\left(45 - \frac{\phi'}{2}\right) + \frac{1}{\tan\left(45 + \frac{\phi'}{2}\right)} \right]$$

(13.5)

which gives

$$P_a = \frac{1}{2}\gamma H^2 \tan^2\left(45 - \frac{\phi'}{2}\right) - 2c'H \tan\left(45 - \frac{\phi'}{2}\right)$$

(13.6)

We can note the following relationships:

$$\tan^2\left(45 - \frac{\phi'}{2}\right) = \frac{1 - \sin\phi'}{1 + \sin\phi'}$$

$$\tan\left(45 - \frac{\phi'}{2}\right) = \frac{\cos\phi'}{1 + \sin\phi'}$$

and Equation 13.6 can be written in the form

$$P_a = \frac{1}{2}\gamma H^2 \left(\frac{1 - \sin\phi'}{1 + \sin\phi'}\right) - 2c'H \left(\frac{\cos\phi'}{1 + \sin\phi'}\right)$$

(13.7)

The above analysis implies some yield of the wall in order to fully mobilize the shear strength of the soil. The soil is tending to push the wall in the horizontal direction, and this case is known as the **active pressure state**.

In some situations, such as anchor blocks in the ground, the situation is reversed; the structure is pushing against the soil, tending to fail the wedge in the upward direction, as illustrated in Figure 13.2. This case is known as the **passive pressure state**. Analysis of the equilibrium situation in the same manner as above for the active case leads to the following expression for the passive force P_p:

$$P_p = \frac{1}{2}\gamma H^2 \left(\frac{1 + \sin\phi'}{1 - \sin\phi'}\right) + 2c'H \left(\frac{\cos\phi'}{1 - \sin\phi'}\right)$$

(13.8)

It will be evident, both intuitively and from the form of the force polygons and Equations 13.7 and 13.8, that the passive force is very much larger than the active force. Equations 13.7 and 13.8 can also be written in the following forms:

$$P_a = \frac{1}{2}K_a \gamma H^2 - 2c'\sqrt{K_a}\,H$$

(13.9)

$$P_p = \frac{1}{2}K_p \gamma H^2 + 2c'\sqrt{K_p}\,H$$

(13.10)

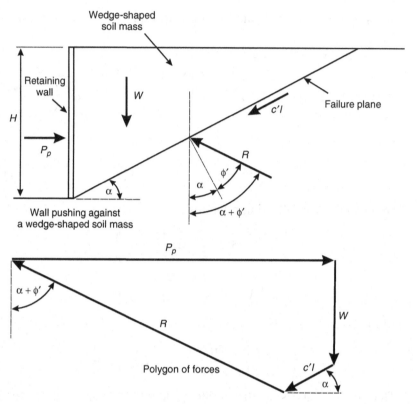

Figure 13.2 Coulomb method for determining passive earth pressure on a retaining wall.

where

$$K_a = \tan^2\left(45 - \frac{\phi'}{2}\right) = \frac{1 - \sin\phi'}{1 + \sin\phi'} \qquad (13.11)$$

$$K_p = \tan^2\left(45 + \frac{\phi'}{2}\right) = \frac{1 + \sin\phi'}{1 - \sin\phi'} \qquad (13.12)$$

and K_a and K_p are respectively termed the active and passive earth pressure coefficients.

The above analysis is for the relatively simple situation of a vertical, frictionless wall, with a horizontal ground surface behind it, and no seepage or pore pressure effects. When the ground surface is not level or when wall friction and seepage effects are included, direct analytical solutions are generally no longer possible. However, the Coulomb method makes possible a simple graphical procedure to obtain a solution for these situations; this is illustrated in Figure 13.3. This figure shows an irregular ground surface behind the wall as well as a phreatic surface and seepage toward the wall.

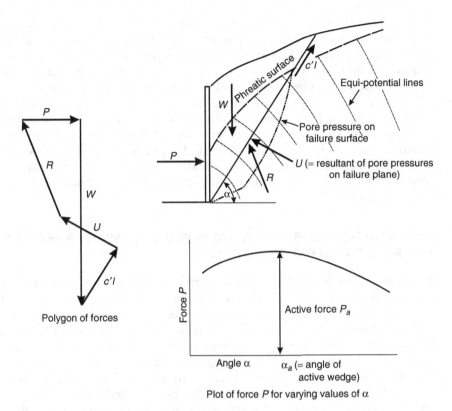

Figure 13.3 Coulomb wedge analysis applied to a more general situation.

The wall is assumed to be sufficiently pervious for water to escape through it or has a thin drainage layer immediately behind it, so that no pore pressure acts directly on the back of the wall. This does not mean that seepage has no effect on the force on the wall, as will become evident from the following analysis.

Except for the influence of seepage within the slope, the forces involved are the same as those in the earlier case illustrated in Figure 13.1. The seepage effect is taken account of by including the total "uplift" force U arising from the pore pressure on the failure plane. This pore pressure force reduces the effective stress on the failure plane and thus the shear strength available on this plane. The polygon of forces is very similar to that in Figure 13.1; the only significant difference is the inclusion of the pore pressure force U.

In this case the angle of the failure plane is uncertain and it is necessary to analyze a series of trial wedges and determine the value of the force P needed to maintain equilibrium for a range of values of α. A graph can then be drawn as indicated in Figure 13.3, and the maximum value of P is the value of the active force P_a occurring when the wedge angle is α_a.

In the above analysis the wall was assumed to be vertical and frictionless. Both nonvertical walls and wall friction can also easily be included in this Coulomb method of analysis. This is the great virtue of the Coulomb method—it allows almost any combination of geometry, soil properties, and seepage conditions to be analyzed following essentially the same procedure. The basic assumption of planar failure surfaces, however, is an approximation and may not lead to mathematically exact solutions. For the active case, however, which is of most significance to geotechnical engineers, the Coulomb analysis is sufficiently accurate for practical design purposes. It should be noted that the Coulomb method does not determine the point of action of the force. For most cases, it is reasonable to assume that it acts at one-third of the height of the wall.

13.2 AT-REST PRESSURE, ACTIVE PRESSURE, PASSIVE PRESSURE, AND ASSOCIATED DEFORMATIONS

In Chapter 4 the concept of "at-rest" earth pressure and the coefficient of at-rest pressure K_o was introduced. For normally consolidated soils that have not been subjected to any unusual external stresses, such as those coming from tectonic forces, the value of K_o is normally around 0.4–0.6. For normally consolidated sands, a reasonable approximation of K_o is given by $K_o = 1 - \sin \phi'$, where ϕ' is the friction angle of the material. The at-rest state is, by definition, one of zero lateral deformation. If lateral deformation occurs, the horizontal stress will change. We can investigate the K_o state and changes due to deformation by setting up an experiment in which a fixed rigid wall initially supports a normally consolidated soil. By measuring the pressure on the wall, we can determine directly the value of K_o. We can then investigate the changes in pressure on the wall if the wall is allowed to move away from the soil or if it is forced to move toward the wall.

The results to be expected are illustrated in Figure 13.4. In the at-rest situation the lateral pressure is the K_o value. If the wall is allowed to yield, that is, move away from the soil, the pressure rapidly falls to a constant value somewhat less than the at-rest value, as illustrated in the figure. This corresponds to the active pressure state. If the wall is pushed toward the soil, the horizontal pressure will steadily increase to reach a value very much greater than either the at-rest value or the active value. The peak value that it reaches is the passive pressure value. For clean, dry sand with a friction angle of $30°$ the K values will be 0.33, 0.50, and 3.0 for the active, at-rest, and passive states, respectively.

13.3 RANKINE EARTH PRESSURES

Rankine used quite a different approach from Coulomb's to determine the stress state on a wall in the active and passive conditions. Instead of considering a possible failure mechanism as the Coulomb method does,

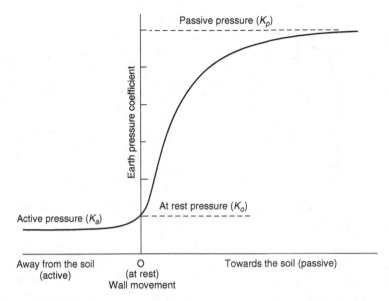

Figure 13.4 Earth pressures on walls and associated lateral deformations.

Rankine investigated the stress state on individual soil elements throughout the soil mass. He did this for the ideal condition of a vertical frictionless wall, a horizontal ground surface, and no pore pressure effects. In both the active and passive conditions, the whole soil mass is considered to be on the verge of failure or, in alternative terminology, in a state of "plastic equilibrium." This is illustrated in Figures 13.5a and 13.5b. In the active state, the vertical stress is the major principal stress and the Rankine active pressure, σ'_a, is the minor principal stress. The failure planes are inclined at $45 + \phi'/2$ to the horizontal. In the passive state, the horizontal stress σ'_p is greater than the vertical stress and thus becomes the major principal stress. The failure planes are now inclined at $45 - \phi'/2$ to the vertical.

Figure 13.5c illustrates the K_o (at-rest) stress state and the two Rankine stress states on the Mohr diagram. In the K_o state, the soil is not at the point of failure and the Mohr's circle is well clear of the failure line. In the Rankine active state, the horizontal stress is decreased until the soil reaches the failure state, and Mohr's circle touches the failure line as indicated. In the Rankine passive state, the horizontal stress is increased until it becomes the major principal stress and the circle again touches the failure line, as indicated in the figure.

In Chapter 9 the following relationship between the principal stresses (Equation 9.9) when the soil is in a failure state was established:

$$\sigma'_1 = \sigma'_3 \left(\frac{1 + \sin \phi'}{1 - \sin \phi'} \right) + 2c' \frac{\cos \phi'}{1 - \sin \phi'}$$

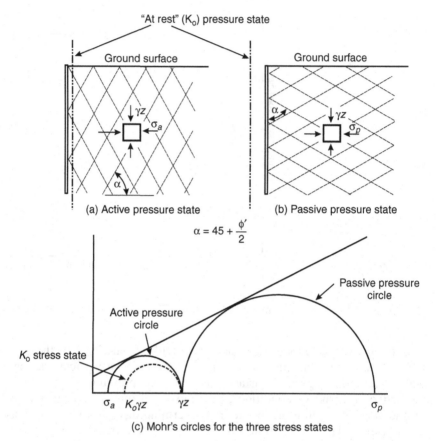

$$\alpha = 45 + \frac{\phi'}{2}$$

(c) Mohr's circles for the three stress states

Figure 13.5 Rankine active and passive pressure states.

Rearranging this gives the following expression for σ'_3:

$$\sigma'_3 = \sigma'_1 \left(\frac{1 - \sin \phi'}{1 + \sin \phi'} \right) - 2c' \frac{\cos \phi'}{1 + \sin \phi'}$$

In the active Rankine state $\sigma'_1 = \gamma z$, where z is the depth, γ is the unit weight of the soil, and σ'_3 is the active earth pressure σ'_a. Thus the active pressure is given by the following relationship:

$$\sigma'_a = \gamma z \left(\frac{1 - \sin \phi'}{1 + \sin \phi'} \right) - 2c' \frac{\cos \phi'}{1 + \sin \phi'} = \gamma z K_a - 2c' \sqrt{K_a} \qquad (13.13)$$

Similarly the passive Rankine state is given by

$$\sigma'_p = \gamma z \left(\frac{1 + \sin \phi'}{1 - \sin \phi'} \right) + 2c' \frac{\cos \phi'}{1 - \sin \phi'} = \gamma z K_p + 2c' \sqrt{K_p} \qquad (13.14)$$

The pressure distributions defined by Equations 13.13 and 13.14 are illustrated in Figure 13.6. It is seen that near the surface, down to the depth z_o, the active pressure is in fact negative, which implies the soil is in a state of tension. This is theoretically correct but is very unlikely to occur in a real soil.

To obtain the total theoretical force on the wall in each case, we can integrate the above expressions:

$$P_a = \int_0^H \left(\gamma z K_a - 2c' \sqrt{K_a} \right) dz$$

$$= \frac{1}{2} K_a \gamma H^2 - 2c' \sqrt{K_a} H \qquad (13.15)$$

$$P_p = \int_0^H \left(\gamma z K_p + 2c' \sqrt{K_p} \right) dz$$

$$= \frac{1}{2} K_p \gamma H^2 + 2c' \sqrt{K_p} H \qquad (13.16)$$

Equations 13.15 and 13.16 are identical to Equations 13.9 and 13.10, so that the Coulomb and Rankine approaches lead to identical solutions

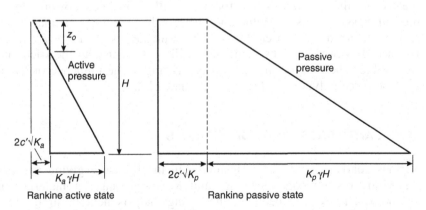

Figure 13.6 Rankine active and passive pressure profiles.

for this case of a frictionless vertical wall with no seepage effects. As explained in Chapter 11, the solution of geotechnical problems involving stability issues is normally done by analyzing possible failure mechanisms, not by investigating the stress state in individual soil elements. The Coulomb method is in keeping with this approach, while the Rankine method is an exception. The Rankine method works in this case because the stress state is identical throughout the soil mass involved in the analysis. The Rankine method, however, does not lend itself to more general situations in the way that the Coulomb wedge method does. The Coulomb method, therefore, is generally of much greater relevance to practical design situations.

We should note also that the term $(1 + \sin \phi')/(1 - \sin \phi')$ is often replaced with $N_{\phi'}$, so that the above expressions become

$$P_a = \frac{1}{2} \frac{\gamma H^2}{N_{\phi'}} - 2c' \frac{H}{\sqrt{N_{\phi'}}} \tag{13.17}$$

$$P_p = \frac{1}{2} \gamma H^2 N_{\phi'} + 2c' H \sqrt{N_{\phi'}} \tag{13.18}$$

13.4 INFLUENCE OF WALL FRICTION

So far we have ignored the influence of wall friction. This has been done purely to simplify the concepts involved. In practice there will always be some friction between the wall and the retained soil. The influence of this friction can be understood by examining Figures 13.1 and 13.2. In the active situation (Figure 13.2) the wedge is tending to move down in relation to the wall so that the wall will exert an upward frictional force on the soil mass. In the passive situation (Figure 13.2) the reverse will be true. This means that the resultant active force P_a will be inclined upward (toward the soil mass), and its magnitude will be less than the value estimated if the wall is assumed to be frictionless. The resultant passive force P_p will be inclined downward and its magnitude will be greater than the frictionless wall value. There is no difficulty in incorporating the influence of friction into the force polygons in Figures 13.1 and 13.2.

13.5 EARTH PRESSURE COEFFICIENTS

Various charts and tables are available in the literature giving values of the earth pressure coefficients K_a and K_p for a range of materials and geometries. These have been obtained using analysis methods similar to the Coulomb method and are generally restricted to cohesionless materials free of any pore pressure effects. They do, however, include sloping walls and wall friction. Jumikis (1962, pp. 570–585.), for example, gives extensive tables of these coefficients. These charts and tables are of limited value and should be used with caution; they are only for cohesionless materials, and they do not take account of any seepage effects.

13.6 TOTAL STRESS ANALYSIS

The above expressions can be modified to cover the undrained case. The friction angle ϕ' becomes zero, and the cohesion intercept becomes the undrained shear strength. The expressions are then as follows:

Earth pressure:

Active pressure $p_a = \gamma z - 2s_u$ (13.19)

Passive pressure $p_p = \gamma z + 2S_u$ (13.20)

Total force:

Active force $P_p = \dfrac{1}{2}\gamma H^2 - 2s_u H$ (13.21)

Passive force $P_p = \dfrac{1}{2}\gamma H^2 + 2s_u H$ (13.22)

These expressions are not of great practical significance, except in special cases where the soil is of very low permeability and only short-term stability is of interest. However, even in this situation, the expressions must be used with caution, as will become evident in the comments made in the following section regarding maximum height of unsupported vertical banks.

13.7 MAXIMUM HEIGHT OF UNSUPPORTED VERTICAL BANKS OR CUTS

The maximum height to which a vertical bank or cutting will remain stable can be determined by theoretical means using the above concepts. Before doing this it must be emphasized that these are theoretical estimates that should not be relied upon in practical situations involving vertical banks. This is an area of considerable theoretical uncertainty and of very important practical significance. Many lives have been lost on construction sites because of the collapse of vertical banks, especially those that form the sides of trenches in which workers carry out pipe or cable laying. *It should be recognized that it is not possible, on the basis of either theory or experience, to say whether a particular trench in soil will remain stable.* There is probably no area of soil mechanics where predictive ability is more limited. *Far too many lives have been lost because people worked in trenches that "appeared perfectly safe" or which analysis showed had adequate safety factors.*

So what does theory suggest? We can use Equation 13.15 and Figure 13.16 to investigate the influence of depth on the force acting on a wall. The pressure is negative down to a depth z_o and then becomes positive. The force on the wall only becomes positive when this positive-pressure zone becomes equal to the negative zone. Thus, if we put $P_a = 0$, this should give us the theoretical depth above which no wall is needed. This gives the "critical" depth as

$$H_c = \frac{4c'}{\gamma\sqrt{K_a}} \tag{13.23}$$

For undrained behavior (the total stress situation) this becomes

$$H_c = \frac{4s_u}{\gamma} \tag{13.24}$$

There is, however, another way of looking at this. As already mentioned, soil is unlikely to sustain negative (i.e., tensile) stress between soil and wall, and it can therefore be argued that the pressure above the depth z_o should not be taken account of in determining the critical depth. Once the depth exceeds z_o, a positive force begins to act on the wall, and hence z_o should be taken as the critical depth. Once the depth exceeds this value, the force becomes positive and support from a wall becomes necessary. We can determine z_o by putting σ'_a in Equation 13.13 equal to zero. This gives the value of $H_c(= z_o)$ as follows:

$$H_c = \frac{2c'}{\gamma\sqrt{K_a}} \tag{13.25}$$

For the undrained case

$$H_c = \frac{2s_u}{\gamma} \tag{13.26}$$

This reasoning thus gives critical depths exactly half those obtained from consideration of the total force rather than the pressure state on the wall. To gain an impression of what Equations 13.25 and 13.26 indicate in a practical situation, we will consider a typical firm to stiff soil. In the author's experience, there are many soils, especially residual soils, in this category, with properties close to the following:

Undrained shear strength $= 100\,\text{kPa}$
Cohesion intercept $c' = 15\,\text{kPa}$
Friction angle $\phi' = 30°$
Unit weight $= 16\,\text{kN/m}^3$

Analysis in terms of effective stress gives $H_c = 3.6$ m and in terms of total stress $H_c = 12.5$ m. The idea that a vertical clay bank could remain stable to a height of 12.5 m is utterly unrealistic, and even a height of 3.6 m is very optimistic for most clays. Simple observation of clay banks shows this to be the case. Agencies concerned with construction safety rightly place limits of 1–1.5 m on the depth of an unsupported trench that workers may enter.

Various explanations are possible as to why the above equations do not give reliable estimates; these include discontinuities in the soil, errors in the determination of soil parameters, the presence of a tensile zone at the top of the bank, and failure to take account of pore pressures in the soil. The last of these factors is probably the most significant, as many, though not all, collapses of vertical banks occur during periods of high rainfall. During dry weather, a large portion of the soil strength may in fact come from negative pore pressures. This is especially likely to be true of residual soils, where water tables are often some depth below the ground surface. This negative pore pressure increases the effective stress and thus the strength of the soil. Periods of rainfall may reduce this pore pressure to zero or very low values, causing a large reduction in soil strength. One of the most important parameters, namely pore pressure, is thus completely ignored by the equations above (Equations 13.4–13.24), and these equations are largely irrelevant to practical situations involving vertical banks.

13.8   CONSTRUCTION FACTORS INFLUENCING EARTH PRESSURES ON RETAINING WALLS

There are a number of practical factors that influence the earth pressure acting on a wall, the most important of which is probably the method used to construct the wall. Several typical ways of building retaining walls are illustrated in Figure 13.7. Figure 13.7a shows how many retaining walls are constructed. A cut is made at a relatively steep angle, with the expectation that the bank will remain stable until the retaining wall is built. The space between the wall and the natural ground is then filled in with backfill material, which could be the excavated original soil or a different material. Granular material is often used for this backfill because of its ease of handling and compaction, and also because it provides a good drainage zone to intercept seepage from the retained ground.

The force acting on the wall can now come from the potential failure of either a wedge within the backfill material or a wedge extending into the natural ground, as shown in Figure 13.7. It is therefore necessary to carry out a Coulomb wedge analysis to determine whether the critical failure plane lies within the backfill or the natural ground. There is an additional factor in this situation which may strongly influence the possible pressure

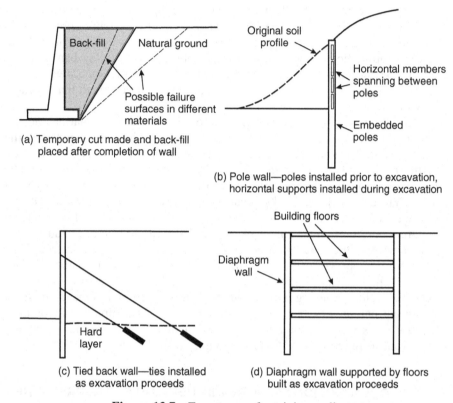

Figure 13.7   Four types of retaining walls.

on the wall; this is the way in which the backfill is placed and compacted. Compaction processes, such as with a steel wheeled roller or a sheepsfoot roller, tend to generate horizontal pressures within the soil, especially if heavy equipment is used close to the wall. These pressures can exceed the active pressures calculated from Rankine or Coulomb analysis. Apart from the weight and type of compaction equipment used, these pressures will depend on the flexibility of the wall behind which the soil is being compacted. For further analysis of this situation see Ingold (1979).

Figure 13.7b shows a retaining wall consisting of embedded poles with horizontal supports spanning between them. Walls of this sort are frequently built when there is no scope available to batter back the slope, as in the case of Figure 13.7a, or there is some risk of a slip occurring if any significant cuts are made into the slope. Holes are drilled and the poles cemented into place prior to any excavation being undertaken. As excavation proceeds, horizontal supports are inserted between the poles to retain the soil. This provides a very safe method of constructing walls in situations where there would otherwise be a danger of collapse. Figure 13.7c shows a "tied-back" wall, that is, a wall that gains its supporting capacity from anchors installed into the ground behind the wall. Such walls are often built in the ground prior to carrying out excavation, somewhat similar to the procedure in Figure 13.7b, and the anchors are installed as excavation proceeds. These tie-back anchors are normally grouted or anchored using mechanical means into suitable firm ground behind the wall. Once they are in place, the anchors are tensioned using suitable jacks. The earth pressure acting on the wall is then dependent on the stress level to which the anchors are tensioned. If there are no tight constraints on deformation of the wall, the stress level need only be sufficient to withstand active earth pressure. However, if deformations are to be kept to a minimum, the anchors can be tensioned to a stress level equivalent to the K_o stress state.

Figure 13.7d shows an excavation carried out between perimeter retaining walls. The perimeter walls are installed before excavation begins and are part of the final structure below ground level. As excavation proceeds, the floors are installed and act as props for the walls. This procedure is often referred to as "top-down" construction. It might appear that such a procedure would prevent any significant horizontal deformation and thus maintain the K_o stress state. This, however, is not the case. Excavation must be done before each floor can be built, and this is sufficient for significant yield of the wall to occur. Also, concrete floors undergo shrinkage as the concrete sets, which also accounts for some movement of the walls. The stress state is thus unlikely to be the original K_o state, but it may still be higher than the active pressure state. Design to the K_o stress level would generally be conservative.

13.9 PROPPED (STRUTTED) TRENCHES

Narrow vertical-sided trenches are frequently used to lay pipelines and various other underground services. These trenches are often supported by horizontal props or "struts" spanning across the trench, as illustrated in Figure 13.8. The struts are installed manually as excavation proceeds.

The procedures used for installing the struts are not particularly "scientific" or even systematic. Wedges are commonly used to tighten up the struts against timber or other material directly supporting the sides of the trench. The extent of tightening depends on the skill and diligence of the workers installing them. Measurements show that the force taken by individual struts varies widely and does not follow any clear pattern. In addition to the installation procedure, the forces in each strut depend also on the geometry of the trench. In some materials, the sides of the trench can be very vertical and planar, and the supporting wall bears evenly against the soil, while in other materials the sides may be irregular and contact between wall and soil is no longer uniform.

Peck (1969) and Terzaghi and Peck (1967) have suggested tentative pressure distributions for the design of struts in deep excavations, as shown

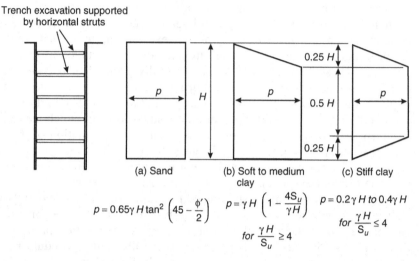

Trench excavation supported
by horizontal struts

(a) Sand

(b) Soft to medium
clay

(c) Stiff clay

$$p = 0.65\gamma H \tan^2\left(45 - \frac{\phi'}{2}\right)$$

$$p = \gamma H\left(1 - \frac{4S_u}{\gamma H}\right)$$

$$for \ \frac{\gamma H}{S_u} \geq 4$$

$$p = 0.2\gamma H \ to \ 0.4\gamma H$$

$$for \ \frac{\gamma H}{S_u} \leq 4$$

Figure 13.8 Suggested pressure diagrams for determining strut loads in braced trench excavations. (From Peck, 1969, and Terzaghi and Peck,1967.)

in Figure 13.8. These pressure distributions are based on a number of measurements of strut loads in various materials and are "envelopes" covering the maximum loads measured. The design of the struts using these charts should therefore ensure that the struts will not be overloaded but does not mean that every strut will be taking these design loads. Some struts may have loads close to these values, while others may have almost no load.

For sand the pressure p is given by

$$p = 0.65\gamma H \tan^2\left(45 - \frac{\phi'}{2}\right)$$

For soft to medium-strength clay the pressure is

$$p = \gamma H\left(1 - \frac{4s_u}{\gamma H}\right) \quad for \ \frac{\gamma H}{s_u} \geq 4$$

For stiff to hard clay the pressure is

$$p = 0.2\gamma H \ to \ 0.4\gamma H \quad for \ \frac{\gamma H}{s_u} \leq 4$$

13.10 RETAINING-WALL DESIGN EXAMPLE

Figure 13.9 shows a stone masonry retaining wall supporting soil with a sloping surface behind it. Walls of this sort rely on the weight of the wall

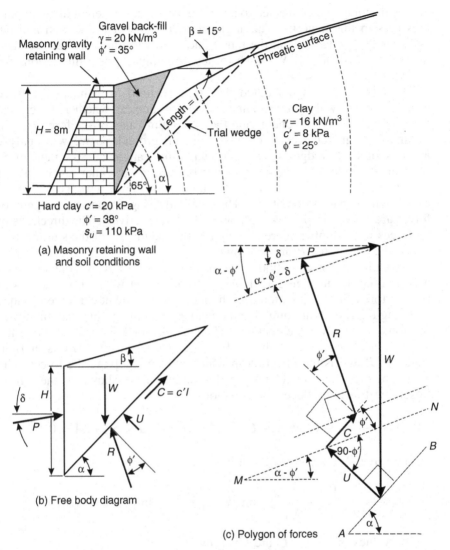

(a) Masonry retaining wall
    and soil conditions

(b) Free body diagram

(c) Polygon of forces

Figure 13.9   Determination of the force P acting on a masonry gravity retaining wall.

itself for their supporting capacity and are commonly referred to as gravity retaining walls. The wall is to be built by first making a temporary cut with a back slope of 65°, then building the wall, and filling in the space behind the wall with free-draining granular fill. Figure 13.9a shows the dimensions of the wall as well as the properties of the retained natural ground and the granular fill to be placed behind the wall. The long-term steady-state seepage condition is also indicated. Because the granular fill is assumed to be free draining and connected to an appropriate outlet, there will be no pore pressures in this material. The angle of friction between the wall and the gravel is designated as δ in the figure and in this example will be assumed to have a value of $10°$. We wish to determine appropriate dimensions for the wall.

The Coulomb wedge method will first be used to estimate the force exerted on the wall by the retained soil. It is unclear whether the critical failure surface (i.e., the one giving the maximum value of the force P) will occur in the gravel backfill or in the natural ground, so we will analyze a series of trial wedges from an initial failure angle of $80°$, then in $5°$ intervals down to $30°$. The analysis can be carried out either graphically or analytically. To demonstrate the method, we will analyze a wedge within the clay, having a base slope of $45°$. The free-body diagram and the polygon of forces are shown in Figures 13.9b and 13.9c, respectively. The directions of all forces in this polygon are known, but the magnitude of the soil reaction R and the wall force P are unknown.

Resolution of forces to find P can be done in several ways, one of which was used earlier in Figure 13.1. An alternative that we will use here is to resolve forces in the direction at right angles to the unknown force R, since by doing this we obtain directly an equation containing only one unknown, the force P that we are seeking. The line AB in Figure 13.9c is at the inclination of the base of the trial wedge, while the line MN is at right angles to R and is the direction in which we are resolving the forces. The angles of the forces in relation to this direction are shown in the figure. Resolving forces produces the equation

$$W \sin\left(\alpha - \phi'\right) + U \sin\phi' - C \cos\phi' - P \cos\left[\left(\alpha - \phi'\right) - \delta\right] = 0$$

The force P is therefore given by

$$P = \frac{1}{\cos\left[\alpha - \left(\phi' + \delta\right)\right]}\left[W \sin\left(\alpha - \phi'\right) + U \sin\phi' - C \cos\phi'\right]$$

The values of W and C are given by

$$W = \frac{1}{2}\gamma H^2 \frac{\cos\alpha \cos\beta}{\sin\left(\alpha - \beta\right)} \qquad C = c'l = c'\frac{H \cos\beta}{\sin\left(\alpha - \beta\right)}$$

The expression for W is only correct if the wedge is within a uniform material. In the present example this is only the case for wedges steeper than $65°$. For wedges in the clay, the value of W must take account of the different unit weights of the gravel and clay.

The value of the pore pressure force U is not amenable to direct calculation. It is given by $U = \Sigma u \, \Delta l$, where u is the pore pressure and l is the length of the failure plane. The parameter U is determined by measuring values of u along the slip plane, taking account of the inclination to the vertical of the equipotential lines. Alternatively, the equipotentials can be assumed to be vertical, which is a reasonable approximation if the phreatic surface is fairly flat.

For the $45°$ wedge shown in the figure, $W = 767.6\,\text{kN}$, $C = 123.2\,\text{kN}$, and $U = 135.8\,\text{kN}$, giving $P = 211\,\text{kN}$. The analysis of those wedges that lie entirely within the gravel backfill ($\alpha = 65°$ to $\alpha = 80°$) is more straightforward than those within the clay, as neither a cohesive component nor a pore pressure component is involved. The results from all wedges are shown in Figure 13.10. As is to be expected, there is a discontinuity in the graph when the angle of the wedge passes the $65°$ boundary between gravel and clay. At this angle, assuming the wedge is within the gravel produces a greater value of P than assuming it is within the clay. However, as the wedge angle is further decreased, the force from the clay wedges increases to reach a maximum when the angle is approximately $42°$. The peak force is about $212\,\text{kN}$.

The second step in the design process is to determine the required dimensions of the wall to ensure it has adequate margins of safety against failure. In Chapter 11 three mechanisms were identified as possible failure modes of the wall:

(a) Sliding horizontally

(b) Overturning

(c) Bearing capacity failure of the foundation soil, which will normally occur near the toe of the wall where the pressure on the soil is greatest

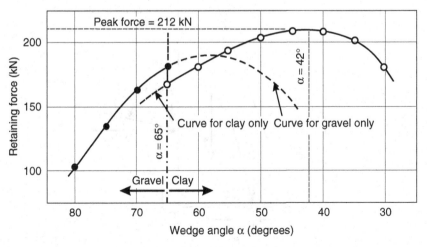

Figure 13.10 Results from Coulomb wedge analysis of retaining wall in Figure 13.9.

The wall dimensions, especially the base width, must be determined to ensure adequate margins of safety against failure by these mechanisms. This is usually done by a trial-and-error procedure. We will adopt a width of 1 m for the top of the wall and determine the necessary base width B. The wall dimensions and the forces involved are shown in Figure 13.11. The force W is the weight of the wall itself and has a center of gravity at a distance z from the inside face of the wall. The force P is the earth pressure force on the wall, 212 kN, as determined above. It acts at one-third of the wall height with an assumed inclination of $10°$. The unit weight of the stone masonry is taken as $22 \, \text{kN/m}^3$. The force R is the resultant of the forces on the wall acting at a distance x from the corner O with an inclination α to the vertical.

Before commencing our analysis, an explanation is necessary with respect to the failure modes of overturning and bearing capacity. These are normally considered distinct failure modes and are analyzed separately. Overturning failure is analyzed by moment equilibrium assuming the wall rotates about the toe, point M in Figure 13.11. However, rotation will only occur about the toe if the wall is built on solid rock. Rotation about the toe implies a very high stress here, so that if the wall is built on soil, bearing capacity failure is likely to occur as the wall overturns. Because of this, the point of rotation will no longer be the toe of the wall. It will be some distance in from the toe and the overturning analysis will no

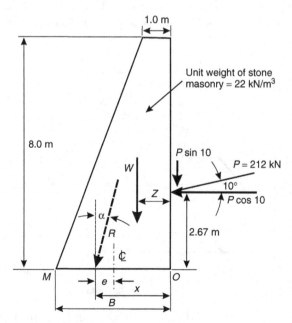

Figure 13.11 Forces involved in stability analysis of the wall.

longer be valid. We should therefore be careful to ensure the wall has a substantial margin of safety against bearing capacity failure and recognize that the safety factor against overturning is an upper limit, except when the foundation material is solid rock or at least very hard soil.

Because this wall is founded on soil, we will begin our analysis by determining the width needed to provide an adequate safety factor against bearing capacity failure. As an inclined and eccentric load is involved, we will use Meyerhof's method for determining the bearing pressure and the bearing capacity. We will demonstrate the procedure by adopting a trial width B of 4 m. The weight of the wall is then 440 kN and the distance z is 1.40 m, and the angle α is $23.7°$. Taking moments about O gives the following equation:

$$212 \cos 10 \times 2.67 + 440 \times 1.4 = (440 + 212 \sin 10)x$$

Solving gives $x = 2.46$ m.

Following Meyerhof's procedure, $e = x - 0.5B = 0.46$ m.

The effective width is given as $B' = B - 2e = 3.08$ m.

The bearing pressure from the wall is $(440 + 212 \sin 10)/3.08$

$$= 155 \text{ kPa.}$$

We will now determine the bearing capacity of the soil. From Equations 12.4 and 12.5, the ultimate bearing capacity is given by $q = s_u i_c N_c + \gamma D$, where

$$i_c = \left(1 - \frac{\alpha°}{90°}\right)^2 = 0.544$$

Taking $N_c = 5.14$ $s_u = 110$ kPa and ignoring the depth factor (since the depth on the toe side of the wall is negligible), we have

$$q = 110 \times 0.544 \times 5.14 = 307 \text{ kPa.}$$

The safety factor is therefore $307/155 = 2.0$. Repeating the analysis for a range of widths produces the results illustrated in Figure 13.12. Analysis has also been carried out using a similar range of widths to determine the safety factors with respect to overturning about the toe of the wall (point M in Figure 13.11). For this example, it is seen that the safety factors are substantially higher than the bearing capacity safety factors. In other words, bearing capacity failure is more critical than overturning failure. For a much harder foundation soil the reverse will be true. The overturning safety factors in Figure 13.12 are therefore "fictional," as the wall does not have this level of security against overturning. Overturning of the wall will result from bearing capacity failure, with respect to which the safety factor is smaller.

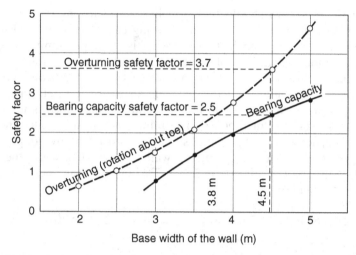

Figure 13.12 Safety factors from stability analysis of the retaining wall in Figure 13.9.

The question of what is an appropriate safety factor receives various answers within the geotechnical profession. It has been traditional to use high safety factors (2.5 or 3) for bearing capacity in designing foundations. This is partly in order to help limit settlements. It can be argued that retaining walls are better able to tolerate deformation than building foundations and a lower safety factor is acceptable. However, it can also be argued that retaining-wall failures are more common than foundation failures, and therefore similar safety factors are appropriate. For this reason we will adopt a safety factor of 2.5 for bearing capacity, which from the graph gives a base width of 4.5 m.

The remaining failure mode to be checked is sliding. The wall is founded on relatively stiff soil, and we will assume that the frictional component of the effective shear strength is fully mobilized between the base of the wall and the foundation soil. With a base width of 4.5 m the weight of the wall is 484 kN, so that the safety factor against sliding is given by

$$SF = \frac{(484 + 212 \sin 10°) \tan 35°}{212 \cos 10°} = 1.75$$

This is not a very high safety factor, but values between 1.5 and 2 are commonly used (for both sliding and overturning). In this case 1.75 is considered appropriate as it has been estimated conservatively by ignoring the cohesive component of strength and the passive resistance at the toe of the wall.

13.11   SHEET PILE (AND SIMILAR) RETAINING WALLS

Sheet piles are flat metal piles used as bracing around the perimeter of excavations. The piles are shaped so that they interlock with each other to form an almost impermeable barrier. Sheet pile walls may be either free standing, in which case they are generally referred to as cantilever walls, or they may be tied back, or propped, near the surface, in which case they are commonly referred to as propped cantilever walls or simply propped walls. In addition to sheet piles, similar walls can be built in a variety of ways, including diaphragm walls and bored "secant pile" walls. The design methods described here are applicable to any wall of this type.

13.11.1   FreeStanding and Propped Cantilever Walls

Figure 13.13 shows a free-standing cantilever retaining wall. Walls of this type, and those using props or anchors, have traditionally been built using sheet piles, but today such walls are frequently built as "diaphragm" walls using slurry trench techniques or as secant bored piles. The capacity of a free-standing wall to resist lateral earth pressure comes from the passive soil pressure acting on that part of the wall embedded in the soil below the base of the excavation. The retained soil will apply active earth pressure over the rear of the wall and tend to cause failure of the wall by rotation around a point O very close to the base of the pile.

Passive pressure then acts on the front of the wall above O and on the rear of the wall below O, as shown in Figure 13.13b. The limiting values of these pressures are the active and passive earth pressures, which are indicated by dotted lines in the figure. The approximate pressure distributions acting on

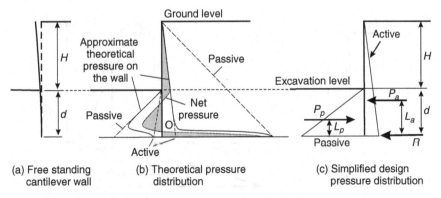

Figure 13.13   Pressure distribution on free-standing "cantilever" retaining wall.

each side of the wall are shown and also the net pressure on the wall. The height of soil that can be supported in this way is not very large.

This pressure distribution is an approximation, and not one that is easily amenable to analytical design methods. To simplify the design process, it is assumed that the passive pressure on the rear of the wall can be concentrated as a single point load at the base of the wall and that full active and passive pressures act on each side of the wall down to its toe. This simplification is shown in Figure 13.13c. The forces P_p and P_a are the resultants of the passive and active pressures, respectively. Moments can then be taken about the base of the wall to determine the required depth d of the wall, that is,

$$P_p L_p - P_a L_a = 0 \qquad (13.27)$$

This analysis produces a cubic equation in d which can be solved by a trial process or an alternative more sophisticated method. The bending moments in the wall can be determined from the equilibrium statics.

Figure 13.14 shows a wall restrained near the surface by a prop or a tie-back. With this arrangement, the wall no longer relies on a restraining moment coming from the embedded depth, so the required embedment depth will be much less than with a free-standing wall. Peak active and passive pressures can be expected to act over the full depth of the wall.

Analysis to determine the necessary embedment depth can now be carried out by taking moments about the point of application of the prop force, that is,

$$P_a L_a - P_p L_p = 0 \qquad (13.28)$$

A cubic equation in d is again produced and can be solved to give the depth d. Force equilibrium can then be used to determine the value of the prop (or tie-back) force P_r.

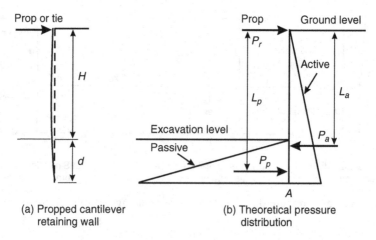

(a) Propped cantilever retaining wall

(b) Theoretical pressure distribution

Figure 13.14   Pressure distribution on a propped or tied cantilever retaining wall.

The above analysis methods are straightforward but do not incorporate any safety factors. Burland et al. (1981) point out that the methods used to apply safety factors in practice vary considerably and can produce widely differing results. In some situations, the methods can produce illogical results, while in others the true safety margin may be much lower than the methods imply. The principal methods for applying the safety factor are the following:

(a) Apply the safety factor to the passive moment so that Equation 13.27 becomes

$$P_a L_a \leq \frac{P_p L_p}{F_p}$$

where F_p is the safety factor.
A similar expression can be derived from Equation 13.28.

(b) Determine the required embedment depth d from Equation 13.27 or 13.28 and increase this by a factor F_d. Various values of F_d are used, ranging from as low as 1.2 up to 1.7.

(c) Apply a safety factor to the shear strength used in determining the magnitude of passive resistance, that is, the value of P_p. This seems the most logical procedure, in keeping with the definition of safety factor given in Chapter 11 (Equation 11.1). However, it is not without its difficulties, especially when the material retained by the wall is not homogeneous.

(d) Another method, not widely used, is to carry out the analysis in terms of the net pressure on the wall and apply the safety factor to the net passive pressure moment.

To overcome the inconsistencies and anomalies inherent in the above methods, Burland et al. (1981) propose a different approach which has general application to all situations. By drawing a comparison with bearing capacity theory, they argue that only that part of the active and passive pressure that arises from the presence of the retained soil should be included in the analysis. The concept is illustrated in Figure 13.15 for a propped wall. The retained soil is defined as that above the excavation level.

Both the active and passive pressures are reduced by the value of the active pressure that would be generated by the self-weight of the soil below the level of the retained soil (i.e., below the excavation level in Figure 13.15). In other words the analysis only involves the active and passive pressures generated or "activated" by the retained soil (the soil above the excavation level). The active pressure is then made up of two components, P_{a1}, which increases linearly with depth, and P_{a2}, which is constant with depth. The passive pressure still increases linearly with depth, but the pressure coefficient is now $K_p - K_a$. The definition of safety factor is given by

$$F = \frac{P_{pn} L_{pn}}{P_{a1} L_{a1} + P_{a2} L_{a2}} \qquad (13.29)$$

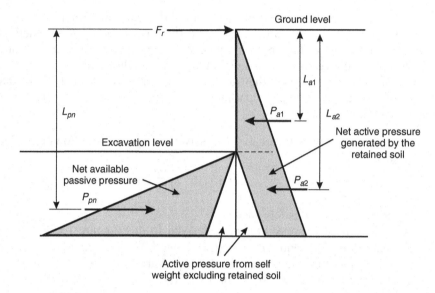

Figure 13.15   Method of Burland et al. (1981) for analysis of a propped embedded retaining wall.

Burland et al. demonstrate that for a variety of situations this method provides safety factors that are very similar to those obtained by applying the safety factor to the soil strength. The argument in favor of their method, rather than applying a safety factor to soil strength [method (c) above], is that it is easier to use when the soil is made up of layers of different properties.

Example Using Different Definitions of Safety Factor   We will consider the simple case of a propped retaining wall of the type shown in Figure 13.16. For simplicity the following assumptions are adopted:

(a) The material is cohesionless ($c' = 0$), and no pore pressures are present.
(b) The wall height $H = 6\,\text{m}$ and is frictionless.
(c) The soil unit weight $\gamma = 20\,\text{kN/m}^3$ and the angle of shearing resistance $\phi' = 30°$.

We can then write:

$$P_a = \frac{1}{2}K_a\,\gamma\,(H+d)^2 \qquad L_a = \frac{2}{3}(H+d)$$

$$P_p = \frac{1}{2}K_p\,\gamma d^2 \qquad L_p = H + \frac{2}{3}d$$

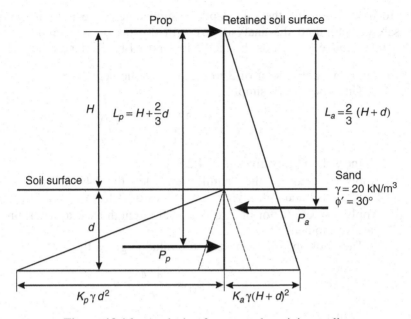

Figure 13.16   Analysis of a propped retaining wall.

The active moment is thus

$$P_a L_a = \frac{1}{3} K_a \gamma (H + d)^3$$

and the passive moment is

$$P_p L_p = \frac{1}{2} K_p \gamma d^2 \left(H + \frac{2}{3} d \right)$$

We will first determine the required depth of embedment with no safety margin involved.

In this case, equating the above expressions gives

$$(H + d)^3 = \frac{3}{2} \frac{K_p}{K_a} d^2 \left(H + \frac{2}{3} d \right) \tag{13.30}$$

with $H = 6\,\mathrm{m}$ and $\phi = 30°$, $K_a = 0.333$, $K_p = 3$, this equation becomes

$$(6 + d)^3 = \frac{27}{2} d^2 \left(6 + \frac{2}{3} d \right)$$

Solving this gives $d = 2.40\,\mathrm{m}$.

To now determine the safe depth of embedment, we must incorporate a safety factor into the analysis. We will do this using methods (a)–(c) described above as well as the method proposed by Burland et al. (1981).

(a) Apply a safety factor of 2 to the passive moment:
   This gives, very simply,

$$(6+d)^3 = \frac{27}{4}d^2\left(6+\frac{2}{3}d\right) \tag{13.31}$$

   and solving this gives $d = 4.25\,\text{m}$.

(b) Apply a factor to the "equilibrium" length, 2.40 m. Factors vary between 1.2 and 1.7, giving new values of d from 2.88 to 4.08 m.

(c) Apply a safety factor of 2 to the shear strength used to determine the passive moment.
   The allowable or mobilized ϕ'_m value is given by

$$\tan \phi'_m = \frac{\tan \phi'}{F}$$

so that

$$\phi'_m = 16.1°,$$

   and $K_p = 1.768$. The equilibrium equation now becomes

$$(6+d)^3 = 7.963\,d^2\left(6+\frac{2}{3}d\right)$$

   and solving this gives $d = 3.66\,\text{m}$.

(d) The Burland et al. method:
   Referring to Figure 13.15 and rearranging Equation 13.29, we have

$$P_{a1}L_{a1} + P_{a2}L_{a2} = \frac{1}{F}P_{pn}L_{pn} \tag{13.32}$$

$$P_{a1} = \frac{1}{2}K_a\gamma H^2 \quad \text{and} \quad L_{a1} = \frac{2}{3}H$$

$$P_{a2} = K_a\gamma Hd \quad \text{and} \quad L_{a2} = H + \frac{d}{2}$$

$$P_{pn} = \frac{1}{2}\left(K_p - K_a\right)\gamma d^2 \quad \text{and} \quad L_{pn} = H + \frac{2}{3}d$$

Inserting these into Equation 13.32 and adopting $F = 2$ lead to the following expression:

$$3\left[12 + d\left(6 + \frac{d}{2}\right)\right] = \left(6 + \frac{2}{3}d\right)d^2 \tag{13.33}$$

Solving this gives $d = 3.90\,\text{m}$.

The depth obtained using Burland et al. is thus close to the value obtained by applying a safety factor to the soil strength. The direct method of applying the safety factor to the passive moment gives a significantly different (although conservative) value. The depth obtained by applying factors to the equilibrium depth gives arbitrary values, which is not surprising as the factors are arbitrary. The lowest factors give an unsafe design, while the highest factors appear to be excessively conservative.

Influence of Seepage and Pore Pressures   If a water table is present or a seepage state exists, these need to be taken into account in the analysis. If a static water table exists and is therefore at the same level on both sides of the wall, then there is no difficulty in taking this into account in the usual way. The total horizontal pressure will be the sum of the effective horizontal stress plus the pore pressure. In this situation, the pore pressure component will be the same on both sides of the wall and will cancel out when considering the moment equilibrium of the wall. If the water level is higher on the retained side than the excavation side, then a seepage condition will exist and the pore pressures are no longer hydrostatic. To determine the pore pressures, we could sketch a flow net and deduce the pore pressures from this in the usual way, as described in Chapter 7. However, an approximation is normally used, as illustrated in Figure 13.17.

Figure 13.17a shows the pore pressure state around the sheet pile wall. It is assumed that the head difference H is lost linearly along the wall. The seepage length along the wall is $h_1 + h_2$, so that the head loss to point P is given by

$$\Delta H = \left(\frac{h_1}{h_1 + h_2} \right) H$$

and the head at P (with respect to P) is given by

$$H_P = h_1 - \left(\frac{h_1}{h_1 + h_2} \right) H = \frac{2h_1 h_2}{h_1 + h_2}$$

and the pore pressure at P is given by

$$u_P = \left(\frac{2h_1 h_2}{h_1 + h_2} \right) \gamma_w$$

The pore pressure distribution is assumed to be linear from the water level on each side of the wall to the maximum value at the toe of the wall. The effect of the seepage state is to raise the pore pressure above hydrostatic on the "downstream" side and to lower it on the "upstream" side. This pore pressure state is then used to determine the effective active and passive earth pressures on each side of the wall. The net pore pressure on the wall is as shown in Figure 13.17b. The forces acting on the wall are therefore made up of the effective active and passive pressure and the net pore pressure forces.

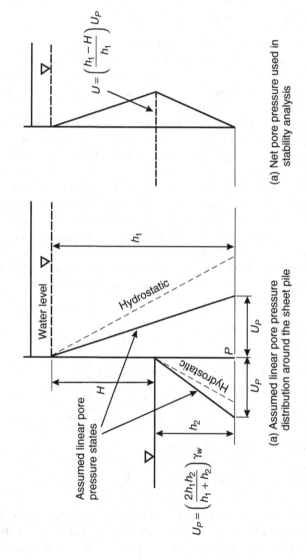

$U_P = \left(\dfrac{2h_1 h_2}{h_1 + h_2}\right)\gamma_w$

Assumed linear pore pressure states

(a) Assumed linear pore pressure distribution around the sheet pile

$U = \left(\dfrac{h_1 - H}{h_1}\right) U_P$

(a) Net pore pressure used in stability analysis

Figure 13.17 Approximate pore pressure state adopted for analysis purposes.

13.12 REINFORCED-EARTH WALLS

13.12.1 Concept and General Behavior

Reinforced earth, shown in Figure 13.18, is a widely used method of ground retention (a form of retaining wall) that involves the use of horizontal strips, or layers, of "reinforcement" embedded in compacted soil to provide tensile strength in the horizontal direction. Reinforced-earth walls are now widely used in place of the gravity and cantilever walls that were used in the past.

The structure is built from the base up, with the reinforcement installed in soil layers at regular intervals as construction proceeds. Facing units, connected to the reinforcement, are placed to form the outside of the retaining wall.

According to the "inventor" of reinforced earth, Henri Vidal (a French engineer), the zone of reinforced soil acts as a "coherent gravity block" and performs like other gravity retaining walls. The reinforcement prevents the development of lateral earth pressure at the face and eliminates the need for an external wall to retain the soil. Vidal developed a systematic design procedure, termed the "coherent gravity block" method, for this gravity block which is still in use, though other more sophisticated methods made necessary by the advent of new reinforcement types have largely taken its place in recent years.

Figure 13.19 illustrates how reinforced-earth walls behave. The possible range of earth pressure within the soil mass, between the K_0 and the active state, and also the pressure that acts on the facing units are shown. Also shown is the way in which the force in the reinforcing strips varies along the strip. Measurements have shown that the point of maximum force occurs some distance away from the wall and these points of maximum force lie

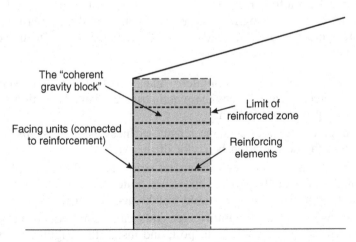

Figure 13.18   Reinforced-earth retaining wall.

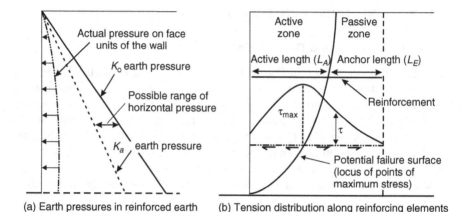

(a) Earth pressures in reinforced earth   (b) Tension distribution along reinforcing elements

Figure 13.19 Earth pressure in the ground and tension force in reinforcing elements.

along a curved line as shown. This potential failure surface is generally not straight as normally assumed for an active Rankine state. The soil zones on each side of this potential failure line are referred to as the active and passive zones. This is a little misleading, as the stress states within these zones do not correspond to the active and passive pressure states as normally understood. The active zone may be in the active state, but the passive zone is not in the passive state.

13.12.2 Reinforcement Types

The Vidal system makes use of metal strips as reinforcement. These are made of galvanized mild steel. Over the last 30 years or so various alternative forms of reinforcement have been developed, the most widely used of which are known as geogrids. These are manufactured from plastic polymers and are quite flexible. Geogrids differ from metal strips in the following respects:

- They are installed as complete "mats" so that they act over the full reinforced area, whereas the metal strips are isolated with a horizontal spacing of about 0.75 m.
- They do not have the same tensile stiffness as metal strips; this influences the deformation behavior of the wall.
- The interaction between the metal strips and soil is a simple function of the area of the strip: This is not quite the case with the geogrids, and empirical factors are used to take account of this.
- Because of their flexibility, geogrids can be rolled up and are thus much easier to handle, transport, and install than rigid metal strips.

13.12.3   Basic Design Procedures

The conventional design procedure assumes that the reinforced mass of soil behaves as a gravity block structure. The width of this block can therefore be determined by analyzing its stability with respect to sliding, bearing capacity, or overturning. This is termed external design. The block can also be analyzed to determine the strength, vertical spacing, and required length of the reinforcement. This is termed internal design. Since the required length of the reinforcement is also the wall width, both external and internal designs determine a minimum width for the wall. The greater of these becomes the design width.

With the Vidal system, internal design normally governs the width of the block rather than considerations of external stability. However, with geogrid reinforcement, the reverse is true. Full coverage of the reinforced area by the reinforcing material means there is a far stronger connection between the soil and reinforcement and the width determined from internal design will normally be less than that determined by external design.

External Design   The external design is no different than the design of other gravity retaining structures and will not be described in detail here. It should be noted that, in addition to meeting sliding, overturning, and bearing capacity requirements, it is generally considered good practice to ensure that the resultant force the block exerts on the foundation soil lies within the middle third of the base. This ensures that there is positive contact between wall and foundation over its full width.

Internal Design   There are two main methods for internal design. First, there is the Vidal coherent gravity block method for walls with metal strip reinforcement. This is essentially a Rankine approach as it focuses on the stress state within the soil mass. Second, there are various forms of wedge analyses, intended for walls with geogrid reinforcement. These are essentially Coulomb methods, involving the analysis of possible failure mechanisms. The latter methods are considered more soundly based and more versatile than the Vidal method. The two-part wedge method described here generally follows the recommendations of the Tensar company for the design of geogrid reinforced walls (Tensar International, 2001). The two methods are described in the following sections.

Vidal Coherent Gravity Block Method   The steps involved in this method are as follows:

1. Determination of the capacity and spacing of the reinforcing elements
2. Determination of the "anchor" length of the reinforcing elements
3. Determination of the total length of reinforcing by adding the "active" and anchor lengths

Determination of Capacity and Spacing of Reinforcing Elements   In the Vidal concept, the function of the reinforcing strips is to prevent the development of horizontal stresses that would otherwise need to be resisted by some form of conventional wall. Hence each reinforcing element must resist the horizontal stress in its vicinity, as indicated in Figure 13.20.

Hence the force T in any reinforcing element is given by

$$T = \sigma_h hb \tag{13.34}$$

where σ_h is horizontal stress and h and b are the vertical and horizontal spacing between elements. The horizontal stress is a function of the vertical stress and can be related to it in the usual way, that is,

$$T = K\sigma_v hb$$

The available strength in the reinforcement is not constant over its full length. Over its central section it is the same as its tensile strength, but it falls off over its anchor sections at each end, as shown in the Figure 13.20, in accordance with Equations 13.35 and 13.36 given below. The value of K might be expected to be the same as K_o if the reinforcement prevents any lateral deformation. This however is not the case, and K has an upper limit of K_o and a lower limit of K_a, the Rankine coefficient of active earth pressure.

With the Vidal system using metal strip reinforcement, empirical rules have been developed, based on experimental work, which show K decreasing from K_0 at the surface to K_a at a depth of 6 m. It appears that little deformation occurs near the surface, but the stress increase with depth

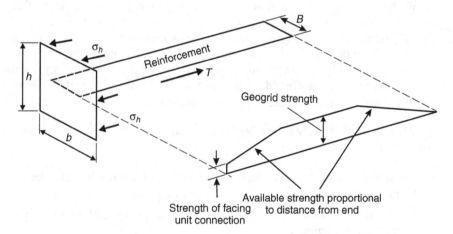

Figure 13.20   Force in a reinforcing element.

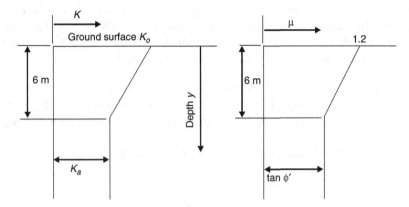

Figure 13.21 Rules for estimation of horizontal pressure and frictional coefficient for Vidal System.

results in yielding beyond a certain depth, and the development of an active Rankine state of stress. This is illustrated in Figure 13.21.

We can note in passing that, when geogrids are used, the reinforcement is normally continuous in the horizontal plane and the term b in Equation 13.34 becomes unity and disappears, so that T is the force per unit length along the wall.

Determination of Anchor Length of Reinforcement   The anchor length is the length in the soil needed to provide the necessary pullout capacity. It is a function of the frictional resistance between reinforcement and soil.

With the Vidal system this length is given by:

$$T_i = \sigma'_v L_E B \frac{\mu}{F} \times 2 \qquad (13.35)$$

where L_E = anchor length
B = width of strip
μ = coefficient of friction between strip and soil
F = safety factor

The value of the coefficient of friction is taken to be 1.2 at the ground surface and to decrease linearly to $\tan \phi'$ at a depth of 6 m, as shown in Figure 13.21.

With geogrid reinforcement the corresponding relationship is

$$T_i = \sigma'_v L_E \alpha_p \frac{\tan \phi'}{F} \times 2 \qquad (13.36)$$

where the symbols have the same meaning as above except for α_p, which is a pullout interaction coefficient. This coefficient accounts for the reduced friction between geogrid and soil (compared to soil alone) and the fact that the geogrid does not occupy the full contact area. The length L_E needs to be calculated at each depth of reinforcement over the full height of the wall.

Two coefficients, α_s and α_p, are commonly used with geogrid reinforcement:

α_s — governing sliding resistance of soil mass along plane of geogrid
α_p — governing resistance of geogrid to pullout from within soil mass

Both coefficients are defined by the relationship

$$\alpha = \frac{\tan \left(\phi' \text{ with geogrid reinforcement} \right)}{\tan \phi'}$$

Typical values for α_s suggested by Tensar are as follows:

Crushed rock and gravel: 0.9–1.0, Sand: 0.85–0.95, Clay: 0.6–0.7

They also suggest that α_p can be taken the same as α_s unless specific tests are done.

Determination of Total Length of Reinforcement   The Vidal method recommends the use of an empirical construction to approximate the boundary of the "active" zone; it is shown in Figure 13.22a. A line is drawn from the toe at an angle of 0.3 horizontal to 1.0 vertical to intersect the ground surface. A vertical line is then taken from this point to a depth equal to half H and connected by a straight line to the toe of the slope.

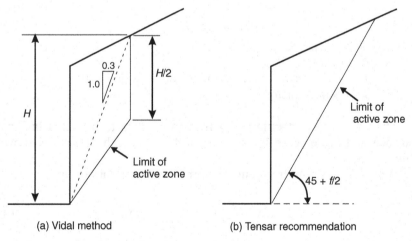

(a) Vidal method   (b) Tensar recommendation

Figure 13.22   Recommendations for determination of the "active" zone. (a) Vidal (metal strip reinforcement). (b) Geogrid reinforcement.

When designing with geogrids, the active zone is normally defined by the Rankine active failure line from the base of the wall, as illustrated in Figure 13.22b. The total length L of the reinforcement is then made up of the "active" length plus the anchor length (see Figure 13.19), that is, $L = L_E + L_A$. The length calculated in this way will vary somewhat over the height of the wall. In practice, a uniform length is normally adopted to avoid the practical complications associated with varying lengths.

Two-Part Wedge Method   The Vidal method suffers from the same limitations as the Rankine method for estimating earth pressure, and as we shall see shortly, it may also lead to unsafe designs if used with geogrids. For this reason, alternative methods have been developed. One of these is the two-part "wedge" analysis, which has become a preferred method for geogrid reinforced walls. It originates from Germany and is essentially the same as the method known as the DIBt (Deutsches Institut fur Bautechnik) method.

The two-part wedge failure mechanism is illustrated in Figure 13.23. It consists of two wedges, one of which is entirely within the reinforced wall and the other within the retained soil. The possible failure mechanisms include rupture or pullout of the reinforcement, sliding on the reinforcement, and sliding on the base soil. If necessary, the method can take account of varying surface shapes as well as surcharge loads acting on the surface. Details of the design process are illustrated in Figure 13.24. Analysis is made of all possible failure modes over the full height of the wall. The objective is to determine the required capacity of the reinforcement and to ensure that no possible failure modes have been overlooked. The great number of possible failure modes means that the method is best used with the aid of a computer program.

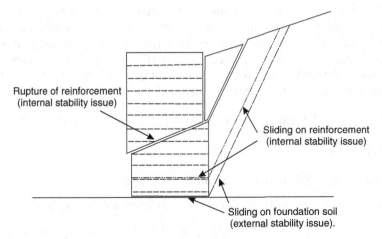

Figure 13.23 Two-part wedge method and possible failure modes.

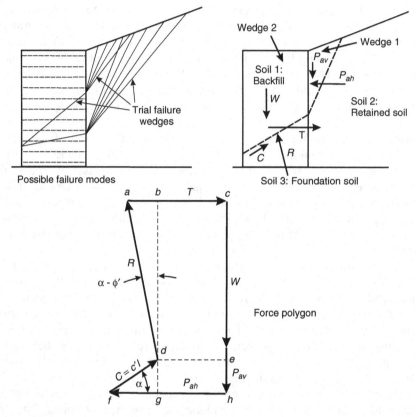

Figure 13.24 Two-part wedge analysis for internal design of a reinforced-earth wall.

Analysis of wedge 1 (within the retained soil) is analyzed in the normal way to determine the magnitude of the active pressure force on the reinforced block This is expressed in terms of its horizontal and vertical components, P_{ah} and P_{av}, respectively. Analysis of wedge 2 is then carried out using the polygon of forces shown in Figure 13.24. The only unknown is the tension T in the reinforcement; this can be obtained as follows:

$$ab = bd \tan(\alpha - \phi'), \text{ i.e., } ac - bc = (ch - eh) \tan(\alpha - \phi')$$

Therefore

$$ac - (fh - fg) = (ch - eh) \tan(\alpha - \phi') \text{ and}$$

$$ac = (fh - fg) + (ch - eh) \tan(\alpha - \phi') \text{ so that}$$

$$T = P_{ah} - C \cos \alpha + (W + P_{av} - C \sin \alpha) \tan(\alpha - \phi')$$

$$(13.37)$$

For cohesionless fill the expression becomes

$$T = P_{ah} + (W + P_{av}) \tan(\alpha - \phi') \qquad (13.38)$$

The force T is the required capacity of the layers of reinforcement intersected by the base of the wedge. Their capacity must be selected or examined to ensure it is greater than the force T. This means checking both the tensile strength and the anchor length of the reinforcement. The design process is illustrated in the following example.

Reinforced-Earth Worked Example The purpose of this example is primarily to illustrate design principles and methods, rather than provide a complete systematic design procedure. Figure 13.25 shows a reinforced-earth retaining wall 9 m high with reinforcement layers at 0.60-m intervals. We will design the wall using both the Vidal method and the geogrid (two-part wedge) method.

Design According to Vidal Method Starting with internal design, the required strength capacity of each layer of reinforcement is given by

$$T = K \gamma d \times 0.6 \, \text{kN/m along wall} = 10.92 K d \ \text{kN/m}$$

Where d is the depth to the reinforcement layer
We can determine the value of K using Figure 13.21, noting the following:

$$K_a = \frac{1 - \sin \phi'}{1 + \sin \phi'} = 0.271$$

$$K_o = 1 - \sin \phi' = 0.426$$

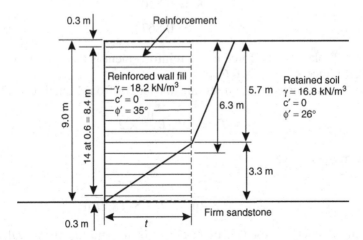

Figure 13.25 Reinforced-earth design example.

Table 13.1 Required reinforcement capacity

Reinforcement Layer Number	Depth	Required Capacity T (kN/m)
1	8.7	25.8
2	8.1	24.0
3	7.5	22.2
4	6.9	20.4
5	6.3	18.6
6	5.7	17.4
7	5.1	16.4
8	4.5	15.2
9	3.9	13.9
10	3.3	12.3
11	2.7	10.5
12	2.1	8.5
13	1.5	6.4
14	0.9	4.0
15	0.3	1.4

Numbering the reinforcement layers from the bottom up gives Table 13.1.

The next step is to determine the required length of the reinforcement and thus the width t of the wall. To do a complete design, we would need to examine the length of reinforcement at every level. However, the length does not normally vary greatly and we will only examine the situation at midheight of the wall (4.5 m). We will assume the reinforcement consists of metal strips 60 mm (0.06 m) wide at 0.60 m spacing horizontally.

From Figure 13.21, by interpolation, we have at 4.5 m

$$K = 0.310$$

$$\mu = 0.867$$

From Table 13.1 the force in the reinforcement is therefore $0.6 \times 15.2 = 9.12$ kN (since spacing is 0.6 m).

We can calculate the required anchor length from Equation 13.23:

$$T_i = \sigma'_v L_E B \frac{\mu}{F} \times 2$$

Rearranging gives

$$L_E = \frac{T_i F}{2\sigma'_v B \mu} = \frac{T_i F}{2 x 4.5 x 18.2 B \mu}$$

Using a safety factor (F) of 2, $B = 0.06$ m, and the above values of T and μ give $L_E = 2.1$ m.

From Figure 13.22, the width of the active zone is $0.3 \times 9 = 2.7\,\mathrm{m}$.

The total length of the reinforcement is thus $2.1 + 2.7 = 4.8\,\mathrm{m}$.

The wall width t must therefore be 4.8 m wide to provide the necessary anchor length of the reinforcement. We should now proceed to the external design to check that this width is adequate. However, for reasons that will shortly become apparent, we will omit this and proceed directly to design the same wall using geogrids.

Design According to Tensar Method for Tensar Geogrids   We will start with the external design and determine the required width of the wall (t) necessary to provide a safety factor of 2.0 with respect to overturning.

The earth pressure force P_{ah} coming from the retained soil is given by

$$P_{ah} = \frac{1}{2}K_a \gamma h^2$$

For $h = 9\,\mathrm{m}$ and $\phi' = 26°(K_a = 0.391)$, the active force $P_{ah} = 265.7\,\mathrm{kN}$.

Assuming the friction between wall and retained soil is fully mobilized, there will be a vertical component of earth pressure given by $P_{av} = P_{ah} \tan 26 = 129.6\,\mathrm{kN}$.

This calculation is not strictly correct as we have used (for simplicity) the Rankine formula in a situation involving wall friction. More accurate values of P_{ah} and P_{av} would be obtained using a Coulomb analysis. However, the Rankine values are conservative and we will use them here.

The weight of the wall (W) is given as $9 \times t \times 18.2 = 163.8t\,\mathrm{kN}$.

We can now take moments about the toe of the wall to determine its width. The wall is founded on firm sandstone so we do not need to consider bearing capacity failure, and rotation can be assumed to occur at the toe.

With a safety factor of 2 and resisting moment equal to 2 times the overturning moment,

$$W \times 0.5t + P_{av} \times t = 2 \times P_{ah} \times 3$$

This gives the quadratic equation $t^2 + 1.582t - 19.47 = 0$ and leads to $t = 3.7\,\mathrm{m}$.

We can immediately see that this is considerably narrower than that obtained using the Vidal method. With this width the weight of the wall (W) is 606.1 kN.

We will check that this width also provides a safety factor of at least 1.5 with respect to sliding:

$$\mathrm{SF} = \frac{(W + P_{av})\tan\phi'}{P_{ah}} = 1.94 \text{ (clearly meets this requirement)}$$

This external design thus shows that the width of 4.8 m obtained using the Vidal design method easily satisfies the external stability requirements.

We now proceed to the internal design, the first step of which is to determine the required capacity of the reinforcement by analyzing a wide range of wedge failures. We will only illustrate the process here by analyzing a single failure mechanism as shown in Figure 13.25. This involves the bottom five layers of reinforcement.

The active force now comes from the external wedge 1, which has a height of 5.7 m, giving

$$P_{ah} = 106.7 \, \text{kN and } P_{av} = 52.0 \, \text{kN}$$

The weight of wedge 2 is 495.0 kN.

The angle α (inclination of the base of the wedge) is given as $\tan^{-1}(3.3/3.7) = 41.7°$.

The total force in the bottom five reinforcement layers can now be calculated from Equation 13.38:

$$T = P_{ah} + (W + P_{av}) \tan(\alpha - \phi') = 171.0 \, \text{kN}$$

Distributing this force over the five layers of reinforcement gives the force in each layer as 34.2 kN. We must now check that the reinforcement layers have adequate anchor length. For this wedge the most critical layer is layer 5 as it has the shortest anchor length, so we will only analyze this layer. Rearranging Equation 13.36, we have for the anchor length

$$L_E = \frac{T_i F}{2\sigma'_v \alpha_p \tan \phi'}$$

Taking $F = 2$, $\alpha_p = 0.7$, this gives $L_E = 0.61 \, \text{m}$. The actual anchor length available is the length on the "passive" side of the failure surface in Figure 13.25. At 6.3 m deep this equals $0.6/\tan 41.7 = 0.67 \, \text{m}$, which is adequate. If this calculation is repeated for other wedges, we find that the maximum force in each layer is slightly higher, 35.5 kN.

We can compare the required capacity of the bottom five layers of reinforcement with the Vidal value by adding up the values in the Vidal table above. This gives the required capacity as

$$\text{Vidal method}: T = 111 \, \text{kN}.$$

$$\text{Two} - \text{part wedge analysis}: T = 171 \, \text{kN}.$$

Thus the value of T from the wedge analysis is clearly much higher than from the Vidal method. At first sight this may seem surprising, because we saw at the start of this chapter that for level ground (and zero wall friction) the Coulomb wedge method and the Rankine method produce identical results. However, the situation with a reinforced-earth wall is different from a conventional wall. With a conventional wall, the restraining capacity of the wall comes from an external source, and the full restraining capacity is available for all possible failure mechanisms. With a reinforced wall, the restraining capacity comes from within the soil mass itself and as such does not remain fully available for all failure mechanisms. In Figure 13.25, the failure mechanism is such that the restraining force comes only from the bottom five layers of reinforcement. The layers above this level do not provide any restraint at all for the wedge failure mechanism involved. The two-part wedge method thus correctly shows that the total required capacity of the reinforcement will be greater than the force required if the soil was retained by a conventional wall. The Vidal method does not take account of this fact.

This does not necessarily mean that the Vidal method is unsafe, provided it is applied only to walls using thin strips of reinforcement. In this case, as we have seen, it leads to a much greater width than the two-part wedge method. If we repeat the above two-part wedge calculation using the Vidal width of 4.8 m, the value of T (force in the bottom five layers) is 101 kN, which is now less than the Vidal value but reasonably close to it. Analysis of other wedges shows that in some cases the wedge method gives greater values than the Vidal method, but the differences between the values are not great. The analysis thus shows that the use of geogrid reinforcement makes possible a narrower wall, but the required strength capacity of the reinforcement is greater than with narrow metal strips. To complete the design, a large number of additional wedges would need analyzing to obtain the maximum strength require in each grid.

13.12.4   Other Matters

Soil Type   Reinforced-earth walls of the Vidal type have traditionally used only granular fill. With the advent of geogrid reinforcement, however, there are no longer any sound reasons (theoretical or practical) for not using clay fill. The much greater contact area between the geogrid and the soil (compared to metal strips) means that there is no difficulty in achieving adequate shear resistance between reinforcement and soil, even if the soil is plastic clay. It is essential that the clay be compacted in a properly controlled manner following procedures described in Chapter 15.

Pore Pressures and Drainage Measures   When granular fill is used, the reinforced-earth wall itself is a drainage material, and provided water can escape from the wall there is little possibility of positive pore pressures developing within the backfill. With clays this is clearly not the case. Clay backfill is of low permeability and will act as a barrier to the flow of any water coming from the soil retained. This is one disadvantage of using clay, but it can be overcome by appropriate measures at the design stage.

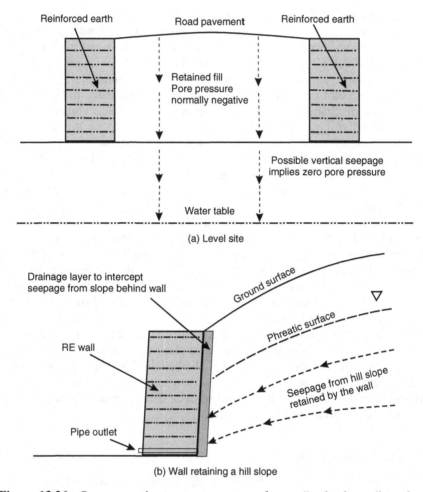

(a) Level site

(b) Wall retaining a hill slope

Figure 13.26 Seepage and pore pressure state for an "embankment" on level ground.

It is necessary to ensure that (a) no pore pressures arise in the RE material itself and (b) the wall does not act as a barrier and cause a rise in the phreatic surface in the material retained behind the wall. Two very different situations can exist, requiring quite different measures. In addition there are a range of conditions between the two, requiring individual judgment.

The first of these conditions is illustrated in Figure 13.26a. This shows reinforced-earth walls retaining a fill constructed on a level site with a water table some distance below the ground surface. If the fill is of similar permeability to the foundation soil, surface water coming from rainfall or any other source will seep vertically downward toward the water table. In this situation the pore pressures will be zero, and there is no real possibility of pore pressures building up in the retained fill or the reinforced earth. If the soil used in the wall and the retained soil is the same as the foundation soil, it is probable that the recompacted soil will be of lower permeability than the undisturbed soil, making it even less likely that there will be pore pressure buildup in the fill. Thus the need for drainage measures is minimal, at least in theory. It may still be prudent to install limited drainage measures for added security.

The second situation is illustrated in Figure 13.26b, which shows a reinforced-earth wall retaining a hillside. In this case there is likely to be a substantial rainfall catchment higher up the slope, causing seepage toward the wall. It is essential therefore to ensure that this seepage is intercepted before it can enter the wall by placing a drainage blanket at the back of the wall. An outlet from this blanket in the form of a pipe is necessary at the base of the wall. Some designers may prefer to also place a drainage layer beneath the base of the wall to make it even more secure from pore pressure buildup. For high walls intermediate drainage layers may be desirable. It is also important to shape and seal the soil at the top of the wall to prevent ponding and minimize entry of water from the surface.

REFERENCES

Burland, J. B, D. M. Potts, and N. M. Walsh. 1981. The overall stability of free and propped embedded cantilever retaining walls. *Ground Eng.*, Vol. 14 No. 5, pp. 28–38.

Ingold, T. S. 1979. The effects of compaction on retaining walls. *Geotechnique*, Vol. 29, pp. 265–283.

Jumikis, A. R. 1962. *Soil Mechanics*. Princeton, NJ: Van Nostrand Company: Princeton, NJ.

Peck, R. B. 1969. Deep excavations and tunnelling in soft ground. In *Proceedings of the Seventh International Conference on Soil Mechanics and Foundation Engineering*, State of the Art Volume, pp. 225–290. Mexico City: Mexican Society of Soil Mechanics and Foundation Engineering.

Tensar International. 2001. *Design Guidelines for Reinforced Soil Structures*. Tensar International, Blackburn, UK: Tensar International.

Terzaghi, K., and R. B. Peck. 1967. *Soil Mechanics in Engineering Practice* (2nd ed.). New York: John Wiley and Sons.

EXERCISES

1. A vertical wall 8 m high retains soil with the following properties:

$$\text{Unit weight} = 18.5\,\text{kN/m}^3,\ c' = 0, \phi' = 28°.$$

The ground surface behind the wall slopes upward at 15°.

Using the Coulomb wedge method determine the active earth pressure force on the wall and the inclination of the plane on which you would expect failure to occur (if the wall had inadequate strength to withstand the force). Ignore any friction between soil and wall and assume no seepage or pore pressure effects are involved. (**261.1 kN, 53°**)

2. A 12-m-high vertical wall supports soil made up of the following layers:

Depth (m)	Soil type	Unit weight (kN/m^3)	c' (kPa)	ϕ' (deg)
0–5	Soft grey clay	18.5	0	28
5–8	Silty sand	21.8	0	35
8–12	Soft gray clay	16.0	0	25

The water table is at a depth of 6 m. Using the Rankine method, determine the total stress profile acting on the wall. Estimate also the total force on the wall and the depth at which it acts. Assume the wall is frictionless. (**601 kN, 8.6 m**)

3. Figure 13.27 shows a concrete gravity retaining wall as well as the properties of the foundation soil and the retained soil. For simplicity, seepage or pore pressure effects can be ignored. Determine the factor of safety of the wall with respect to failure by each of the following:

   (a) Sliding, assuming sliding resistance is given by the frictional component only of the foundation soil (**1.71**)

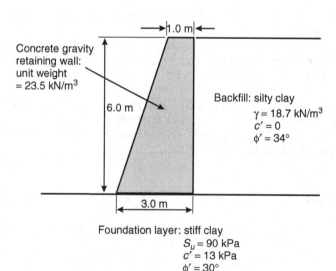

Figure 13.27   Concrete gravity retaining wall (exercise 3).

(b) Overturning about the toe of the wall **(2.83)**

(c) Bearing capacity failure, calculating the bearing capacity using the undrained shear strength of the foundation soil **(2.56)**

4. A long trench 12 m deep is to be excavated in soft to medium clay having an undrained shear strength of 25 kPa and a unit weight of 17.5 kN/m^3. Props are to be installed to support the excavation at 2.0-m intervals vertically and 3.0-m intervals horizontally, starting from a depth of 1.0 m. Determine the required strength capacity of the props at depths of 1, 7, and 11 m. **(440 kN, 660 kN, 660 kN)**

5. A reinforced-earth wall is to be erected to retain a soil height of 8 m. The ground to be supported is level, and no seepage or pore pressure effects are expected. The foundation of the wall will be on firm rock. It is planned to use geogrid reinforcement at a spacing of 0.5 m; the top and bottom geogrids will be 0.25 m, respectively, from the top and bottom of the wall. The properties of the retained soil and the wall fill are as follows:

Soil	Unit Weight (kN/m^3)	Cohesion Intercept c' (kPa)	Friction Angle ϕ' (deg)
Retained soil	15.8	0	24
Wall fill	17.3	0	32

Determine the following:

(a) The required width of the wall, adopting safety factors of 2 and 1.75 with respect to overturning and sliding failure, respectively. **(3.5 m)**

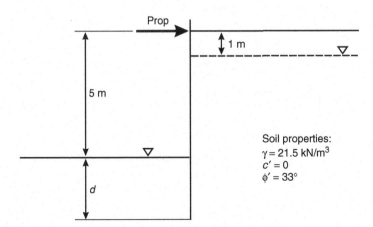

Figure 13.28   Sheet pile wall (exercise 6).

(b) The total strength capacity of the bottom two geogrids and also of the bottom 5 geogrids. Do this using the two-part wedge analysis method. **(19.9 kN, 28.0 kN)**

6. Figure 13.28 shows a sheet pile retaining wall installed into a cohesionless soil having the properties shown.

Determine the required embedment depth d and the prop force F for the following cases:

(a) With no water table, seepage, or pore pressure effects present **(2.60 m, 44.6 kN, 2.66 m, 45.8 kN)**

(b) With the water level 1 m deep in the retained soil and at the new ground level within the excavation **(6.0 m, 107.9 kN, 6.5 m, 135.9 kN)**

Do this by two methods, first by applying a safety factor of 2 to the soil strength (used in calculating the passive resistance) and second by using the Burland et al. method with a safety factor of 2.

CHAPTER 14

STABILITY OF SLOPES

14.1 INTRODUCTION

Geotechnical engineers are concerned with the following three types of slopes:

(a) Natural slopes, that is, slopes found in nature and formed by natural processes such as erosion and/or tectonic movement

(b) Slopes created by excavation or cutting into natural soils

(c) Slopes constructed of soil, such as embankments for highways or earth dams

The possibility of slope failure is present in all such slopes, arising primarily from gravity forces but strongly influenced by seepage effects in the soil and any external forces. Various ways in which slopes may fail are illustrated in Figure 14.1.

The most common types of slope failure approximate to rotational slides, meaning their failure surfaces are close to circular arcs in shape, as shown in Figures 14.1a, c, and d. In some situations, the presence of a harder soil layer near the surface may dictate the form of failure. In particular, when a hard layer is found close to the ground surface with the same inclination as the surface, the slide is likely to be relatively shallow and of constant depth. Movement in this case is parallel to the ground surface, and the slide is known as a translational slide. This form of failure is shown in Figure 14.1b.

Whether a slide is deep or shallow, as illustrated in Figures 14.1c and d, respectively, depends on the nature of the soil and the seepage conditions.

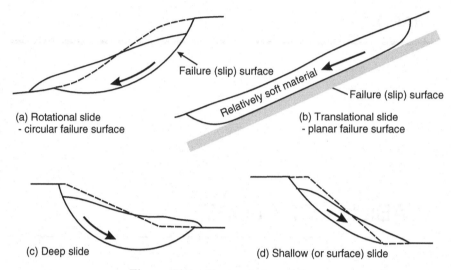

(a) Rotational slide
- circular failure surface

(b) Translational slide
- planar failure surface

(c) Deep slide

(d) Shallow (or surface) slide

Figure 14.1 Types of slope failure.

Soft soils tend to fail with deep-seated slip surfaces, as in Figure 14.1c, while stronger soils tend to fail on shallower surfaces, as in Figure 14.1d. High pore pressures within the slope tend to produce deeper slips.

In the following sections, analytical methods are presented for estimating the stability of slopes. It should be clearly appreciated that these methods have many limitations and are not the sole tools available to geotechnical engineers for assessing the stability of slopes. This is especially true of natural slopes. A full list of the methods available for assessing the stability of natural slopes would include the following:

(a) Visual inspection of the slope
(b) Geological appraisal of the slope and surrounding area
(c) Inspection of aerial photos
(d) Inspection of existing slopes in similar materials to the slope in question
(e) Slip circle analysis and other forms of analytical analysis

The first four are very important but are outside the scope of this book, which covers only analytical methods. Analytical methods are best suited to constructed slopes made of compacted soil, such as highway embankments and earth dams, where the quality and uniformity of the embankment soil can be assured by appropriate monitoring and control testing during construction. For natural slopes, (a) to (d) above are generally of more importance than analytical methods.

14.2 ANALYSIS USING CIRCULAR ARC FAILURE SURFACES

As already observed, the shear surface on which slips and landslides occur often approximates to a circular arc, and for this reason the most commonly used method of analysis makes the assumption that the surface has this form; this method is called slip circle analysis. The basis of the method is illustrated in Figure 14.2.

The actual surface on which failure is most likely to take place is unknown, and hence it is necessary to analyze many different potential failure surfaces. The surface having the lowest safety factor is the one on which failure is most likely to occur and is termed the "critical failure surface" or "critical circle." Figure 14.2 shows one possible failure surface—a circular arc with centre at point O. The equilibrium of the block of soil bounded by this failure surface is investigated by considering moments about the point O.

Moment tending to cause failure is given as W_x, where W = total weight of soil mass

$$x = \text{lever arm}$$

The moment resisting failure is given as

$$S_m R = s_m L R$$

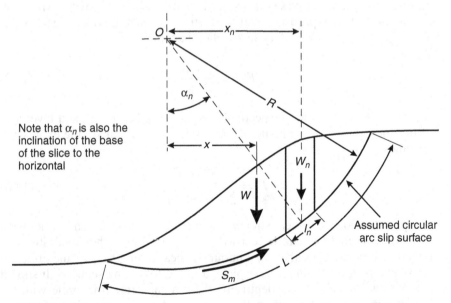

Figure 14.2 Circular arc (slip circle) method of slope stability analysis.

where S_m = total shearing resistance along slip surface needed to maintain stability (i.e., mobilized shear resistance)
$\quad s_m$ = mobilized shear strength of soil
$\quad L$ = total length of slip surface.
$\quad R$ = radius of this surface

These moments must be equal:

$$Wx = S_m R = s_m LR$$

and

$$s_m = \frac{Wx}{LR}$$

The safety factor F is defined as the ratio of the available strength to the mobilized strength; thus

$$F = \frac{s}{s_m} = \frac{sLR}{Wx} \tag{14.1}$$

This is a very simple expression, and in principle this method of analysis is straightforward. However, there are a number of practical considerations that complicate its application in practice. The first complicating factor is that soils are seldom homogeneous. To overcome this, the soil mass is divided into a number of segments, termed slices, one of which is shown in Figure 14.2. The appropriate soil properties are then applied to each slice. The expression for the safety factor then becomes

$$F = \frac{\sum sl_n R}{\sum W_n x_n} \tag{14.2}$$

where l_n, W_n, and x_n are respectively the arc length, weight, and lever arm of the slice shown (the nth slice). Noting that $x_n = R \sin \alpha_n$ the expression becomes

$$F = \frac{\sum sl_n}{\sum W_n \sin \alpha_n} \tag{14.3}$$

Up to this point the shear strength term (s) in the expression has not been defined. The analysis can be carried out in terms of either total stress or effective stress using either the undrained shear strength S_u or the effective stress parameters c' and ϕ', as explained in the following sections. It should be noted at this stage that identifying the critical circle (the circle with the lowest safety factor) requires the analysis of a large number of circles and is a very tedious undertaking if done manually. Fortunately, computer programs are readily available for such analysis.

14.2.1   Circular Arc Analysis Using Total Stresses

In this case the shear strength used in the analysis is the undrained strength of the soil and Equation 14.3 becomes

$$F = \frac{\sum s_u l_n}{\sum W_n \sin \alpha_n} \qquad (14.4)$$

where s_u is the undrained shear strength. The method is thus very easy to apply to undrained situations provided reliable measurements can be made of the undrained strength of the soil. The undrained case, however, is of limited practical relevance, as will be discussed later.

It is possible to produce the solution to Equation 14.4 in the form of a stability chart for the case of a uniform soil overlying a hard layer and simple slope geometry. This is done by introducing two dimensionless parameters, $s_u/\gamma H (= c/\gamma H)$, known as the stability number, and a depth factor D, which is the ratio of the depth of the hard layer to the slope height. A common form for such a stability chart is shown in Figure 14.3. If we

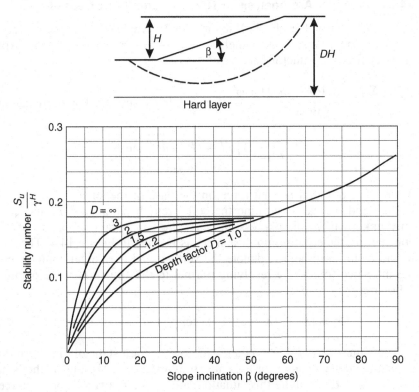

Figure 14.3   Stability chart for the total stress (undrained) case for clay of uniform strength.

know the inclination of the slope and the depth to the hard layer, then we can read off the value of the stability number and thus determine S_u, the value of the undrained shear strength needed to maintain stability, in other words the mobilized shear strength. Comparing this with the actual shear strength of the soil gives us the value of the safety factor.

The following two points can be noted from Figure 14.3:

(a) For slopes steeper than 54°, the depth of the hard layer no longer influences the stability number. This is because the critical circles in this case are all contained within that part of the slope above the toe. No circles penetrate into the soil deeper than the toe level.

(b) For slopes flatter than 54° on uniform soil ($D = \infty$), the maximum height of the slope is independent of the slope inclination, since the value of $S_u/\gamma H$ is no longer dependent on the slope angle β. This observation is primarily of theoretical interest since no clays are of infinite depth and their strength normally increases with depth.

14.2.2 Circular Arc Analysis in Terms of Effective Stresses

In this case the value of s in Equation 14.3 is replaced with the strength in terms of effective stresses namely, $s = c' + (\sigma - u)\tan \phi'$, and the expression for the safety factor becomes:

$$F = \frac{\Sigma\left[c'l + (\sigma l - ul)\right]\tan \phi'}{\Sigma W \sin \alpha} \quad \text{(for simplicity suffix } n \text{ has been omitted)}$$

or

$$F = \frac{\Sigma\left[c'l + (P - ul)\right]\tan \phi'}{\Sigma W \sin \alpha} \tag{14.5}$$

where P is the total normal force on the base of the slice.

Provided we can determine the soil strength parameters c' and ϕ', the only unknown in this expression is P. We can obtain the value of P by considering the equilibrium of a slice, as is done in Figure 14.4. The forces X and E are used here to denote the vertical and horizontal components of the interslice forces. By resolving normal to the slip surface we obtain the following expression:

$$P = \left[W + (X_n - X_{n+1})\right]\cos \alpha - (E_n - E_{n+1})\sin \alpha \tag{14.6}$$

The values of $(X_n - X_{n+1})$ and $(E_n - E_{n+1})$ are statically indeterminate, so that the determination of P is not straightforward. Various methods have been proposed to handle this situation, all of which involve some degree of approximation. The simplest method is to ignore the value of these interslice forces and take $P = W \cos \alpha$. However, it can be shown that this method

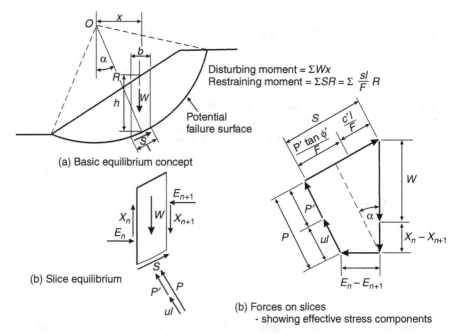

Figure 14.4   Effective stress slip circle analysis and force equilibrium of a typical slice.

may seriously underestimate the value of F, especially where the central angle of the failure arc is large or where the pore pressure is high.

Bishop (1955) showed that a much better estimate of F is obtained if it is assumed that the resultant of the interslice forces acts in the horizontal direction. When this assumption is made, we can determine the value of $P'(= P - ul)$:

$$P' = \frac{W - l\left(u \cos \alpha + \dfrac{c'}{F} \sin \alpha\right)}{\cos \alpha + \dfrac{\tan \phi'}{F} \sin \alpha} \tag{14.7}$$

Substituting this into Equation 14.5 gives the following expression for the factor of safety:

$$F = \frac{1}{\Sigma W \sin \alpha} \Sigma \left[\{c'l \cos \alpha + (W - ul \cos \alpha) \tan \phi'\} \frac{\sec \alpha}{1 + \dfrac{\tan \phi' \tan \alpha}{F}} \right] \tag{14.8}$$

It is often convenient, and common practice, to express the pore pressure in terms of a dimensionless parameter r_u, defined as

$$r_u = \frac{u}{\gamma h}$$

where γ is the unit weight of the soil and h is the height of the slice. Thus r_u is the ratio of pore pressure to total vertical stress. Noting also that $b = l \cos \alpha$ and $W = \gamma \, bl$, the above expression can be written as

$$F = \frac{1}{\Sigma W \sin \alpha} \Sigma \left[\{c'l \cos \alpha + (W - ub) \tan \phi'\} \frac{\sec \alpha}{1 + \dfrac{\tan \phi' \tan \alpha}{F}} \right]$$

(14.9)

$$F = \frac{1}{\Sigma W \sin \alpha} \Sigma \left[\{c'b + \tan \phi' \, (W \, \langle 1 - r_u \rangle)\} \frac{\sec \alpha}{1 + \dfrac{\tan \phi' \tan \alpha}{F}} \right]$$

(14.10)

Equation 14.9 is the "simplified" Bishop equation, but it is commonly referred to as just the Bishop equation. The term simplified is used because it ignores the resultant of the vertical component of the interslice forces. Bishop illustrated how these forces can be taken into account, but their influence on the safety factor is very small. Equation 14.9 is thus the form normally used in the Bishop method and is generally recognized as giving the most reliable value of safety factor. It is undoubtedly the most widely used circular arc method worldwide.

The safety factor F appears on both sides of the Bishop equation, so it is necessary to use an iterative process to determine F. The rate of convergence is generally very rapid, and only two or three iterations are needed. Because of the number of circles that need to be analyzed and the iterative nature of the solution, manual determination of F is an extremely tedious process. As noted earlier, computers have long been used for solutions.

14.2.3 Example Calculation Using Bishop Method

Figure 14.5 shows a slope of simple geometry and one possible failure circle. The application of the Bishop method to this circle is illustrated in Table 14.1. For convenience of presentation, only six slices are used, which is much less than used by most modern computer programs. Slices 1–5 are 20 m wide and slice 6 is 25 m wide. The pore pressure has been estimated taking into account the actual shape of the equipotential lines.

The safety factor has been initially "guessed" to be 1.50, and at completion of the first iteration a safety factor of 1.40 is obtained. Repeating the

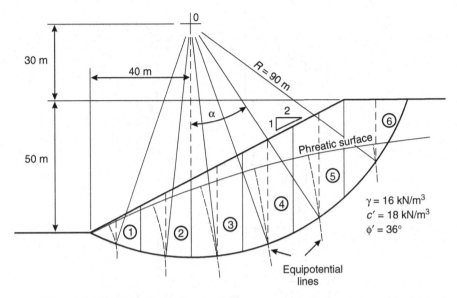

Figure 14.5   Example using the Bishop method of slip circle analysis.

analysis using this value gives a new value of 1.38. Further iterations may lower the value slightly more, but it is evident that the value converges rapidly toward its true value.

14.2.4   Bishop's Method for Submerged Slopes

Some slopes are partially submerged in water, either permanently or during part of their design life. Embankment dams clearly belong in this category. The influence of submergence is twofold. First, the presence of a body of external water provides a buttressing effect against the submerged slope and increases the resisting forces helping maintain the stability of the slope. Second, the internal seepage pattern will be affected by the changed boundary condition at the submerged slope. Submergence will normally raise the phreatic surface within the slope and increase the pore pressures. This in turn will lower the shear strength of the soil and lower the stability of the slope. The two effects are in opposing directions and the more dominant of the two will determine whether the safety factor rises or falls.

Bishop proposed a convenient method for taking account of the external buttressing force; this is illustrated in Figure 14.6. The disturbing moment about O is reduced by the moment arising from water pressure on ABC. The simplest method to take account of this moment is to imagine a section of water bounded by $ABCDA$, and to consider the equilibrium of this segment with respect to moments about O.

Table 14.1 Bishop Method Applied to Circle in Figure 14.5

1	2	5	6	7	8	9	10	11	12	13		14	
										$\dfrac{\sec\alpha}{1+\dfrac{\tan\phi'\tan\alpha}{F}}$		12×13	
Sl No	W $kN \times 10^3$	α	$W\sin\alpha$ $kN \times 10^3$	$c'b$ $kN \times 10^3$	U kPa	Ub $kN \times 10^3$	$(W-ub)$ $kN \times 10^3$	$(W-ub)$ $\tan\phi'$ $kN \times 10^3$	$7+11$ $kN \times 10^3$	$F = 1.50$	$F = 1.40$	$F = 1.50$	$F = 1.40$
1	2.69	−19.5	−0.90	0.36	61	1.22	1.47	1.03	1.39	0.911	0.902	1.27	1.25
2	7.68	−6.4	−0.85	0.36	192	3.84	3.84	2.69	3.05	0.954	0.951	2.91	2.90
3	11.0	6.4	1.22	0.36	250	5.00	5.98	4.19	4.55	0.954	0.951	4.34	4.33
4	12.8	19.5	4.26	0.36	281	5.62	7.18	5.03	5.39	0.906	0.896	4.88	4.83
5	12.7	33.8	7.05	0.36	243	4.86	7.81	5.47	5.83	0.909	0.893	5.30	5.21
6	7.71	53.7	6.22	0.45	63	1.26	6.45	4.52	4.97	1.081	0.989	5.06	4.92
Σ			17.00									$\Sigma 23.76$	$\Sigma 23.44$
												$F = 1.40$	$F = 1.38$

$F = 1.50$ $F = 1.40$

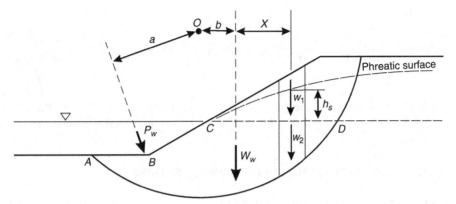

Figure 14.6 Method for taking account of partial submergence in stability analysis.

The normal pressures on AD all passes through point O and thus exerts no moment. Hence the moment equilibrium of $ABCDA$ is maintained by the external water pressure on ABC and the mass of $ABCDA$ itself. Therefore the moment $W_w b = P_w a$, where W_w is the weight of the mass of water $ABCDA$ and P_w is the force resulting from the water pressure on ABC. The disturbing moment in our slip circle analysis must therefore be reduced by the moment about point O of the mass of water $ABCDA$. To do this, we can simply use the submerged unit weight of the soil for those sections of the soil mass below the *external* water level, that is, below CD. By doing this we are reducing the disturbing moment by $P_w a$. The disturbing moment is now given by $\Sigma(W_1 + W_2)x$, where W_1 and W_2 are the weights above and below the line CD, determined using the unsubmerged and submerged unit weights, respectively

Note that the boundary CD implies nothing about the magnitude of the pore pressures inside the slope and is used only to obtain a statically equivalent disturbing moment. Pore pressures inside the slope must be treated in the normal way. The above procedure is simply a device to calculate the moment of the external water force on ABC. Bishop's expression becomes

$$F = \frac{1}{\Sigma(W_1 + W_2)\sin\alpha} \Sigma \left[c'b + (W - ub)\tan\phi' \right] \frac{\sec\alpha}{1 + \dfrac{\tan\phi'\tan\alpha}{F}}$$

$$(14.11)$$

In Equation 14.11, the use of the submerged density only occurs in the expression for the disturbing moment, as it is this moment that is reduced by the buttressing effect of the external water pressure. However, in order to avoid having to calculate W as well as $W_1 + W_2$, the above expression can be written as

$$F = \frac{1}{\Sigma \, (W_1 + W_2) \sin \alpha} \Sigma \left[c'b + (W_1 + W_2 - u_s \, b) \tan \phi' \right] \frac{\sec \alpha}{1 + \dfrac{\tan \phi' \tan \alpha}{F}}$$

(14.12)

where $u_s \, (= h_s \gamma_w)$ is the pore pressure calculated with respect to the level of submergence. Equations 14.11 and 14.12 are essentially identical.

14.3 STABILITY ANALYSIS OF INFINITE SLOPES

Many slopes are very large and the slips that occur in them approximate to translational slides with a failure plane parallel to the surface of the slope. A slip of this type is shown in Figure 14.1b. The stability in this situation can be examined reasonably accurately by ignoring the end effects (i.e., the upper and lower ends of the sliding mass) and evaluating the equilibrium of an element of the soil above the failure plane. The equilibrium situation is illustrated in Figure 14.7. The slope angle is β and the depth of a potential failure plane is H. The water table depth is H_w. The soil element is of height H, width b, and base length along the slip plane l.

From the geometry and static equilibrium of the element (parallel to the failure plane) we can write:

Weight of soil element $W = \gamma \, Hb$

Pore pressure on failure plane $u = \gamma_w h_s = \gamma_w (H - H_w) \cos^2\beta$

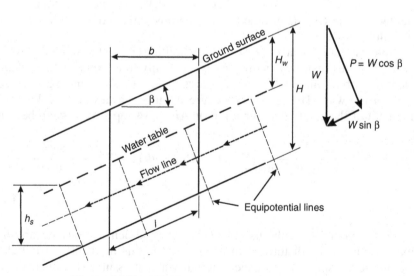

Figure 14.7 Stability analysis of an infinite slope.

Base length $l = b/\cos\beta$

Disturbing force $= W\sin\beta$

Total force normal to failure plane $P = W\cos\beta$

Note that the forces acting on the ends of the soil element are identical and cancel out in equilibrium analysis:

Resisting force $= c'l + (P - ul)\tan\phi' = c'l + (W\cos,\beta - ul)\tan\phi'$

Taking the safety factor (F) as the ratio of resisting force to disturbing force gives

$$F = \frac{c'l + (W\cos\beta - ul)\tan\phi'}{W\sin\beta} \tag{14.13}$$

Substituting W, l, and u with the expressions above leads to the following equation:

$$F = \frac{c'}{\gamma H\cos\beta\sin\beta} + \left[1 - \frac{\gamma_w}{\gamma}\left\{1 - \frac{H_w}{H}\right\}\right]\frac{\tan\phi'}{\tan\beta} \tag{14.14}$$

For the case of limiting equilibrium (SF $= 1$), this becomes

$$\frac{c'}{\gamma H\cos\beta\sin\beta} = 1 - \left[1 - \frac{\gamma_w}{\gamma}\left\{1 - \frac{H_w}{H}\right\}\right]\frac{\tan\phi'}{\tan\beta} \tag{14.15}$$

Equation 14.15 defines the relationship between the shear strength parameters needed to maintain the stability of the slope. For a given slope angle, water table depth, soil unit weight, and friction angle ϕ', the value of c'/H is constant. To maintain stability, the minimum value of the cohesion intercept c' must increase linearly with depth, a fact pointed out by Taylor (1948).

For the case of a slope in which no water table is present, the expression becomes

$$\frac{c'}{\gamma H\cos\beta\sin\beta} = 1 - \frac{\tan\phi'}{\tan\beta} \tag{14.16}$$

And for the case of a slope with a water table at the ground surface the expression becomes

$$\frac{c'}{\gamma H\cos\beta\sin\beta} = 1 - \left[1 - \frac{\gamma_w}{\gamma}\right]\frac{\tan\phi'}{\tan\beta} \tag{14.17}$$

where γ and γ_w are the unit weights of the soil and water, respectively

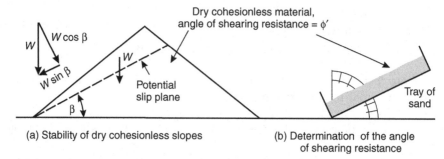

(a) Stability of dry cohesionless slopes

(b) Determination of the angle of shearing resistance

Figure 14.8   Stability of dry cohesionless material (clean sand or gravel).

These equations are essentially the same as those given by Taylor (1948), in a slightly different form. We can note also that if the soil is cohesionless and no pore pressures are present, Equation 14.16 becomes $\tan \beta = \tan \phi'$. This means that the steepest slope possible in a dry cohesionless material (a clean sand or gravel) is the angle of shearing resistance (ϕ') of the material. This fact can be demonstrated directly by considering the equilibrium of the sand heap shown in Figure 14.8a.

The slip plane of a potential wedge failure is shown. The safety factor of this wedge is given by

$$F = \frac{W \cos \beta \tan \phi'}{W \sin \beta} = \frac{\tan \phi'}{\tan \beta} \qquad (14.18)$$

Therefore F is only greater than unity if $\tan \beta \leq \tan \phi'$, so that the steepest angle at which the material remains stable is given by $\beta = \phi'$. This relationship provides a simple method of measuring the friction angle ϕ' of dry sand, as illustrated in Figure 14.8b. If we place some of the sand in a tray (to a suitable depth, say 5–10 cm) and slowly raise one end of the tray, then the sand will remain stable until a certain inclination is reached. At this point a shallow depth of sand will slump downward, indicating failure of the slope; the sand has reached its equilibrium inclination. Measurement of the angle of the tray when this occurs is a measure of the ϕ' value of the sand. This value is accurate for low stress levels. The Mohr–Coulomb failure line is not always linear and the ϕ' value at higher stress levels may be less than the value at the stress levels in the tray.

14.4   SHORT- AND LONG-TERM STABILITY OF BUILT SLOPES

In Section 9.3, a description was given of the changes in shear strength that occur with time when the load on a soil is reduced, as, for example, when an excavation is made, and also when the load is increased, as, for example, during construction of an embankment. It is important to understand these processes and the relevance of total stress and effective stress analysis to these situations. If the construction period is short, it is likely that little or no movement of water will occur into or out of the soil, and analysis of stability can be carried out on a total stress basis. This is often referred to as the **short-term** stability, the **end-of-construction** case, or the **undrained case**. Changes in stability will occur over time that may be positive or negative, depending on whether construction will cause water to seep toward the slope or away from it. Detailed accounts of the significance of these changes and how they can be taken account of are given in the following sections.

14.4.1 Excavated Slopes

We will illustrate the short-term (end-of-construction) and long-term stability situation for a 30-m-deep excavation in clay as illustrated in Figure 14.9. This shows half of a symmetrical excavation. The properties of the clay are indicated in the figure. The water table is at a depth of 3 m and the clay is assumed to be fully saturated. Figure 14.9a shows the situation immediately on completion of the excavation and Figure 14.9b shows the situation in the long term.

Provided the excavation is made rapidly and the soil is of low permeability, the stability at the completion of construction can be estimated using a total stress analysis and the undrained strength.

Using Figure 14.3 and taking $D = \infty$:

$$s_u/\gamma H = 0.18,$$

$$s_u \text{ (mobilized shear strength)} = 0.18 \times 17 \times 30 = 92 \text{ kPa.}$$

$$\text{Safety factor} = 140/92 = 1.52$$

From this point onward the effect of reducing the load on the soil strength must be considered, and this can only be done reliably using an effective stress analysis. However, as discussed in Chapter 9, effective stress analysis does not normally produce the same safety factor as total stress analysis, so that to estimate the change in safety factor over time, we must use a consistent method. We will therefore reestimate the end-of-construction safety factor using effective stresses, still assuming the behavior to be undrained. The procedure for estimating the effective stresses is illustrated by considering points A and B.

Point A: This is at the center of the excavation where the edge effects of the slopes can be ignored. The total stress on the soil here is reduced (as a result of the excavation) by

$$\Delta\sigma = h\gamma = 30 \times 17 = 510 \text{ kPa.}$$

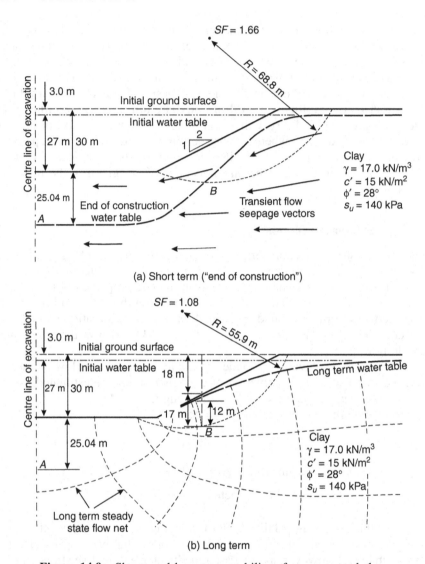

$SF = 1.66$

$R = 68.8\,\text{m}$

3.0 m

Initial ground surface

Initial water table

27 m | 30 m

2
1

Clay
$\gamma = 17.0\ \text{kN/m}^3$
$c' = 15\ \text{kN/m}^2$
$\phi' = 28°$
$s_u = 140\ \text{kPa}$

Centre line of excavation

25.04 m

End of construction
water table

B

Transient flow
seepage vectors

A

(a) Short term ("end of construction")

$SF = 1.08$

$R = 55.9\,\text{m}$

3.0 m

Initial ground surface

Initial water table

18 m

Long term water table

27 m | 30 m

17 m

12 m

B

Clay
$\gamma = 17.0\ \text{kN/m}^3$
$c' = 15\ \text{kN/m}^2$
$\phi' = 28°$
$s_u = 140\ \text{kPa}$

Centre line of excavation

25.04 m

A

Long term steady
state flow net

(b) Long term

Figure 14.9 Short- and long-term stability of an excavated slope.

The pore pressure will be reduced by exactly the same amount, since a fully saturated clay has a B parameter of 1 (Section 6.3). Thus $\Delta u = 510$ kPa. This means the reduction in the head of water in the soil is given by

$$\Delta h_w = u/\gamma_w = 510/9.8 = 52.04\,\text{m}$$

The water table thus drops by 52.04 m, resulting in a new depth of 25.04 m (27.0 − 52.04) below the base of the excavation. The pore pressure will be negative above this level.

Point B: Close to the slope and within the slope itself, the changes of pore pressure are influenced by changes in both total stress and shear stress, and exact estimates of these changes are not possible. An approximate estimate can be made by assuming, first, that the change in total vertical stress is given by the weight of the excavated soil column directly above the point and, second, that the pore pressure coefficient A is small. This means the change in pore pressure is given by the change in total vertical stress, as was the case at point A. Table 14.2 shows the calculations for each point.

An approximate new water table based on the above method of estimation is shown in the figure. Stability analysis of this end-of-construction situation using Bishop's method (and the SlopeW computer program) results in a safety factor of 1.66, taking into consideration the influence of negative pore pressure above the water table. The reduced pore pressures in the vicinity of the cutting will cause seepage toward the excavation, as shown in Figure 14.9a. Seepage during this phase is transient, and the pattern changes over time. It is not possible to draw a flow net during this phase as seepage is entering the vicinity of the excavation and is being absorbed by the soil itself. There is no "downstream" end to the seepage regime. We can only draw vectors of flow direction at specific times.

In the long term, a steady-state flow net will develop as water seeps out of the ground into the base of the excavation. An approximate steady-state flow net is shown in Figure 14.9b for this long-term condition. Using this flow net, analysis shows that the long-term safety factor is then only 1.08, which is clearly inadequate if we are designing this cutting to be permanent. The changes in pore pressure, effective stress, and safety factor that occur as a result of making the excavation are shown in Table 14.2 and illustrated in Figure 14.10. The pore pressure and effective stress curves in this figure are for point B.

It is thus not surprising that many cuttings in clay remain stable for a considerable length of time but eventually fail. The time for the pore pressure change to take place depends on the permeability of the soil, or more correctly on the coefficient of consolidation of the soil during unloading (i.e., during swelling). In some clays, especially heavily overconsolidated sedimentary clays, it may take decades for the steady-state seepage pattern to develop, while in others, especially some residual soils, it may be only a matter of weeks or even days.

14.4.2 Embankments on Soft Clays

The effect of building an embankment is the opposite to making a cutting. The increased load on the soil results in an increase in pore pressure in the soil. This causes drainage away from the soil, accompanied by consolidation and an increase in strength in the vicinity of the embankment. If analysis or design is based on total stresses using the undrained shear strength of

Table 14.2 Stress Changes Resulting from Excavation

	Initial	Change During Excavation	End of Construction	Long Term
			Point A	
Total stress σ (kPa)	$55.04 \times 17 = 935.7$	$30 \times 17 = 510.0$	$25.04 \times 17 = 425.7$	425.7
Pore pressure u (kPa)	$52.04 \times 9.8 = 510.0$	510.0	zero	284.9
Effective stress σ' (kPa)	425.7	0	425.7	$425.7 - 284.9 = 140.8$
			Point B	
Total stress σ (kPa)	$35 \times 17 = 595.0$	$18 \times 17 = 306.0$	$17 \times 17 = 289.0$ (or $595.0 - 306.0$)	289.0
Pore pressure u (kPa)	$32.0 \times 9.8 = 313.6$	306.0	$313.6 - 306.0 = 7.6$	$12 \times 9.8 = 117.6$
Effective stress σ' (kPa)	281.4	0	281.4	$289.0 - 117.6 = 171.4$

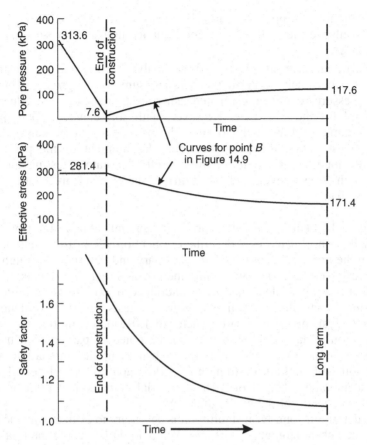

Figure 14.10   Pore pressure and effective stress (at point B) and safety factor changes over time.

the soil, this will be a safe (conservative) approach. For this reason, small safety factors may be accepted during construction on the assumption that they will increase over time.

To examine the change in stability over time of the cutting in the previous example, we used an effective stress analysis. While it is also possible to do this with embankments, there are practical difficulties that generally make a total stress approach based on the undrained shear strength more appropriate. The reasons for this are as follows:

(a) Most embankments that require careful evaluation of stability are those built on soft, normally consolidated clays. It is generally easier to measure the undrained shear strength of a soft clay than it is to measure the effective strength parameters c' and ϕ'. This is because the stress levels involved are generally quite small, and many laboratories are not equipped to carry out tests at these low stress levels. It is also difficult to obtain good-quality undisturbed samples of soft clays.

(b) To undertake an effective stress analysis at the end of construction, it is necessary to know the pore pressures in the ground. These pore pressures could be estimated using an approximate procedure as we did for the cut slope. However, with soft clays the pore pressure parameter A is likely to be high and needs to be taken account of in estimating the pore pressure. Measurement of the parameter A is not easy, since it requires careful laboratory testing and varies with stress level and position in the ground in relation to the embankment.

Figure 14.11 illustrates the stability of an embankment built on the soft normally consolidated clay described in Section 9.2.2. It is found at Mucking on the north coast of the Thames estuary and described by Pugh (1977) and Wesley (1975). The initial undrained shear strength of the clay is shown in Figure 14.11a, determined from undrained triaxial tests on high-quality undisturbed samples, as well as in situ vane tests. Using these data, a total stress analysis produces a safety factor of 1.01, indicating that the embankment is only marginally stable. It can be noted in passing that using the charts in Figure 14.3 and treating the embankment and foundation as a uniform soil with undrained strength of 15 kPa and unit weight of 21 kN/m^3 give a maximum height of 4.08 m. A trial embankment built at the site failed at a height of 4.3 m (Pugh, 1978).

To determine long-term stability, it is necessary to estimate the increase in undrained shear strength over time. We could do this using the relationship given in Chapter 9, namely,

$$\frac{s_u}{\sigma'_c} = \frac{\sin \phi'}{1 + \sin \phi'} \tag{14.19}$$

This relationship is really only valid for remolded soils that have been normally consolidated (i.e., in the laboratory), and as explained in Chapter 9, the relationship does not entirely agree with measurements on undisturbed soils. In order to obtain a reliable estimate of the expected increase in undrained strength, it is preferable to undertake laboratory measurements. Results of such measurements on Mucking clay samples from a depth of 3.3 m are shown in Figure 14.12. The undrained strength rises rather slowly at first and then increases to approach a value directly proportional to the consolidation pressure given by $s_u/\sigma'_c = 0.30$. This is in good agreement with Equation 14.19, which gives $s_u/\sigma'_c = 0.34$ for $\phi' = 32°$, the approximate value for the Mucking clay.

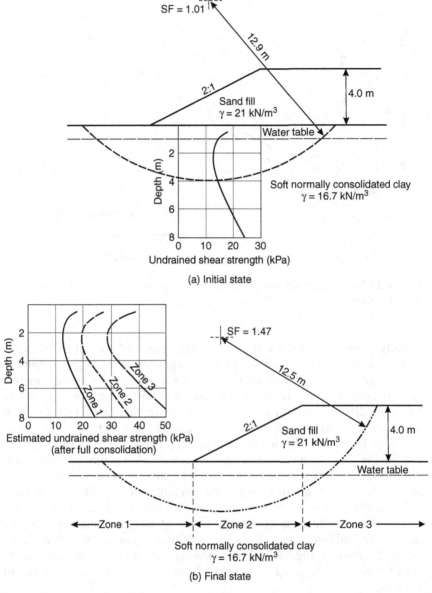

Figure 14.11 Short- and long-term stability situation of an embankment on soft clay.

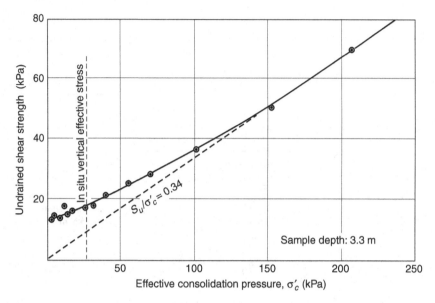

Figure 14.12 Undrained shear strength versus consolidation pressure for the Mucking clay.

Only those parts of the soft clay that experience additional stress from the embankment will show an increase in strength. Away from the embankment no change will occur, and under the central part of the embankment the full effect will occur. To allow for this, the clay beneath the embankment has been divided into three zones as shown in Figure 14.11b. In the first zone (zone 1) nothing changes, in the second zone an increase in strength resulting from half the height of the embankment is assumed to occur, and in the third zone the full effect occurs. The graph in Figure 14.12 was obtained from samples where the in situ vertical effective stress was about 25 kPa. Using this graph, the increase in strength that occurs is 122 percent in zone 3 and 54 percent in zone 2. Assuming that similar increases occur throughout the full depth of the clay and applying these ratios to the original strength lead to the two new curves of undrained shear strength shown in Figure 14.11. Repeating the undrained stability analysis with the new strengths gives a safety factor of 1.47, which is a substantial increase.

The above procedures are approximate only, and more sophisticated approaches are possible. For example, the relationship between undrained strength and consolidation pressure was obtained from isotropic consolidation on samples from one depth (3.3 m) only. More accurate relationships would be obtained by testing samples from several depths and by using anisotropic (K_o) consolidation. In particular, the increase in strength near the ground surface, where the soil has been "hardened" by surface drying, is unlikely to be as great as that assumed in Figure 14.11. The increases in stress in the ground could also be determined more accurately using a finite element study rather than adopting the simplistic division of the clay into three zones.

In practice, this increase in strength with consolidation is frequently used as a basis for increasing the height of embankments on soft clays. When this is done, it is essential to monitor the strength increase to ensure that design assumptions and expectations are fulfilled. In the author's view, the undrained strength should be monitored directly rather than relying on settlement or pore pressure records as indicators that the desired effect is being achieved. It is preferable to do this by suitable field methods, such as the cone penetrometer test (CPT), or the field vane. Measurements should be made before construction starts and at appropriate times during construction and after completion.

14.5   STABILITY ANALYSIS FOR EARTH DAMS

The methods used to determine safe slopes for earth dams are essentially the same as for other slopes. However, the pore pressures in the slopes vary during the life of the dam, and slope design must therefore focus on estimating these varying pore pressures. There are three critical conditions for which analyses must be made in the design of an earth dam:

(a) At the end of construction
(b) When the dam is full—steady state condition
(c) After sudden drawdown

Effective stress analysis using Bishop's method or any other reputable method may be used for all these cases. The challenge is to determine pore pressures under the three conditions.

14.5.1   Estimation of Pore-Water Pressures During or at End of Construction

As the embankment is built, the steadily increasing vertical load induces pore pressure rise in the compacted soil, and these pore pressures will continue to rise unless the soil is highly permeable and the pressure dissipates as construction proceeds. It is convenient to express the change in pore pressure Δu as a function of $\Delta\sigma$, the increase in total vertical stress, that is, $\Delta u = \overline{B}\Delta\sigma_1$. For earth dams with relatively gentle slopes, it is a reasonable approximation to take $\Delta\sigma_1 = \gamma d$, where γ is the unit weight of the soil and d is the vertical distance below the soil surface. We can use the expression presented earlier in Chapter 9 (Equation 9.14) to derive a relationship between the change in pore pressure and the change in total stress:

$$\Delta u = B \left[\Delta \sigma_3 + A \left(\Delta \sigma_1 - \Delta \sigma_3\right)\right] \qquad (14.20)$$

Dividing by $\Delta \sigma_1$ gives

$$\frac{\Delta u}{\Delta \sigma_1} = B \left[\frac{\Delta \sigma_3}{\Delta \sigma_1} + A \left(1 - \frac{\Delta \sigma_3}{\Delta \sigma_1}\right)\right]$$

$$= B \left[1 - (1 - A)\left(1 - \frac{\Delta \sigma_3}{\Delta \sigma_1}\right)\right] = \overline{B} \qquad (14.21)$$

For earth dam materials, A is usually small and well below unity; hence $\overline{B}$ is always less than B. Taking $\overline{B} = B$ is therefore a conservative assumption.

If we assume that the pore pressure in the compacted fill is initially zero and no dissipation occurs during construction, then the pore pressure at the end of construction is given by $u = \overline{B}\gamma d$, where d is the depth below the surface. The assumption that the initial pore pressure is zero is not normally true but is a conservative assumption. Well-compacted fill can be expected to have a high negative pore pressure at the time of compaction so that a considerable thickness of fill will need to be placed above it before the pore pressure becomes positive.

The value of $\overline{B}$ (or B) can be measured in a triaxial cell. It is very simple to measure B, but it is a little more complicated to measure $\overline{B}$. The horizontal stress (cell pressure) needs to be increased as the test proceeds to maintain the expected σ_1'/σ_3' ratio that will apply in the actual embankment.

The value of B or $\overline{B}$ is strongly influenced by the compaction water content. This is illustrated in Figure 14.13, which shows typical results of B or $\overline{B}$ measurements plotted against water content in relation to the optimum water content. The optimum water content is that at which compaction of the soil will be most effective and is explained more fully in Chapter 15.

Pore pressures can be controlled during construction by various methods, the main ones being the following:

I. Controlling the compaction water content
II. Altering the rate of construction to allow time for dissipation
III. Using drainage layers to accelerate the rate of pore pressure dissipation

For a given soil, the most important factor influencing pore pressures during construction is the compaction water content. As Figure 14.13 demonstrates, the value of the parameter B is very dependent on the water content in relation to the optimum water content. To limit pore pressures, it is desirable that the soil be compacted no wetter than the optimum water content.

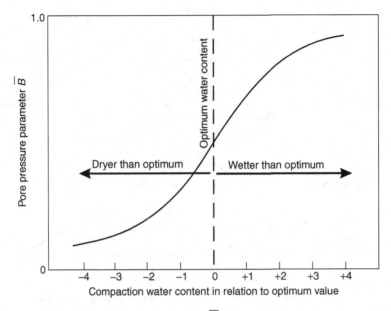

Figure 14.13   Pore pressure parameter $\overline{B}$ versus compaction water content.

This may not be practical in some situations, where the natural water content of the soil is high and the climate does not make possible adequate drying before compaction.

A possible pore pressure state in a homogeneous clay embankment built on a foundation of moderate permeability is illustrated in Figure 14.14a. This is only likely to be the situation if the soil is of low permeability and the compaction water content is on the high side of the optimum water content. Pore pressures during construction, and also on a long-term basis during the life of the dam, are normally monitored by installing piezometers in the embankment as construction proceeds.

Altering the rate of construction to allow time for pore pressures to dissipate may be possible in some situations but would involve costly delays to the completion of the project which may be unacceptable. A measure sometimes used to limit pore pressures is to install horizontal sand layers in the slopes to provide drainage paths for the escape of pore pressures, as illustrated in Figure 14.14b. For dams on clay foundations, where the shear strength of the foundation may be of concern, sand drains may also be used to accelerate pore pressure dissipation in the foundation.

14.5.2 Full-Reservoir Steady-State Seepage Condition

Throughout most of its life a dam is likely to be operating full or nearly full. At the design stage, flow nets can be established from which to estimate pore pressures for use in stability analysis. For simple cross sections, hand

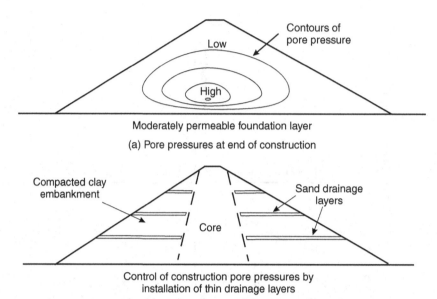

Moderately permeable foundation layer

(a) Pore pressures at end of construction

Control of construction pore pressures by
installation of thin drainage layers

(b) Use of sand layers to accelerate rate of pore pressure dissipation

Figure 14.14 Possible pore pressures state in a clay embankment and measures to control them.

sketching is satisfactory, but for complex cross sections, computer programs are more appropriate. After completion and filling of the reservoir, the design flow net can be checked from the piezometer records and the stability analysis rechecked if necessary.

14.5.3 Rapid Drawdown Pore Pressures

Stability against rapid drawdown is often the controlling factor determining the inclination of upstream slopes in dams. Drawdown is "rapid" if little pore pressure dissipation occurs during the drop in water level. For large dams of low-permeability soil, a drawdown rate of 0.2 m/day may be "rapid." Drawdown has two effects:

1. The removal of the water takes away the buttressing effect of the water pressure, so that the sliding moment and shear stress on a potential sliding surface are increased.
2. The removal of the water pressure also results in a reduction in the internal pore pressures; this needs to be accounted for when carrying out stability analysis using effective stresses.

The net result of rapid drawdown is almost invariably a reduction in the safety factor. The pore pressure change can be expressed as

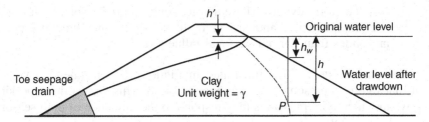

Figure 14.15   Estimation of rapid drawdown pore pressures.

$\Delta u = \overline{B}\Delta\sigma_1 = \overline{B}\gamma_w h_w$, where h_w is the reduction in water level. When considering pore pressures during construction, we had previously seen (Equation 14.21) that

$$\overline{B} = \frac{\Delta u}{\Delta\sigma_1} = B\left[1 - (1 - A)\left(1 - \frac{\Delta\sigma_3}{\Delta\sigma_1}\right)\right]$$

Under rapid drawdown $B = 1$ (or nearly 1) while $\Delta\sigma_3 < \Delta\sigma_1$, so that $\overline{B} > 1$. Taking $\overline{B} = 1$ is safe as it means the estimated reduction in pore pressure is less than the true reduction. Taking $\overline{B} =$ unity, the pore pressure at point P after drawdown can be calculated as follows, making use of Figure 14.15.

The initial pore pressure prior to drawdown is given by $u_o = \gamma_w\left(h - h'\right)$. The change in pore pressure during drawdown is $\Delta u = \gamma_w h_w\left(reduction\right)$. Therefore after drawdown the pore pressure is

$$u = u_o - \Delta u = \gamma_w\left(h - h' - h_w\right) \tag{14.22}$$

The influence of drawdown needs to be checked for various levels of drawdown. In some unusual situations it may be that the drawdown of $^1\!/_2$ or $^3\!/_4$ of total height gives a lower safety factor than full drawdown.

14.6   INFLUENCE OF CLIMATE AND WEATHER ON STABILITY OF SLOPES

Rainfall is undoubtedly the most common trigger for failure of slopes, especially natural slopes in residual soils. Rainfall causes an increase in pore pressures and a rise in the water table, leading to reduced shear strength and possible failure. The influence of rainfall is made up of two components:

1. Regular seasonal influence. This is cyclical in nature and for many climates is reasonably predictable, as described in Chapter 4.

2. Isolated storm events. These are generally unpredictable, both in timing and intensity, and are more likely to be the direct trigger of landslides than normal seasonal influences.

Seasonal influence was illustrated in Figure 4.6. An amended version of that figure is presented here as Figure 14.16. This shows three possible ways in which pore pressures may respond to the influence of both seasonal and storm events:

(a) Response to both seasonal change and storm events. This can be expected to be the case close to the surface.
(b) Response to seasons but no response to storms. Because storm events are of much shorter duration than seasons, their influence is likely to fade away at a much shallower depth than seasonal influence.
(c) No response to either seasons or storms.

The trends shown in Figure 14.16 are to be expected in clay slopes on the basis of theoretical considerations, assuming the soil conditions are uniform. However, actual records from piezometers installed in slopes in Hong Kong (*Geotechnical Manual for Slopes*, 1984) show a considerably more complex picture. Some piezometers show no response to seasons but react sharply to storm events. Others show seasonal effects but very little response to storms. The explanation for this behavior is probably to be found in nonuniform soil conditions. The soils involved in the Hong Kong measurements were predominantly weathered granites, which are relatively coarse grained (silty

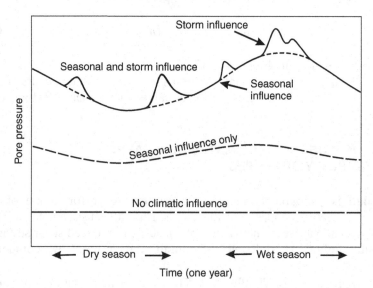

Figure 14.16 Pore pressure response to climatic effects in clay slopes.

sands) and involve major variations in properties depending on the degree of weathering.

Theoretical Example A theoretical analysis of the way rainfall can influence the stability of a natural clay slope is illustrated in Figures 14.17 and 14.18. This analysis examines the effect of continuous rainfall on a particular slope for a period of 20 time intervals. The time intervals used here are days, but they could be hours, weeks, or years, depending on the permeability and compressibility of the soil. The soil properties used in the analysis are $\gamma = 17 \, kN/m^3$, $c' = 16 \, kPa$, $\phi' = 35°$, $k = 0.05 \, m/day$, and $m_v = 0.0001 \, kPa^{-1}$. The slope is shown in Figure 14.17. The water table is initially assumed to be at a low level, only a little higher than the toe of the slope. Analysis of the pore pressure change has been carried out using the computer program Seep/W (which can handle transient flow situations). The seepage states determined in this way have then been introduced into stability analysis using the program Slope/W.

The way the pore pressure changes is illustrated in Figure 14.18a for one representative location, namely line $a - b$ of Figure 14.17. The assumption at the start of the analysis is that the pore pressures above and below the water table are those satisfying hydrostatic equilibrium. This is a crude approximation, as the actual state will be governed by the preceding weather

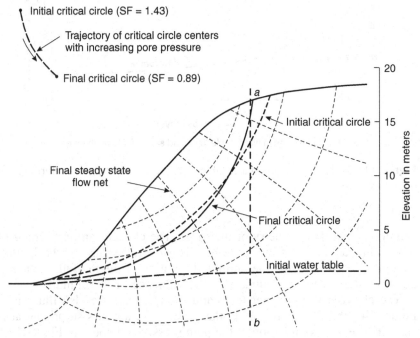

Figure 14.17 Influence of direct rainfall on the stability of a slope.

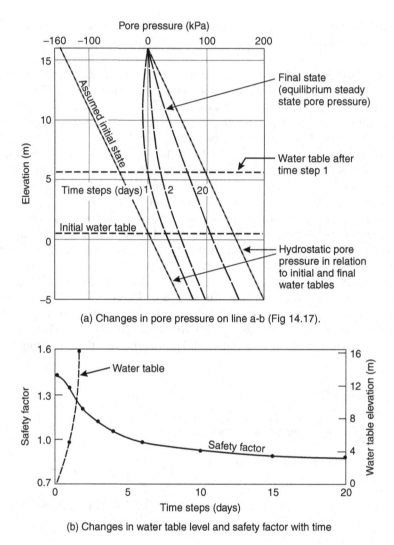

(a) Changes in pore pressure on line a-b (Fig 14.17).

(b) Changes in water table level and safety factor with time

Figure 14.18 Pore pressure, water table, and safety factor changes during rainfall on a clay slope.

conditions and the geometry of the slope. The changes are in keeping with those illustrated earlier in Figure 4.5. However, because of the boundary conditions used here and the unlimited duration of the rainfall, the analysis does not show a lower boundary to the influence of the rainfall.

The changes in the water table and safety factor are also illustrated in Figure 14.18b. The sharp rise in the water table appears surprising at first sight. However, examination of the pore pressure contours in Figure 14.18a shows it to be quite logical. Rapid changes in the water table occur initially, and in the first time step it rises by almost 6 m and by the second time step it has already reached the ground surface. A small change in the position of the pore pressure profile, as occurs between time steps 1 and 2, results in a very large rise in the water table.

The pore pressure continues to rise and the safety factor continues to fall long after the water table reaches the ground surface. The safety factor is 1.43 at the start and declines to unity shortly after 5 time steps, at which point the slope will theoretically fail. An equilibrium condition is reached after about 20 time steps; the flow net for this condition is shown in Figure 14.17. It should be noted that in the steady-state condition the pore pressures are still well below the hydrostatic values in relation to the water table at the ground surface. This is because the equipotential lines are far from vertical.

In practice, it is most unlikely that the seepage state will ever reach an equilibrium situation during a rain storm. Pore pressures will be transient, rising to a certain level, and then declining again, as indicated in Figure 14.16. The storm duration in this example could be only one or two time steps, so that in theory the slope will not fail. Figure 14.18a also shows the position of the critical failure circles at the start and at the end. It is seen that there is not a great change in their positions

In concluding this section on the influence of rainfall on natural slopes, the following observations are made:

(a) The soil properties used in the above example do not appear to be very representative of most soils, although they can be found in some residual soils. More representative values would produce much longer time steps and would suggest that very long storm durations would be required to initiate failure. Despite this, field evidence shows clearly that rainfall is the principal trigger for most slips in natural slopes, especially those consisting of residual soils. The explanation may be that many residual soils tend to be heterogeneous and contain numerous discontinuities, so that the true influence of rainfall may depend much more on these properties than on the parameters of the intact soil.

(b) The analysis above is not intended to suggest that it might be possible to predict slope failure from a knowledge of the slope and soil properties together with rainfall data. Neither the slope properties nor the rainfall and storm characteristics are ever likely to be known with the degree of reliability needed to make realistic predictions.

14.7   STABILITY ANALYSIS USING NONCIRCULAR FAILURE SURFACES

The use of circular arc stability analysis is very widespread and has proved its usefulness as an analytical tool for assessing the stability of slopes and as a design tool for man-made slopes. However, there are certain natural situations where failure is more likely to take place on noncircular surfaces. The presence of either significantly harder or softer layers is an example of such situations. An approximate analysis may still be possible using the circular arc method, but it is also possible that such an analysis will not give a reliable answer. Because of this, a number of methods have been developed for analyzing noncircular failure surfaces. It is not the intention here to give a detailed account of these methods, only to make some general observations.

Figure 14.19 shows failure of a soil mass on a circular and a noncircular failure surface. Failure on a circular arc surface is always kinematically feasible. The sliding soil mass can move and rotate on such a surface without suffering any distortion, at least in theory. This is illustrated in Figure 14.19a. The success of the Bishop method of analysis is partly due to this fact. Bishop's routine method assumes that there are no vertical interslice forces, only horizontal ones. This is clearly a reasonable assumption, since there are no obvious physical phenomena acting that are likely to cause vertical interslice forces. However, if the failure surface is not circular, then movement of the soil mass cannot occur without causing distortion of the mass. This is illustrated in Figure 14.19b by imagining the soil is divided into discrete slices as is done in stability analysis using the method of slices.

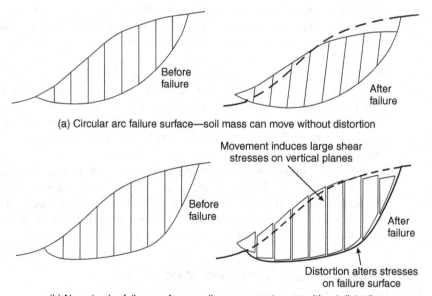

(a) Circular arc failure surface—soil mass can move without distortion

(b) Non-circular failure surface—soil mass cannot move without distortion

Figure 14.19 Circular and noncircular failure surfaces and influence on interslice forces.

Once movement commences, the slices will tend to move relative to each other and shear stresses will build up at the vertical boundaries. This effect will be minor over those sections of the failure plane where the curvature is constant (or is linear) but will be very significant where the curvature of the failure plane changes rapidly. Shear failure is likely to be occurring in some zones within the soil mass as well as on the slip plane. It is worth noting also that the interslice shear force may not always act in the same direction at all slice boundaries. In Figure 14.19b the interslice forces all act in the same direction; at the right side of each slice the adjacent slice is causing a downward drag on the boundary. This may not always be the case, especially if curvature of the failure plane becomes convex rather than concave.

It is clear from the above considerations that inclusion of the interslice shear forces is essential in any method of analysis using noncircular failure surfaces. This has been done in various ways by the methods that have been developed to date. The best known methods are probably those by Morgenstern and Price (1967) and Sarma (1973).

REFERENCES

Bishop, A. W. 1955. The use of the slip circle in the stability analysis of slopes. *Geotechnique*, Vol. 5 No 1, pp. 7–17.

Geotechnical Manual for Slopes, 2nd ed. (1984). Geotechnical Engineering Office, Civil Engineering Department, Government of Hong Kong.

Morgenstern, N. R., and V. E. Price. 1967. The analysis of the stability of general slip surfaces. *Geotechnique*, Vol. 15, No. 1, pp. 79–93.

Pugh R. S. 1978. The strength and deformation characteristics of a soft alluvial clay under full scale loading conditions. PhD thesis, Imperial College, University of London.

Sarma, S. K. 1973. Stability analysis of embankments and slopes, *Geotechnique*, Vol. 23, No. 3, pp. 423–433.

Taylor, D. W. 1948. *Fundamentals of Soil Mechanics*. New York: John Wiley and Sons.

Wesley, L. D. 1975. Influence of stress path and anisotropy on the behaviour of a soft alluvial clay. PhD thesis, Imperial College, University of London.

EXERCISES

To solve most of the questions in this chapter, it is necessary to make use of a slope stability computer program. Many such programs are readily available, and most universities will have a slope stability program available for use by students. A convenient program used by the author is Slope/W;—a student version is available free of charge from the maker's website (http://www.geo-slope.com).

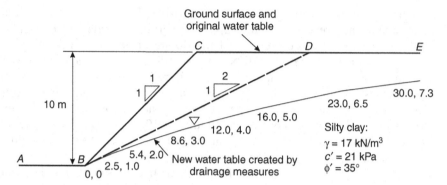

Figure 14.20 Slope details for stability analysis (exercises 1 and 2).

1. Figure 14.20 shows a slope 10 m in height with an inclination of 1:1 defined by the lines *ABCDE*. The slope consists of silty clay with the properties shown in the figure. Determine the safety factor of the slope using the Bishop method for the following conditions:

   (a) The water table (under worst conditions) is at the ground surface. Assume the seepage is horizontal and equipotential lines are vertical. (**1.04**)

   (b) The slope is entirely free of any seepage or pore pressure effects. (**1.98**)

2. In its worst condition, the slope in exercise 1 (Figure 14.20) is close to failure (SF = 1.04). It is desirable to increase the safety factor of the slope. Determine the new safety factor if the measures described in (a), (b), or (c) are taken:

   (a) Drainage measures are installed that ensure the water table does not rise above the level shown. Assume equipotentials are vertical. (**1.89**)

   (b) The slope is flattened to have a new shape defined by lines *ABDE* and thus an inclination of 2:1. Assume the water table is again at the new ground surface, with seepage in the horizontal direction and vertical equipotential lines. (**1.65**)

   (c) Both measures in (a) and (b) are implemented. (**2.18**)

3. A cutting for a new highway is to be made in level ground to a depth of 15 m. The soil consists of stiff to hard clay having the following properties: $\gamma = 16.8\,\text{kN/m}^3$, $c' = 23\,\text{kPa}$, $\phi' = 30°$.

   Determine suitable slope angles for the cutting, adopting a safety factor of 1.5 and making the following assumptions:

   (a) There is no seepage or pore pressure in the slope ($r_u = 0$).(**47°**)

   (b) The value of r_u is 0.45. (**2H:1V**)

4. Figure 14.21 illustrates a slope that is showing signs of developing instability; in other words its safety factor is close to unity. The drawing

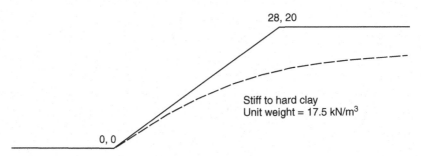

28, 20

Stiff to hard clay
Unit weight = 17.5 kN/m³

0, 0

Figure 14.21   Slope for back analysis (exercise 4).

also shows the phreatic surface in the slope established from stand-pipe piezometers. The coordinates (x, y) of the phreatic surface are as follows:

$$0, 0; \; 5, 3; \; 11, 6; \; 15, 8; \; 20, 10; \; 25, 12; \; 29, 13; \; 35, 14; \; 45, 15; \; 60, 15$$

The unit weight of the soil is $17.5 \, \text{kN/m}^3$.

By a back-analysis procedure and assuming the soil is homogeneous, determine the combinations of c' and ϕ' that correspond to a safety factor of unity. Plot a graph of c' versus tan ϕ'. This can be done by assuming values of ϕ', say at $10°$ intervals, and then by a trial-and-error process adjusting the value of c' to make SF $= 1.0$.

(No solution is given here as this would eliminate the trial-and-error method.)

5. A very large uniform slope has an angle of inclination of $25°$ to the horizontal. A layer of hard rock is found at a depth (vertically) of 8 m below the ground surface and its surface is parallel to the ground surface. Assuming the soil has a unit weight of $16.5 \, \text{kN/m}^3$ and an angle of shearing resistance of $20°$, determine the value of the cohesion intercept c' at the soil–rock interface needed to maintain equilibrium (SF $= 1$) for the following conditions:

(a) No seepage or pore pressures are present in the slopes. **(1.1 kPa)**

(b) The water table is 1.0 m deep (vertically) and parallel to the ground surface.

(Assume seepage and equipotential lines are respectively parallel to and perpendicular to the ground surface.) **(31.6 kPa)**

CHAPTER 15

SOIL COMPACTION

15.1 EARTHWORKS AND SOIL COMPACTION

Earthworks, which involve the excavation, transport, and recompaction of soil, are carried out for a variety of reasons, including reshaping of the ground to make it more suitable for suburban or commercial use, the construction of embankments for highways or railways, and the construction of earth dams. As mentioned in Chapter 8, the term **compaction** should not be confused with consolidation. Compaction is a mechanical process in which dynamic energy is used to make the soil more compact; it squeezes air out of the void spaces and thus pushes the soil particles closer together. It does not involve the removal of any water from the soil during the compaction process itself, although drying or wetting the soil may be done prior to commencement of the compaction operation.

Compaction can be carried out using various devices. In the field, compaction of clay can be done using various types of equipment, including smooth wheeled rollers, pneumatic-tyred rollers, "sheepsfoot" rollers, and grid rollers. Compaction of granular materials is best done using vibrating rollers, which are normally smooth wheeled. In the laboratory, soil is normally compacted using falling-weight hammers (or rammers) of known dimensions.

15.2 COMPACTION BEHAVIOR OF SOILS

The behavior of soils during the compaction process is best understood by considering the laboratory tests normally used to measure compaction characteristics. These tests are generally referred to as the standard Proctor test

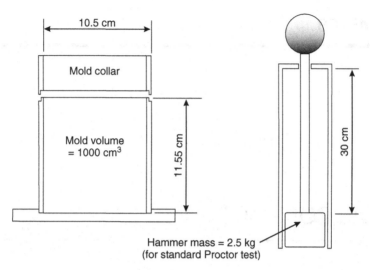

Figure 15.1   Apparatus used for Proctor compaction tests.

and the modified (or heavy) compaction test. The tests involve compacting the soil in a cylindrical container (compaction mold) of known dimensions using a specified number of layers and hammers of known weight and fall height. The apparatus is illustrated in Figure 15.1.

The tests are carried out on a series of samples of the soil prepared at different water contents with approximately equal increments between them. Each sample is tested in turn. The soil is systematically compacted in layers of equal thickness, carefully selected so that the top of the final layer is just above the top of the mold, extending into a collar that sits on top of the mold. The collar is then removed and the soil trimmed level with the top of the mold. The soil is weighed and a sample taken for moisture content measurement. The mold used for the two tests (standard and modified) is identical, but the hammer used for the modified test is heavier with a greater drop height, and the number of layers in the modified test is five instead of the three layers used in the standard test. Details of the mold, hammer, and number of layers in the two tests are given in Table 15.1.

The modified test involves a substantially greater energy input (termed the compactive effort) than does the standard Proctor test. The results of the tests are plotted as graphs of dry density versus water content. Typical test results for a clay are shown in Figure 15.2. Also plotted on the graph is the zero air voids line; this indicated the maximum value the dry density can have at any particular water content.

It is clear from the shape of the graphs that for a given compaction effort there is a water content at which the density is maximum. This water content is termed the optimum water content, and the corresponding dry density is known as the maximum dry density. These parameters are used

Table 15.1   Details of the standard (Proctor) and modified compaction tests

	Standard Proctor Test	Modified (or Heavy) Test
Mold diameter (cm)	10.5	10.5
Mold height (cm)	11.55	11.55
Mold volume (cm^3)	1000	1000
Number of soil layers	3	5
Weight of hammer (kg)	2.5	4.5
Fall height of hammer (cm)	30	45
Blows per layer	27	27

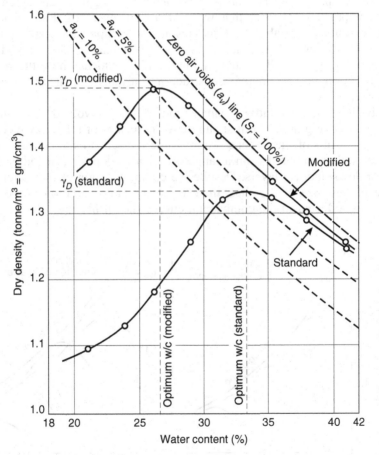

Figure 15.2 Compaction tests on clay—standard Proctor test and modified (heavy) test.

in the traditional method for controlling the compaction of soil in the field (see Section 15.3). It should be clearly appreciated that the optimum water content is not a unique property of the soil; it is simply the preferred water content at which to compact the soil using a particular method of compaction or particular compactive effort. The optimum water content of the soil is different for each compactive effort. For the soil in Figure 15.2, the optimum water content from the modified compaction test is about 8 percent below that for the standard compaction test. This is normal for moderate- to high-plasticity clays. It is worth noting also that for moderate- to high-plasticity clays the optimum water content from the standard Proctor compaction test is normally close to the plastic limit.

Before describing the traditional compaction control method based on these compaction tests (and possible alternatives), there are other aspects of compaction behavior that we need to be aware of. First, many soils are extremely variable, and the results of compaction tests of the sort described above on samples from these soils may produce a wide range of results. This is illustrated in Figure 15.3, whichs show the results of compaction tests from two construction sites in the Auckland region of New Zealand.

Both sites are relatively small, and the soil involved is of the same geological origin. At the first site the soil consists of relatively recent sedimentary soils of Pleistocene origin, while the second site consists of much older soils weathered from a range of volcanic deposits, including basaltic lava flows and ash layers. Despite the common origin of the materials and the limited size of the sites, there is a very large variation in the type of soil, as reflected in the compaction curves shown in Figure 15.3.

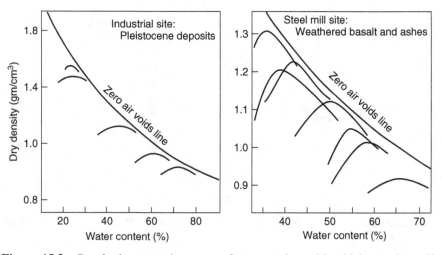

Figure 15.3 Standard compaction curves from two sites with widely varying soil types. (After Pickens, 1980.)

Second, some unusual soils do not conform to the behavior illustrated in Figures 15.2 and 15.3. Their compaction curves do not show clear peaks of dry density and thus do not indicate optimum values of water content. One such soil is clay containing a high proportion of the clay mineral allophane. As mentioned in Chapter 1, allophane is a very unusual mineral and clays with a high allophane content display some rather unique properties. One of their distinctive characteristics is the very flat nature of their compaction curves. Figure 15.4 shows typical results of compaction tests on two samples of allophane clay.

The tests have been carried out by progressively drying the soil from its natural water content, which was 166 and 195 percent for samples A and B, respectively. Fresh soil was used for each point on the compaction curve. This procedure is essential as repeated compaction may progressively soften the soil, and excessive drying before testing can irreversibly alter its properties (see Wesley, 2002). Standard compactive effort was used for the tests. Sample A shows an optimum water content around 135 percent, although it is not well defined, and sample B does not show an optimum value at all. The highest dry density is achieved by progressively drying the soil to very low water contents, but such a procedure is unlikely to be feasible in practice.

Third, some soils become softer during the compaction process. As we have seen, many soils, especially residual soils, are "structured," and when remolded this structure is broken down and the soil becomes softer. It is important to recognize that compaction of a soil can have two important but very different effects, one of which is not actually compaction at all. These effects are:

(a) "Densifying" the soil, that is, pushing the particles closer together and squeezing out air entrapped between the particles.

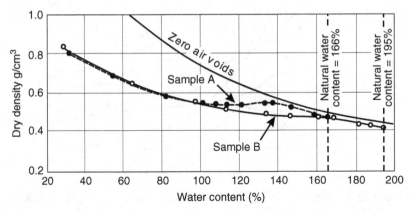

Figure 15.4 Standard Proctor compaction test on an allophane clay.

(b) Remolding the soil, causing it to soften. This is usually accompanied by release of water trapped within or between the particles, adding to the softening process.

Figure 15.5 illustrates the softening effect that compaction has on volcanic ash soils in Japan. Tests have been done on a range of volcanic ash soils in which strength measurements have been made after the soils have been compacted using varying compactive effort. The strength has been measured using a cone penetrometer test and the compactive effort varied by changing the number of blows of the compaction hammer. It is seen that nearly all of the samples become softer as the blow count increases. Only sample A shows a consistent strength increase until the blow count reaches about 70, beyond which a small decrease in strength occurs. The softening effect produced by higher compactive effort is often referred to as "overcompaction."

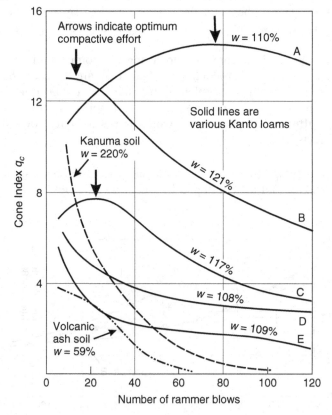

Figure 15.5 Influence of repeated compaction on the strength of volcanic ash soils. (After Kuno et al., 1978.)

15.3   CONTROL OF COMPACTION

15.3.1   Traditional Method of Compaction Control

The normal method for controlling compaction is to carry out a laboratory compaction test, either the standard Proctor or the modified test, and use the results to specify limits of water content and dry density. Common practice would be to specify water content limits within several percentage points on each side of the optimum water content and minimum dry density not less than 90 or 95 percent of maximum dry density. The choice between the standard Proctor test and the modified test should be made taking into consideration the following factors:

(a) The purpose for which the soil is being compacted. If it is a relatively low embankment for a road, then a very high-quality material is unlikely to be required and the standard Proctor test would be appropriate. However, if the objective is a high earth dam, then a higher quality fill may be desirable, in which case the modified test would be more appropriate.

(b) The equipment available for compaction in the field. If only light equipment is available, then it would be unlikely that the density corresponding to the modified test could be achieved.

(c) The natural water content of the soil and the weather conditions at the site. If the natural water content is high in relation to the optimum water content from the standard Proctor test and climatic conditions are such that drying of the soil may be difficult, then it would clearly be impractical to adopt the modified optimum water content, which would require much more drying than the standard Proctor optimum value.

It will be apparent from the compaction characteristics described in Section 15.2 that the traditional method for controlling compaction is not always easy to apply. In particular, when soil characteristics vary as much as in Figure 15.3, use of the traditional method is simply not practical, and alternative methods are desirable. In the author's experience, the method described in the following section is a very appropriate and practical alternative. It was developed in New Zealand to cope with the rapid variations that occur in many local residual soils and is described in detail by Pickens (1980). An outline only of the method is given in the next section.

15.3.2   Alternative Compaction Control Based on Undrained Shear Strength and Air Voids

The principal objectives in compacting soil are normally to create a fill of high strength and low compressibility and, in the case of water-retaining fills, of low permeability. It is also desirable that the fill will not significantly soften with time as a result of exposure to rainfall. In adopting the traditional control method it is assumed that by aiming for maximum density the above objectives will be achieved. This is not automatically true, and there is no reason why other parameters will not achieve the intended objectives equally well. Undrained shear strength and air voids appear to be suitable alternative parameters and are more directly related to the intended properties of the fill.

Figure 15.6 illustrates the basis for using undrained shear strength; it shows the results of a standard Proctor compaction test on clay, during which measurements of undrained strength have been made, in addition to density and water content. The measurements have been made using both a hand shear vane and unconfined compressive tests on samples of the compacted soil. The two strength measurements give significantly different results.

It is seen that at the optimum water content the undrained shear strength is about 150 kPa from the unconfined tests and about 230 kPa from the vane tests. Conventional specifications may allow water contents 2 or 3 percent greater than optimum, in which case the comparable shear strength values would be about 120 and 180 kPa. Thus to obtain a fill with comparable properties to those obtained with conventional control methods, specifying a minimum undrained shear strength in the range of about 150–200 kPa

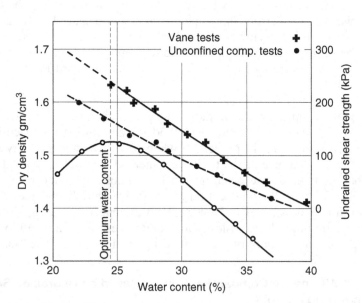

Figure 15.6 Standard Proctor compaction test on clay, including measurements of undrained shear strength.

would be appropriate. This would put an upper limit on the water content at which the soil could be compacted. Since the undrained shear strength steadily rises with decreasing water content, the required shear strength could be achieved by compacting the soil in a very dry state, which would generally be undesirable, as dry fills may soften and swell excessively when exposed to rainfall. To prevent the soil from being too dry, a second parameter is specified, namely the air voids in the soil.

Figure 15.2 indicates that at optimum water content the air void in the soil is generally about 5 percent. If the soil is compacted 2–3 percent drier than the optimum water content corresponding to the compaction effort being used, the air voids may be as much as 8 or 10 percent. Thus to prevent the soil from being compacted too dry an upper limit is placed on the air voids, normally in the range of 8–10 percent. Figure 15.7 illustrates how this method of controlling compaction relates to the traditional method. The zero-air-voids line is always the upper limit of the dry density for any particular water content and thus applies to both methods. The traditional method involves an upper and lower limit on water content and a lower limit on dry density and thus encloses the area shown in the figure. The alternative method involves an upper limit on water content, corresponding to the minimum shear strength, and a line parallel to the zero-air-voids line representing the upper limit of air voids. There is no specific lower limit of water content, but the air voids limit prevents the soil from being too dry.

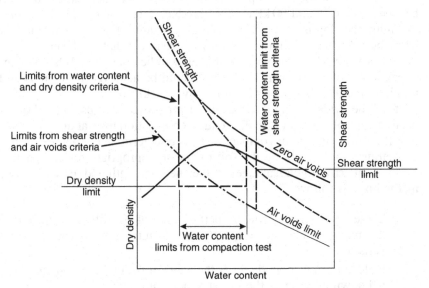

Figure 15.7   Compaction control using alternative specification parameters.

Experience has shown that suitable limits for the two control parameters are as follows:

Undrained shear strength (hand vane values): Not less than 150 kPa (average of 10 tests)

Minimum single value: 120 kPa

Air voids (for "normal" soils): Not greater than 8 percent

These values have been found to be very satisfactory in producing firm, high-quality fills. The undrained shear strength can be measured in situ by a hand shear vane or by taking samples for unconfined compression tests. The hand shear vane is the much simpler of the two methods. The air voids can only be determined by measuring the density and water content in the usual way. The author's experience has been mainly in temperate or wet tropical climates, where it is often the case that the soil is too wet and the undrained shear strength criterion is difficult to meet while the air voids requirement is easily achieved. This means that the quality control consists essentially of checking the shear strength. With the hand shear vane this checking can be done as the compaction operation proceeds.

While the criteria above are suitable for a wide range of compaction operations, there are some situations where other properties may be important, and the criteria can be adjusted accordingly. For example, the core of an earth dam built on compressible foundations or in a seismic zone may need to be plastic, or ductile, to allow for possible deformations in the dam. This can be achieved by adopting a lower undrained shear strength; a value between about 70 and 90 kPa would produce a reasonably plastic material, assuming the clay is of moderate to high plasticity. For a clay embankment being built for a new highway, it may be desirable that the layers closest to the surface (on which the pavement itself will be constructed) have a higher strength than those deeper down. This could be achieved by increasing the required undrained shear strength to, say, 200 kPa.

It will be evident from the account given above that this method of compaction control does not actually require compaction tests at all. However, it is still useful to carry out compaction tests to determine the degree of drying or wetting needed to bring the soil to a state appropriate for compaction.

To summarize, the advantages of the shear strength and air voids control method are as follows:

1. Large variations in soil properties present no difficulty in applying the method. The same specification limits apply regardless of the variations.
2. Field control is more direct as the value of the undrained shear strength is known as soon as the measurements are made.
3. The specification is easily varied to produce fills with particular properties needed in special situations.

15.4   DIFFICULTIES IN COMPACTING CLAYS

15.4.1   Soils Considerably Wetter Than Optimum Water Content

It is not infrequent that the natural water content of a soil is much greater than the value suitable for effective compaction, and considerable drying is needed before compaction can commence. If the soil is of low plasticity, drying is not particularly difficult, but drying soils of high plasticity can be very difficult. There are no easy ways to overcome this problem, but the following are essential requirements:

- Adequate spells of fine sunny weather. This can be a very uncertain expectation in some countries.
- Plenty of wide open space to spread out the soil for drying. Such a space should be created or obtained so that it is exposed to maximum direct sunlight and also to maximum wind.
- Good site management. This means organizing the whole operation to maximize fine spells of weather for drying and also being ready to "seal" the surface of any uncompacted material or stockpiles if rain is approaching. This "sealing" can be done by shaping all exposed surfaces so that rainfall cannot pond on them and rolling the surface with a smooth wheeled roller to create a tight impermeable surface layer.

15.4.2   Soils That Soften During Compaction

As we have seen, many residual soils are "structured"; that is, they have some form of bonds or weak cementation between their particles. When remolded by the compaction process, this structure is broken down and the soil becomes softer, as shown in Figure 15.5. When dealing with soils of this type, it is therefore important to understand their properties and plan the compaction criteria accordingly. It may be necessary to choose between two options as follows:

1. Drying the soil to its optimum water content and using the normal compactive effort to produce a high-quality fill
2. Accepting that substantial drying is not feasible because of weather conditions and adopting a much lower compactive effort so that the soil can be effectively compacted at (or close to) its natural water content

To determine the feasibility of the second option, it is desirable to conduct trials involving the excavation, transport, and compaction of the soil. Excavation, transport, and spreading should be carried out in such a way that disturbance and remolding of the soil is kept to a minimum. In other words the natural structure and strength of the soil should be retained as much as practical. The compaction operation should similarly be conducted so that remolding the soil is minimized. Light, tracked, equipment is likely to be most appropriate for this purpose, and the compaction process consists essentially of "squeezing" intact fragments of soil together to form a uniform fill. For this purpose only a few passes of the compaction equipment is likely to be preferable to a large number of passes, which may progressively soften the soil and not make it more compact or stronger.

15.5   COMPACTION OF GRANULAR AND NONPLASTIC MATERIALS

Granular materials show rather different compaction characteristics to cohesive soils. Conventional Proctor compaction tests may or may not show clear maximum dry densities and optimum water contents. Also, with clean granular materials, water drains from the material rather rapidly and in many cases the water content is "arbitrary" and does not greatly influence the dry density. Despite these factors, the test can be used to give a reasonable indication of the density that ought to be achieved by conventional compaction methods in the field. However, it may turn out that a considerably higher density can easily be achieved in the field. For this reason it is generally preferable to carry out compaction trials to determine a reasonable "target" density to be specified for field compaction.

For controlling compaction of granular materials, it is generally preferable to use a strength test rather than a density measurement. This is easily done using simple field penetrometers. These penetrometers are usually hand-operated dynamic cone penetrometers, and the parameter measured is blows per distance penetrated. By conducting such tests on trial compaction fills, appropriate values can be established for controlling the rest of the project. Figure 15.8a shows the principle of such a penetrometer. For cohesive fills it is also possible to gain an empirical measure of undrained shear strength by using static cone penetrometers of the type also shown in Figure 15.8b. These are pushed into the soil at a steady rate and the cone resistance measured on the dial gauge. This can be correlated with the undrained shear strength of the soil.

Examples of the two types of penetrometers are the following:

Dynamic: Scala Penetrometer

Tip: 20 mm diameter, $30°$ point angle.
Hammer weight: 9.0 kg
Height of fall: 51.0 cm
Parameter measured: S = number of blows/300 mm

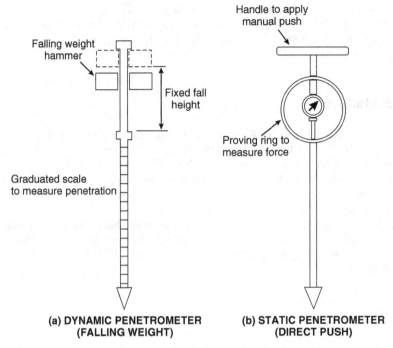

Figure 15.8   Hand penetrometers that can be used to control compaction.

This parameter can be correlated with other material properties of parameters. The correlation with the SPT test is SPT N value = 1.5 times N value from Scala penetrometer.

A common criterion used for compaction of granular fill with the Scala penetrometer is the following:

> The number of blows to drive the Scala penetrometer from a depth of 50–150 mm below the compacted fill surface should not be less than 10–12.

Static: U.S. Army Corps of Engineers Cone Penetrometer

Tip: 12.8 mm diameter, $30°$ point angle

Parameter measured: Force in units of 0.5 lb (= 1.102 kg). This was the original parameter used by the Corps of Engineers for calibrations with Californian Bearing Ratio (CBR) values. For general use it would be better to simply measure force so this could be converted to a simple cone resistance value (q_c) in pressure units such as kg/cm^2 or kPa.

The cone resistance can be calibrated with CBR values or with undrained shear strength (S_u). A reasonable correlation with the latter is the same as that for the Dutch static cone penetrometer test, namely, $S_u = q_u/N_k$ where $N_k = 15$ to $N_k = 20$.

REFERENCES

Kuno, G., R. Shinoki, T. Kondo, and C. Tsuchiya. 1978. On the construction methods of a motorway embankment by a sensitive volcanic clay. In *Proc. Conf. on Clay Fills*. London (November 1978), pp. 149–156.

Pickens, G.A. 1980. Alternative compaction specifications for non-uniform fill materials. *Proc. Third Australia-New Zealand Conf. on Geomechanics*, Wellington (May 12–16, 1980), Vol. 1, pp. 231–235.

Wesley, L.D. 2003. Geotechnical characterization and behaviour of allophane clays. *Proc. International Workshop on Characterisation and Engineering Properties of Natural Soils*. Singapore (December 2002), Vol.2, 1379–1399. Rotterdam: A. A. Balkema.

CHAPTER 16

SPECIAL SOIL TYPES

16.1 GENERAL COMMENTS

This book has been written based primarily on the author's own experience as a geotechnical engineer, which has been in the relatively wet, temperate climate of New Zealand and the wet, tropical climate of Southeast Asia. The material presented reflects this background and is primarily relevant to fully saturated soils. Other parts of the world, including some parts of Southeast Asia, have quite different climates and soil conditions, which give rise to soil groups not covered so far in this book and which are indeed not often covered in textbooks at all. This chapter therefore gives a brief introduction to three soil groups of considerable importance to geotechnical engineers, namely partially saturated soils, expansive or swelling soils, and collapsing soils.

These soils tend to belong to the residual soil group rather than the sedimentary group, and, as with residual soils, understanding their behavior generally involves two basic factors:

(a) Understanding the properties of the material
(b) Understanding the seepage and pore pressure state in which they exist, especially that above the water table

16.2 PARTIALLY SATURATED SOILS

16.2.1 Occurrence

The principal factors that govern the existence of partially saturated soils are:

- Particle size (grading) of the soil
- Climate
- Depth of the water table
- Topography

With fine-grained soils, especially those made up mostly of silt and clay-sized particles, water cannot freely enter or freely drain from the soil under gravity because it is prevented from doing so because of surface tension effects at the air–water boundary. Soils, like porous stones, have an "air entry value," that is, the pressure needed for air to enter the soil. This is very high for clays. The boundary between fine- and coarse-grained soils, 0.06 mm, appears to be the particle size limit below which water cannot easily drain from the soil under gravity. Once fine-grained soils become saturated, they generally only become unsaturated by evaporation at the ground surface, or any other exposed surface, such as a cut slope. The evaporation process overcomes the inability of water to drain from the soil in the liquid phase.

The influence of climate is twofold. If the climate is warm and wet, it promotes intense weathering of parent rocks and therefore favors the formation of fine-grained soils. It also means a plentiful supply of rain, inhibiting evaporation. Both these factors mean the soil is likely to be fully saturated, except very close to the ground surface. If the climate is dry, the opposite applies; regardless of temperature, the weathering process will be less intense and residual soils are likely to be relatively coarse. This fact plus the reduced availability of water means the soils are likely to be partially saturated.

The depth of the water table and the topography are closely related. Clearly, soils are much more likely to be fully saturated in low-lying, poorly drained areas (with shallow water tables) than in steep hilly terrain where drainage can take place much more readily and the water table is likely to be deep.

These comments are generalizations and do not take account of the influence of the parent rock or soil type. Hong Kong, for example, is relatively warm and wet, and it might be reasonable to expect that soils here would be fully saturated. However, the parent rock in Hong Kong is granite, which has a large quartz component that is relatively resistant to weathering. The result is that the residual soils in Hong Kong are fairly coarse (silty sands) and therefore unlikely to be fully saturated. In addition, Hong Kong is a very hilly place, which contributes to the fact that many of its soils are partially saturated. However, in more tropical countries where granite is commonly found, such as Malaysia, weathering is more intense and the resulting soils are finer and consist of clays or silty clays. They are therefore less likely to be partially saturated than the weathered soils of Hong Kong.

The nonuniformity of residual soils also has an influence on their degrees of saturation. Cracks, joints, and even coarse bedding planes may provide "paths" along which evaporation can occur, inducing the adjacent soil to be less than fully saturated. They also provide channels by which water can reenter and allow pore pressures to build up during periods of heavy rainfall.

16.2.2 Measurements of Degree of Saturation

There are surprisingly few comprehensive measurements of degree of saturation in natural soils, especially residual soils where it is likely to be of most significance, and not many measurements of how it varies with seasonal or storm events.

Measurements made in tropical red clays and allophane clays by the author (Wesley, 1973) show the soils to be fully saturated except in the top 1–2 m. Lumb (1962) gives some limited data for the weathered granite soils of Hong Kong, shown in Figure 16.1, which illustrates a different situation. The degree of saturation in the weathered Hong Kong granite at the two sites investigated averages about 40 percent at the site where no water table exists and 90 percent at the site with a water table at 8 m (25 ft). Particle

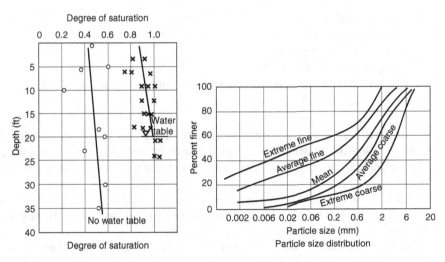

Figure 16.1 Degrees of saturation and particle size curves for Hong Kong weathered granite. (After Lumb, 1962.)

size curves for Hong Kong weathered granites, also given in Figure 16.1, show that the soil is a silty sand, with an average silt content of only about 10 percent. This is a very different material to the red clays and allophane clays described by Wesley (1973). We should note that with the water table at 8 m the degree of saturation is still high, averaging about 90 percent.

Not all weathered granites are similar to the Hong Kong granites. In Malaysia, which has a wet, tropical climate, the weathered granite soils are more clayey and better described as silty clays than as silty sands. Consequently their degree of saturation is likely to be much higher than the weathered granites of Hong Kong.

16.2.3 Mechanics of Partially Saturated Soils

The application of conventional soil mechanics concepts, especially the principle of effective stress, to partially saturated soils, is not straightforward, because partially saturated soils are no longer two-phase systems. There are clearly three phases—solids, water, and air. Because of surface tension effects at the air–water interface, the air pressure is not the same as the pore water pressure. This complicates the relationship between total stress, pore pressure, and effective stress, and the conventional expression for effective stress is no longer valid. Various forms of an equivalent effective stress equation for partially saturated soils have been proposed; see, for example, Fredlund and Morgenstern (1977). The first and probably the best known is that of Bishop, which has the form

$$\sigma' = (\sigma - u_a) + \chi(u_a - u_w) \qquad (16.1)$$

where σ' = effective stress
σ = total stress
u_a = pore air pressure
u_w = pore water pressure
χ = parameter depending on degree of saturation, with values from 0 to 1

The term $u_a - u_w$ is normally used to designate the "suction" (commonly called the matric suction) in the soil. In soil slopes u_a will usually approximate to atmospheric pressure, while the pore water pressure will be negative.

To help understand this equation, consider a natural soil close to the ground surface. The air phase is considered to be continuous throughout the soil and therefore at atmospheric pressure.

The air pressure (atmospheric) is thus zero, so the equation becomes

$$\sigma' = \sigma - \chi u_w$$

Normally the equation would be

$$\sigma' = \sigma - u_w$$

The effect of air in the soil is therefore to reduce the influence of the pore pressure on the effective stress in the soil (since χ is less than 1). This is to be expected because the pore pressure no longer acts on the total cross section of the soil mass. Figure 16.2 illustrates the way the pore pressure relates to the effective stress above the water table, depending on the degree of saturation and therefore the value of the parameter χ.

Attempts to apply equations such as 16.1 to analyze practical problems in terms of effective stress have not met with much success. There are clearly two main difficulties in applying this equation to practical situations. The first is that partially saturated soils are very unlikely to exist in a constant state of saturation. Their degree of saturation is likely to rise and fall with climatic effects, and relating degree of saturation to weather effects is very difficult. The second difficulty is the measurement of the parameter χ and relating it to degrees of saturation. Difficult and time-consuming tests are needed to do this. The difficulty of analyzing unsaturated soils using rigorous procedures has led to the development of simpler "ad hoc" (semiempirical) procedures, some of which appear to be finding limited use in engineering applications.

The principal interest in the analytical treatment of partially saturated soils appears to arise in relation to slope stability, in particular the wish

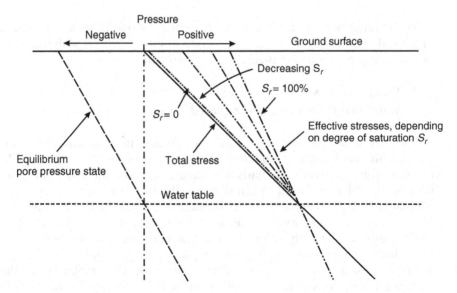

Figure 16.2 Relationship between pore pressure and effective stress for varying degrees of saturation.

to be able to take account of the varying pore pressure state with time, as governed by weather conditions. Two behavior "models" for handling partially saturated soils are therefore required—one to relate the pore pressure state in the ground to weather and hydrological conditions and the other to relate the soil strength to the pore pressure state. The first of these clearly involves a transient analysis of the seepage state and the second a model that takes account of variations in the degree of saturation.

Lam, Fredlund and Barbour (1987) have proposed a model for dealing with seepage in saturated and partially saturated soils which indirectly incorporates changes in the degree of saturation over time. They make use of the conventional transient form of the continuity (Equation 7.9), expressing it as follows:

$$k \left(\frac{\partial^2 h}{\partial x^2} + \frac{\partial^2 h}{\partial y^2} \right) + Q = m_w \gamma_w \frac{\partial h}{\partial t} \qquad (16.2)$$

where Q = rate of flow into element from an external source
m_w = slope of volumetric water content with change in pore pressure u
θ = volumetric water content, that is, volume of water per unit volume of soil

Hence

$$m_w = \frac{\partial \theta}{\partial u}$$

The parameter θ is simply a measure of the volume of water contained in the soil. This changes as water flows into or out of the soil with a change in the pore pressure u and is made up of two components:

1. Water that flows out as a result of compression of the soil element
2. Water that flows out and is replaced by air

For fully saturated soils m_w becomes identical to the well-known parameter m_v, the coefficient of one-dimensional compressibility of the soil. In practice, when analyzing partially saturated soils, according to Lam et al. (1987), the volume of water from compressibility of the soil is very small compared with the volume flowing out and being replaced by air. Examples of the parameter θ for several soils are shown conceptually in Figure 16.3.

The soils remain fully saturated until the pore pressure becomes negative. Once the pore pressure becomes negative, the possibility arises of air entering the soil. With sand this occurs relatively easily—water leaves the void space and the volumetric water content falls rapidly. As the particle size becomes smaller, it becomes more difficult for air to enter the sample and the volumetric water content no longer drops so rapidly. With true clays

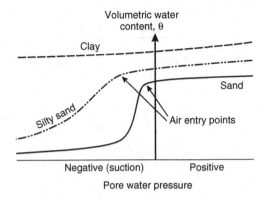

Figure 16.3 Examples of volumetric water content curves for several soils.

the air entry value (the suction needed to draw air into the soil) is so large that air can no longer be drawn into the soil by negative pore pressure (in most practical situations) and there is no sharp change of gradient of the volumetric water content graph. The soil remains fully saturated unless it is exposed to evaporation, in which case it may become partially saturated. In many wet tropical and temperate climates, intense weathering produces very deep fine-grained residual clays. These generally remain fully saturated except during hot dry seasons, when surface evaporation may result in partial saturation close to the ground surface.

The parameter m_w can be considered as taking the place of the soil compressibility in familiar situations, such as the Terzaghi consolidation theory. It is worth noting that the Terzaghi 1-D equation is a simple case of transient flow and is easily derived from Equation 16.2:

$$k\left(\frac{\partial^2 h}{\partial x^2} + \frac{\partial^2 h}{\partial y^2}\right) + Q = m_w \gamma_w \frac{\partial h}{\partial t}$$

For one-dimensional flow in the vertical direction the x term disappears. Also there is no external source of water so Q also disappears. This leaves

$$\frac{\partial h}{\partial t} = \frac{k}{m_w \gamma_w} \frac{\partial^2 h}{\partial y^2}$$

which is essentially the Terzaghi equation, since h is directly related to the pore pressure ($u = \gamma h$).

The second soil parameter (in addition to the m_w or θ parameter) is the coefficient of permeability k (or hydraulic conductivity as it is commonly called in groundwater studies). In a steady-state, fully saturated situation, k is unlikely to change very much with stress level or changes in pore pressure. However, with partially saturated soils, the situation is quite different as the permeability rapidly decreases as the degree of saturation falls.

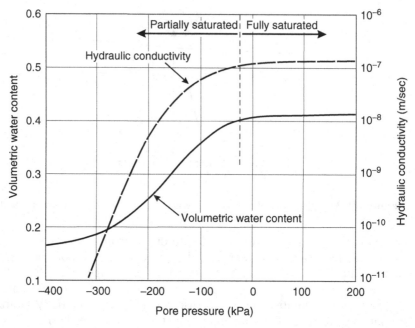

Figure 16.4 Volumetric water content and hydraulic conductivity versus pore pressure.

Figure 16.4 illustrates the typical way both θ and *k* may vary with pore pressure in a clayey silt.

It should be recognized that in the above method of analysis the air in the soil is assumed to be in a continuous state, so it has only one value, which is normally zero (atmospheric). The above procedure provides a solution for the value of the head (*h*) in the soil, to which the pore pressure (*u*) is directly related. The analysis incorporates the influence of partial saturation by relating both the volumetric water content and permeability to the pore pressure. These relationships are governed by the degree of saturation. This method of Lam et al. (1987) is incorporated into the commercial seepage computer program Seep/W, which forms part of a group of geotechnical programs known as GeoStudio.

With respect to the second behavior model required, namely a way of taking account of the influence of soil suction on the shear strength of the soil, the work of Fredlund and co-workers appears to be the best known, and their approach is incorporated into the Slope/W computer program, which is part of GeoStudio.

Fredlund and Rahardjo (1993) advocate treating the two stress state variables $\sigma - u_a$ and $u_a - u_w$ separately and independently evaluating their contribution to soil behavior. In other words, their approach avoids determination of the true effective stress from the values of $\sigma - u_a$ and $u_a - u_w$, which was Bishop's approach. Fredlund has proposed the following expression for shear strength:

$$\tau = c' + (u_a - u_w)\tan\phi^b + (\sigma - u_a)\tan\phi' \qquad (16.3)$$

where ϕ^b is the angle of cohesion intercept increase with increasing suction.

Equation 16.3 provides a semiempirical means of taking account of the influence of suction when the soil is not fully saturated. In effect, the frictional component of shear strength is being divided into two components, one arising primarily from the total stress and one from the soil suction. The latter is being expressed as an increase in the cohesive component of the shear strength. The concept is illustrated in Figure 16.5.

According to Fredlund and Rahardjo, the value of ϕ^b is usually found to be between 15° and 20° but theoretically could equal 45°. It varies with degree of saturation and would be 45° at full saturation (i.e., the effective stress would relate to the pore pressure in the usual way). Fredlund and Rahardjo advocate the use of Equation 13.2 for slope stability analysis of unsaturated slopes—to take account of the contribution coming from suction above the water table. This procedure forms part of the computer program Slope/W to enable it to be applied to partially saturated soils.

The Fredlund expression seems less than satisfactory from a theoretical viewpoint because it implies that the increase in shear strength from the negative pore pressure is a cohesive contribution rather than a frictional component. The Fredlund and Bishop approaches are illustrated in Figure 16.6.

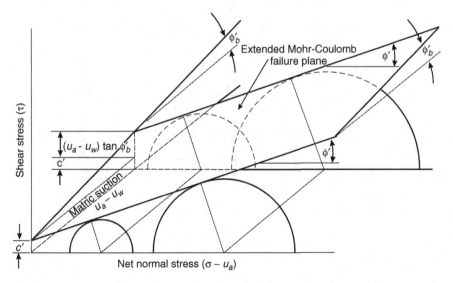

Figure 16.5 Extended Mohr–Coulomb failure envelope for unsaturated soils. (From Fredlund and Rahardjo, 1993.)

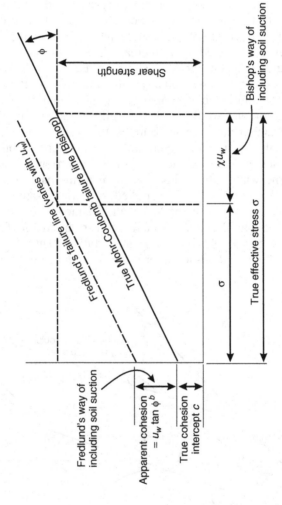

Figure 16.6 Comparison of shear strength expressions of Fredlund and Bishop.

As indicated, the value of ϕ^b varies with degree of saturation, so that in any real situation it is likely to vary with depth in the ground as well as over time depending on weather variations. To properly take account of this variability would be difficult. The computer program Slope/W does not currently incorporate a function that automatically changes the value of ϕ^b as the degree of saturation or the soil suction changes.

16.3   EXPANSIVE OR SWELLING CLAYS

In some parts of the world, buildings can suffer quite severe damage from a group of soils known as expansive or swelling clays. The structures most commonly damaged are houses and other types of low-rise buildings such as warehouses. This is because their foundations are normally close to the surface and the buildings have little structural rigidity. The clay on which the building is founded takes up water and swells unevenly, causing distortion of the building and associated damage. The damage can be quite severe. Expansive clays are usually associated with semiarid or arid climates, particularly South Africa, Australia, parts of North and South America, and the Middle East. However, the phenomenon is probably found in many other countries to a lesser extent, but damage to buildings is not great enough to be a matter of real concern.

16.3.1   Basic Concepts of Expansive Behavior

Literature on expansive clays sometimes implies that this problem arises purely from the type of soil, but this is not true. Two basic factors must be present for the problem to arise:

1. The existence of clayey soils containing a significant proportion of clay minerals, especially those of the smectite family (i.e., active clay minerals)
2. A fairly dry climate, with long dry seasons and some periods of rainfall

In other words "expansive clays" reflect both their composition and the environment in which they exist. It is wrong to think of expansive clays as simply a particular type of clay. There are plenty of similar clays in other parts of the world that do not give rise to swelling problems because they exist in different climatic conditions. It is better to think in terms of expansive clay conditions rather than expansive clays. Expansive clays tend to be found in relatively flat areas, where drainage conditions favor weathering processes leading to the formation of minerals belonging to the smectite family (e.g., montmorillonites), which are prone to large volume changes with changing water content. As discussed earlier in Chapter 3, soils in the CH group on the plasticity chart are those most likely to give rise to shrink/swell problems.

The mechanism or "explanation" for expansion is that during long dry spells the soil dries out and shrinks due to high suction ("tension") that develops in the pore water. When water becomes available to the soil, it literally "sucks" up water and swells. Taking up water means the suction decreases and the effective stress in the soil decreases, accompanied by a positive volume change (i.e., an increase in volume). In most expansive clay situations, the soil is partially saturated, and the degree of saturation increases as swelling takes place. It is not however necessary for the clay to be partially saturated—the problem could still arise with a fully saturated soil, though it would be unlikely to be very severe. Expansive clay behavior is essentially a more severe case of the situation discussed earlier in Chapter 4 (Figures 4.4 and 4.5). When foundations are involved, the influence of pore pressure changes may be greater than the influence of applied loads. In a number of countries or areas around the world, the climate is essentially dry most of the time, so that the soil exists permanently in a "high suction" state, and only external intervention in some form alters this state.

The depth to which drying out of the soil occurs can be very great. Blight (1997) states that soil profiles can dry out to depths of 15–20 m during a long dry season. Heave of up to 70 mm has been measured at markers embedded in the soil at a depth of 7.5 m, which means that the soil must have undergone drying to a considerable greater depth than this.

The cause of the increased availability of water to the soil is usually a change in the land use at the surface. This change may involve construction of pavements or structures, conversion of land to livestock farming or crop production, or development of gardens and parks. These activities can have the effect of providing additional water from external sources, such as irrigation or leaking pipes, or they may simply cut off avenues for water escape, such as surface evaporation.

Construction of a house or building, or simply the construction of a sealed car-park or airport runway, may cause severe expansion simply because it cuts off evaporation and upsets the natural water balance. A possible situation where this could occur is illustrated schematically in Figure 16.7.

It is thus not necessary that there be an external (man-made) water source to cause an increase in the water content of the soil and consequent swelling. There may of course be additional wetting of the soil from irrigation, leaking water supply pipes, and so on.

16.3.2 Estimation of Swelling Pressure and Swell Magnitude

Various methods have been proposed to estimate the "swelling pressure" of an expansive clay and more importantly the magnitude of swell that may occur in a particular situation. It is the potential magnitude of swell that is

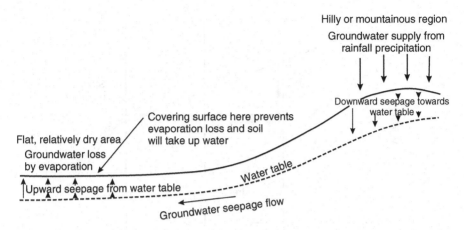

Figure 16.7   Groundwater system where construction covering the ground surface can upset the existing equilibrium situation—and cause swelling.

of concern to foundation designers. Procedures that have been proposed for estimating swell range from completely empirical methods based on simple indicator tests to methods making use of the principle of effective stress.

Methods making use of the principle of effective stress involve the following:

1. Estimation of the initial and final effective stresses and hence the change in effective stress throughout the soil layer
2. Determination of the swell characteristics of the soil
3. Calculation of the swell of sublayers throughout the soil profile, in much the same way as is done for estimating settlements in fully saturated soils

Figure 16.8 illustrates the total and effective stress states in a soil when the swelling pressure or heave magnitude is measured, starting with the soil at its in situ stress state. Consider that two identical undisturbed samples have been obtained and set up in conventional odometers and the existing overburden stress applied (using dry porous stones). Point A represents this state in terms of void ratio and total stress. The effective stress in the soil is unknown.

Two tests can then be carried out. In the first test the vertical stress is kept constant, water is allowed access to the sample, and the magnitude of the swell is measured. The soil swells from point A to point C. The pore pressure at point C is now zero and the total and effective stresses are the same.

In the second test, swelling of the soil is prevented when water is added, and measurement is made of the pressure needed to prevent swelling. This can be done either by adding weights to the apparatus as soon as any swell is

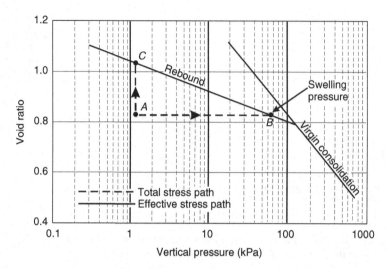

Figure 16.8 Basic soil-swelling model. (After Frydman, 1992.)

detected or by using some form of rigid restraint and a rigid load-measuring system. The pressure thus measured is given by point B. In moving from point A to point B, no volume change has occurred, which means that there has been no change in the effective stress acting on the soil. At point B the pore pressure is now zero and the total stress and effective stress are again equal.

The vertical stress can now be reduced in a controlled manner and the swell measured. The soil will then follow the line BC if the vertical stress is reduced to that which originally acted on the sample. The line BC is therefore the normal swell or "rebound" line obtained when loads are reduced in a conventional odometer test, and its slope is given by the swell parameter C_S.

This procedure can be used for both partially and fully saturated soils, and the same concept of behavior applies. However, with a fully saturated soil, the initial suction will exactly equal the pore pressure, and the stress state is not complicated by the presence of air. When no vertical strain is permitted and water is added to the odometer, the pore pressure will increase from its suction value to zero, and the total stress will increase by exactly the same amount. In other words, in a fully saturated soil, the negative pore pressure in the soil and the swell pressure are identical. In a partially saturated soil, the suction no longer induces an effective stress in the soil of the same magnitude as the suction itself. Thus the stress to be applied to prevent the soil from swelling will be correspondingly less than the suction.

These concepts are illustrated in Figure 16.9. First this illustrates the stress state of the soil when it is in its natural state—unrestrained and no

	Fully saturated	**Partially saturated**
In natural state, no external stress, no water available to soil (undrained).	Pore pressure $= u$ $\sigma' = \sigma - u = -u$ $\sigma' = -u$	Pore pressure $= u$ $\sigma' = \sigma - \chi u = -\chi u$ $\sigma' = -\chi u$
Confined in apparatus, external stress applied, to prevent volume change, water freely available to soil (drained).	σ_s ↓↓↓↓↓ Pore pressure $= 0$ $\sigma' = \sigma_s - u = \sigma_s - 0 = \sigma_s$ $\sigma' = \sigma_s$ Therefore $\sigma_s = -u$	σ_s ↓↓↓↓↓ Pore pressure $= 0$ $\sigma' = \sigma_s - u = \sigma_s$ $\sigma' = \sigma_s$ Therefore $\sigma_s = -\chi u$

Figure 16.9 Relationship between swelling pressure and pore pressure for fully and partially saturated soil.

water is available to it. Second it illustrates the stress state after the sample is set up in an apparatus such as an odometer, and water is allowed to freely enter it while the volume is kept constant. The stress needed to maintain this constant volume is measured. Because no volume change is permitted, the effective stress state must be the same in each condition. This leads to the expressions given for the swelling pressure σ_s.

There are various other ways to measure the swell pressure. The above procedure is not as straightforward as it appears, as it requires an extremely rigid system to ensure no vertical movement occurs. An alternative method is illustrated in Figure 16.10.

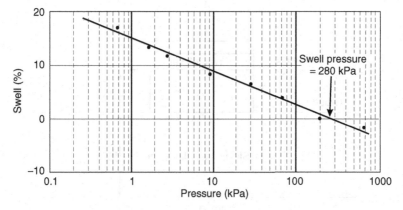

Figure 16.10 Measurement of swell pressure and potential swell magnitude by applying different stresses and measuring swell.

A series of identical samples are prepared and differing vertical stresses are applied before any water is allowed in the apparatus. This can conveniently be done using the conventional odometer apparatus. After the loads are applied, water is added to the apparatus and the swell of each sample is measured. If the loads added exceed the swell pressure, then the sample will compress, and the addition of water will have no effect.

By drawing a "best-fit" line, the swell pressure can be determined. The line established in this way, if a log scale for pressure is used, should be the swell line corresponding to the swell index C_S.

16.3.3 Estimation of Swell Magnitude

An experimental result of the form shown in Figure 16.10 can be used directly to estimate the potential magnitude of swell in a soil if water becomes freely available to it.

Example   Consider a layer of potentially expansive clay 8 m thick overlying hard rock. A storage tank 10 m in diameter is to be constructed on it to store water to a depth of 4 m, as illustrated in Figure 16.11. There is concern that the tank could be damaged by swell of the soil. Samples taken from the center of the clay layer are tested using the procedure described above and give the result shown in Figure 16.10. The unit weight of the soil is measured at $17.5 \, \text{kN/m}^3$.

Estimate the potential heave of the ground surface at the center of the tank. For simplicity we will consider this layer as a single layer. Dividing it into sublayers would give us a more accurate answer:

Existing total stress at center of clay layer $= 4 \times 17.5 = 70.0 \, \text{kPa}$

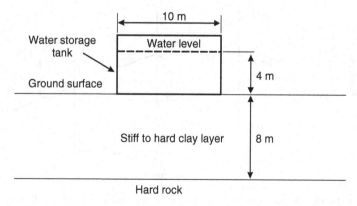

Figure 16.11   Example for estimation of swell magnitude.

To estimate the stress increase at the center of a circular tank, we can use the appropriate chart from Chapter 5:

Radius of tank $(a) = 5\,\text{m}$

Depth to center of layer $(z) = 4\,\text{m}$

$z/a = 0.8$

Influence factor $I_\sigma = 0.74$

Applied stress from tank at surface $= 4 \times 9.8 = 39.2\,\text{kPa}$

Stress increase from tank $= 0.74 \times 39.2 = 29.4\,\text{kPa}$

Thus total applied stress $= 70.0 + 29.4 = 99.4\,\text{kPa}$

Swell in percent (from graph) $= 2.7\%$

Therefore total ground heave $= \dfrac{2.7}{100} \times 8 \times 1000\,\text{mm} = 216\,\text{mm}$

This calculation is straightforward and provides a simple procedure for estimating "potential" ground heave. However, it involves rather simplistic assumptions which cannot be justified on a general basis.

These assumptions are as follows:

1. The graph of the type shown in Figure 16.10 is representative of the soil on a general basis. This is not the case, as the graph will normally depend on the time of the year when the soil is sampled. The graph will be different if the soil is sampled at the end of a long dry period to what it would be after a period or wet weather.
2. The "wetting" of the soil is sufficient to cause the suction to entirely disappear; that is, the pore pressure becomes zero. In some cases this may be true; however, it is probable that in most practical situations the suction does not reduce to zero. In situations where the water table is deep and the climate rather dry, it is unlikely that the soil suction would be completely destroyed by water availability.

If the soil is sampled in its driest state, then the heave calculated using the above method will be an upper limit and thus conservative. It will be an estimate of "potential" heave, and not necessarily a realistic estimate of what may occur in practice.

16.4 COLLAPSING SOILS

The term "collapsing" soil is used rather loosely in the soil mechanics literature but is generally taken to mean a soil that undergoes a sudden decrease in volume when water is made available to it. Generally, the term is used in relation to foundation performance, that is, a situation where an external load has been applied to the soil prior to water becoming available and causing collapse to occur. However, in some cases at least, collapse may occur under the existing overburden pressures. The following soil types may result in collapsing behavior:

1. Loess soil types. These are wind-blown deposits consisting of predominantly silt- and sand-sized particles. The deposition method results in a very loose open structure, and subsequent weathering may produce small quantities of clay minerals, which tend to act as weak bonding agents between the coarser particles. These deposits generally exist in dry or semidesert conditions.

2. Highly weathered residual soils. In some environments, weathering of rock leads to very open, porous soils. This occurs because of the leaching processes associated with weathering. It appears that this process can occur in a wide variety of rocks. Blight (1997) associates this with granites, but Vargas (1976) mentions a wide variety or rock types, from sandstones and basalts to granites and metamorphic rocks, found in various parts of Brazil. In Brazil, the weathering process has produced a surface layer with high void ratio and low water content, known locally as "porous clay."

3. Saline soils. In some arid environments, including parts of northern Chile, concentrations of salt occur that act as cementing agents between soil particles. This occurs in situations where groundwater is being recharged by aquifer flow from distant sources and steady evaporation is occurring, leading to a slow but steady increase in salt in the soil.

The above soils generally have reasonable strength and stability in their natural environment. However, if subject to loading and the addition of water, they are likely to undergo a sudden decrease in volume, which is referred to as collapse.

The behavior of a collapsing soil is illustrated diagrammatically in Figure 16.12. This shows the results of three odometer tests carried out on undisturbed samples of a collapsing soil. Curve (a) is for the soil tested at its natural water content without access to any water. Curve (b) is for the same soil tested after water has been added to the odometer cell, and the time allowed for the sample to take up water before the first load is added. Curve (c) is for a test started on the natural soil, but water is added at a particular stress level while the load is maintained constant at that value.

Curve (a) for the natural soil shows that the soil is essentially unstable in this state with respect to the influence of water. Curve (b) illustrates conventional "stable" behavior, as expected, after water is added. The sample is no longer susceptible to collapse or expansion due to water addition. It

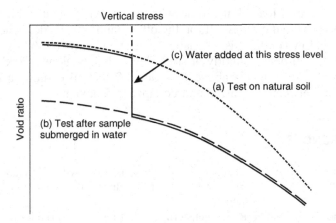

Figure 16.12 Behavior of a collapsing soil.

is probable that the degree of saturation is quite high after the addition of water.

The volume changes that occur in collapsing soils *cannot be predicted or explained on the basis of effective stress considerations.* In the diagram above, the soil at its natural water content is likely to be subject to high pore water suction and thus to a high effective stress. The addition of water will reduce this suction and thus reduce the effective stress in the soil. Despite this, the volume decreases rather than increases.

Some partially saturated soils show both expansive and collapse behavior, as illustrated in Figure 16.13.

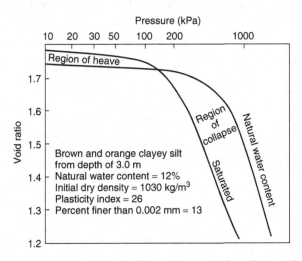

Figure 16.13 Behavior of a clayey silt (residual weathered quartzite) to loading at natural water content and after saturation. (After Blight, 1997.)

If this soil is loaded at a time when its water content is high, it would compress without collapse. If, on the other hand, it is loaded when its water content is low (in this case its natural water content) and then subsequently wetted, it will heave slightly at stress levels below about 100 kPa and will collapse at stress levels above this value. This soil exhibits a continuous spectrum between expansive and collapsing behavior.

REFERENCES

Blight G., ed. 1997. Mechanics of residual soils. Prepared by Technical Committee 25, International Society for Soil Mechanics and Foundation Engineering, Rotterdam: A. A. Balkema.

Fredlund, D. G., and N. R. Morgenstern. 1977. Stress state variables for unsaturated soils. *J. Geotech. Eng. Div. ASCE*, May 1977, pp. 447–467.

Fredlund, D. G., and H. Rahardjo. 1993. *Soil Mechanics for Unsaturated Soils*. New York: John Wiley and Sons.

Frydman, S. 1992. An effective stress model for swelling of soils. In *Proceedings, 7th International Conference on Expansive Soil*, Vol. 1, pp. 191–195. Lubbock: Texas Technical University Press.

Lam, L., D. G. Fredlund, and S. L. Barbour. 1987. Transient seepage model for saturated—unsaturated soil systems: A geotechnical engineering approach. *Canadian Geotechnical Journal*, Vol. 4 No 24, pp. 565–580.

Lumb, P. 1962. The properties of weathered granite. *Geotechnique*, Vol. 12 No. 3, pp. 226–243.

Vargas, M. 1976. Structurally unstable soils of Southern Brazil. In *Proceedings, 10th International Conference on Soil Mechanics and Foundation Engineering*, Moscow, 239–246.

Wesley, L. D. 1973. Some basic engineering properties of halloysite and allophone clays in Java, Indonesia. *Geotechnique* Vol. 23, No 4, 471–494.

INDEX

active earth pressure, 309
  concept of, 313, 314
  coefficient of, 310
activity, 4, 36, 37
aging of soils, 3
air voids, 15
allophone, 4, 5
allowable bearing capacity,
  262
allowable settlement, 170–172
angle of shearing resistance
  ("friction angle"), 58, 186,
  226–227
angular distortion, 171
anisotropy, 216, 218, 219
area ratio, 240
at rest earth pressure, 57
Atterberg limits, 5, 31, 41
  definition of, 31, 32
  test for, 31–33
  use in soil classification, 34, 35,
    38, 39

bearing capacity, 259, 261, 262,
  267–274
  allowable, 262
  eccentric and inclined loads,
    270–272

equations, 268–273
of clay, 272–274
of sand, 276–278
ultimate, 262, 267
Bishop method of slices, 360–366
block sampling, 241
boreholes, 235
braced excavations, 321–322
bulk density, 15, 22
Burland-Burbidge sand settlement
  method, 176–178

Casagrande construction, 166
classification methods, 37, 38, 40
  for residual soils, 44–47
  Unified soil classification system
    (USCS), 38
clay, 27
  definition of, 28
  particle size, 28
  plasticity of, 40
clay fraction, 29
clay minerals, 2–6, 47
coarse-grained soils, 27, 28, 39, 42,
  52,
  classification of, 38
  definition of, 27
  particle size of, 28

Printed in the United States
By Bookmasters